Christophe Wyon and Marcel Van de Voorde
Semiconductors and Nanoelectronic Chips

Also of interest

Semiconductor Spintronics
2nd editon
Thomas Schäpers, 2021
ISBN 978-3-11-063887-5, e-ISBN (PDF) 978-3-11-063900-1

Quantum Technologies for Engineers
Quantum Computing, Quantum Communication, Quantum Sensing
2nd revised and extended edition
Rainer Müller and Franziska Greinert, 2026
ISBN 978-3-11-914611-1, e-ISBN (PDF) 978-3-11-222443-4

Nanoelectronics
From Device Physics and Fabrication Technology to Advanced Transistor Concepts
2nd Edition
Joachim Knoch, 2024
ISBN 978-3-11-105424-7, e-ISBN (PDF) 978-3-11-105442-1,
e-ISBN (EPUB) 978-3-11-105501-5

Christophe Wyon and Marcel Van de Voorde

Semiconductors and Nanoelectronic Chips

Artificial Intelligence, Technology, Sustainability, and Supply Chain

DE GRUYTER

Authors
Dr. Christophe Wyon
CEA-Leti
CEA Grenoble
17 rue des martyrs
38054 Grenoble Cedex
France
E-mail: cwyon@cea.fr
ORCID Number: 0009-0000-4509-060X

Prof. Dr. Marcel Van de Voorde
Rue du Rhodania, 5
BRISTOL - A, App. 31
3963 CRANS - MONTANA
Switzerland

ISBN 978-3-11-224794-5
ISBN 978-3-11-224795-2 (PDF)
ISBN 978-3-11-224796-9 (EPUB)
DOI https://doi.org/10.1515/9783112247952

Library of Congress Control Number: 2026943681

Bibliographic information published by the Deutsche Nationalbibliothek
The Deutsche Nationalbibliothek lists this publication in the Deutsche Nationalbibliografie; detailed bibliographic data are available on the Internet at https://dnb.dnb.de.

De Gruyter and Walter de Gruyter GmbH are part of De Gruyter Brill.
www.degruyterbrill.com

Questions about General Product Safety Regulation:
productsafety@degruyterbrill.com

Cover illustration: Vertigo3d/E+/Getty Images
Printing: CPI books GmbH, Leck

FSC www.fsc.org
MIX
Papier | Fördert gute Waldnutzung
FSC® C083411

"The world is moving so fast these days that the man who says it can't be done is generally interrupted by someone doing it".

Elbert Hubbard (1856–1915)

Foreword from Luc Van den hove

Artificial intelligence is rapidly becoming a defining force in economic growth, industrial competitiveness, and geopolitical influence. As AI applications expand from the digital world into physical systems such as autonomous vehicles, robotics, healthcare, telecommunications, defense and industrial automation, semiconductors have become a strategic foundation of future prosperity. The ability to design, manufacture, and deploy advanced computing systems will increasingly determine which regions capture the value created by AI.

Europe has a unique opportunity to grow its position in the AI era, not by replicating the approaches of the United States or Asia, but by building on its own distinctive strengths. Europe possesses the strongest world-leading semiconductor research capabilities, strong universities, manufacturing expertise, equipment and materials leadership, and globally competitive industries that are poised to become major adopters of AI technologies. These assets provide a foundation for a new model of innovation in which AI, semiconductor technology, and industrial applications evolve together.

The central challenge is that the future of AI will be shaped by the close interaction between algorithms, computing architectures, and semiconductor technologies. AI innovation cycles are measured in weeks, while semiconductor development requires years of investment and coordination. At the same time, growing computing demands are creating unprecedented energy challenges, making efficiency a critical competitive advantage. Future breakthroughs will depend on the ability to co-optimize software, hardware, and manufacturing technologies across the entire technology stack.

To seize this opportunity, Europe must pursue three strategic objectives. First, it must preserve and strengthen its leadership in semiconductor research and innovation by supporting advanced R&D infrastructure and pilot lines that bridge scientific discovery and industrial deployment. Second, it must translate research excellence into commercial success by expanding chip design capabilities, fostering a stronger fabless ecosystem, increasing access to growth capital, and accelerating the development of AI-specific semiconductor solutions. Third, to address the critical demand for semiconductors, Europe must secure access to advanced manufacturing capacity while maintaining strengths in specialty and mature-node technologies that remain essential for many industrial sectors.

However, none of these objectives can be achieved in isolation. Success requires a coordinated ecosystem that links research institutions, startups, established industry, investors, and policymakers. The challenge Europe faces is not internal competition between member states or organizations, but intense global competition for technological leadership. As a result, European initiatives in semiconductors, AI, defense, automotive, and industrial policy must be aligned within a common strategic framework across the Member States.

 | https://doi.org/10.1515/9783112247952-202

The overarching message is optimistic but urgent. Europe possesses many of the capabilities required to become a leader in the next generation of AI-driven innovation. By combining research excellence, design leadership, manufacturing capacity, and industrial deployment into a coherent strategy, Europe can build a competitive and resilient technology ecosystem that creates economic growth, strengthens strategic autonomy, and ensures a significant role in shaping the future of AI.

This book arrives at an important moment. It offers readers a comprehensive overview of a field that increasingly influences every aspect of modern life. More importantly, it encourages reflection on how technology, industry, policy, and society must work together to shape the future.

Luc Van den hove
President IMEC
www.imec-int.com
July 2026

Foreword from Françoise Chombar

Semiconductors currently experience explosive market growth and have bolstered further strategic importance driven mainly by AI and data centers demand, and the electrification and automation of everything. The semiconductor sector already occupies the central stage globally—not just in trade, but as the wellspring of innovation, product development, and added value. It is entirely appropriate that the European Union and its Member States dedicate their focus to this vital industry. This book by Dr. Christophe Wyon and Prof. Marcel Van de Voorde is an insightful contribution to this debate.

It is from a perspective shaped by decades at the heart of Europe's semiconductor industry that I'm writing this foreword. As co-founder, former CEO, and current chairwoman of Melexis, a fabless mixed-signal semiconductor company, as co-founder of X-Fab, Europe's largest pure play foundry, having taken up board membership at multiple European technology companies, and as investor in semiconductor related start-ups and scale-ups, my journey has afforded me a unique vantage point. I write this as a European globalist and STEMinist, driven by the conviction that our continent's future hinges on the decisions we make today.

The world stands at a pivotal juncture. In my forty years immersed in this industry, I have never witnessed such a converging storm of change, uncertainty, and importantly: huge opportunity. Technological advances unfold at breakneck speed against a backdrop of increasing global fragmentation. What remains clear is that Europe must close the gap on some critical technologies; failing to do so would leave us exposed not only in terms of economic growth and competitiveness, but also in our security and strategic positioning, all necessary to retain our democratic, inclusion and peace dividends. While I believe interdependencies are not in se a problem, they should at the very least be mutual in nature and partnerships with like-minded countries should be consciously pursued and nurtured.

Europe's semiconductor strengths reside in areas such as "More than Moore," compound semiconductors, photonics, equipment, and advanced materials. Our research and development capabilities are robust, yet we must acknowledge and address a deficit in scaling, speed, industrialization, and commercialization. Europe should not try to just copy paste others. Rather, while systematically mitigating the effects of our weaknesses, our policies and decisions should first and foremost reinforce what we do best, with a focus on future oriented end applications and their value chain robustness.

In my humble opinion, following themes emerge as priorities, some of which are more deeply explored in this book: STEM (science, technology, engineering and mathematics) talent and general STEM literacy; knowledge institutions that enable ecosystems, engrain an entrepreneurial spirit and treat spin-offs as joint ventures for sustainable societal impact rather than trophies they own (aka 4th generation institutions), success being measured by how well the spin-offs and ecosystems thrive;

a coherent and right-sized regulatory framework that stimulates entrepreneurship and innovation; a focus on developing European domestic demand for microelectronic hardware, because supply follows demand, not the other way round; proven, non-subsidy based, public financing models that encourage intra-European collaboration, stop infighting between member states and reduce wasting our scarce resources; energy security and affordability; and the free flow of critical materials.

More than ever, Europe needs to become more integrated, yet also more agile, forward-looking, and pragmatic. Our continent's future progress and prosperity depend on a strategic approach to semiconductor policy—one that targets investments to align with Europe's unique industrial fabric, leverages our inherent strengths, and nourishes the renewal of our innovation ecosystem.

As you read this book, I invite you to consider above mentioned themes not merely as policy challenges, but as true opportunities to shape the future of our European continent. Let us move forward with clarity, cooperation, and resolve. "Move fast and mend things" could well serve as our guiding principle.

Françoise Chombar
Chairwoman Melexis nv
https://www.melexis.com/
https://www.linkedin.com/in/fchombar/
June 2026

Contents

Acknowledgements

Pr. Van de Voorde acknowledges to his son Marc-Philip Van De Voorde for the managerial advice on making the right decisions during the preparations for this book.

Dr. Wyon is grateful to his colleagues Marc Aid and François Perruchot for their help in numbering figures and references of this book.

Abstract

This book highlights the ubiquitous and vital role of semiconductors in the AI era, where demand for high-performance logic chips and memories is booming. Due to the tremendous expansion of AI, and to their strategic value, semiconductor supply chains, from the raw materials to advanced chip packaging, have become geopolitical assets and critical national security infrastructure. Facing global vulnerabilities, governments are now working to reshore chip design and production. Finally, the document reviews Europe's current position and outlines a long-term vision to strengthen its supply chain resilience.

 | https://doi.org/10.1515/9783112247952-204

Introductory Remarks

Semiconductor integrated circuits, or chips, are silicon wafers imprinted with transistors. Modern devices contain chips ranging from millions to tens of billions of transistors, with the most advanced AI chips reaching hundreds of billions. Chips are now embedded across consumer electronics, automotive, industrial, medical, agricultural, and military sectors, making semiconductors essential to economic competitiveness, national security, and everyday life. Due to the chip ubiquity (Figure 1), the semiconductor industry is often called as the "Industry of Industries," since semiconductors rank first as the most traded goods in global trade statistics since 2020.

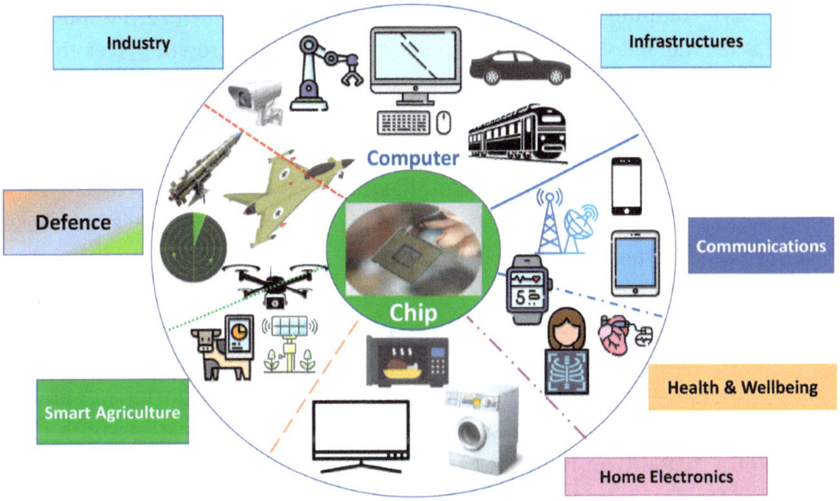

Figure 1: Ubiquity of nanoelectronic chips.

Chip manufacturing spans a complex global value chain. It begins with extracting and refining raw materials or recycling electronic waste to obtain high-purity Si, Ge, W, In, etc., and specialty gases and chemical precursors. Equipment suppliers provide wafers, lithography masks, gases, and fabrication tools. Designers create chip layouts that foundries process on silicon or alternative substrates such as SiC and GaAs. After fabrication, wafers are diced into chips, electrically tested, packaged, and assembled onto printed circuit boards for end customers.

The dramatic spread of chips stems from an enormous cost reduction by a factor 10^7 per transistor over decades, driven by exponential increases in transistor density and shrinking transistor dimensions – from micrometer scales in the 1980s to nanometer, and even Angstrom scales today. Smaller transistors yield better performance and energy efficiency, enabling more powerful and compact devices. "Leading-edge" chips use the most advanced process technologies for microprocessors and high-performance memory,

 | https://doi.org/10.1515/9783112247952-205

while "legacy" chips rely on older technology nodes and serve automobiles, aircraft, consumer electronics, medical devices, and many embedded systems. Because the industry evolves rapidly, today's cutting-edge nodes become tomorrow's legacy nodes.

However, the cost of research, design, and manufacturing for leading-edge chips has grown exponentially over the last three decades. Rising development and capital expenses had driven many Japanese, European, and some US companies to outsource leading-edge production to Taiwanese and Korean foundries and to rely on Asian providers for chip packaging. As a result, critical elements of the semiconductor supply chain have become highly fragmented and globally dispersed.

The AI revolution, accelerating since about 2015, has intensified demand for specialized AI chips such as digital processors and memories. Training large AI models consumes massive compute resources, hardly pushing demand toward the most advanced process technologies and costly chip designs. AI's rapid growth drives the demand for logic devices and memories that are smaller, faster, and more energy-efficient, while semiconductor advancements, such as the move to very advanced technology nodes (≤ 2nm), enable breakthroughs in AI capabilities. The huge demand of AI-based data centers for memories recently induced a dramatical increase in their average selling cost and a shortage for other applicative sectors requesting these memories, such as smartphones, PC, etc.

On one hand, the demand for smaller, faster and more energy-efficient AI chips will drive the semiconductor R&D for, at least, the 5 to 10 forthcoming years. On the other hand, the development of AI can greatly improve the semiconductor manufacturing thanks to a better and real-time monitoring of the process parameters, to an enhanced defect capture rate and its classification in killer defects or not, and to the implementation of a digital twin of the semiconductor manufacturing line. AI can also reduce the time for designing the most advanced chips and semiconductor systems, through finding the design solution leading to the optimal power, performance and area of semiconductor chips.

Driven by the tremendous demand AI chips, semiconductor market forecasts expect semiconductor demand to expand significantly, with projections to double within a few years and approach or exceed $1 trillion by 2030 or even by 2026 or 2027, according to more recent forecasts.

AI is also reshaping global geopolitics. Countries increasingly view strategic control over compute, data, and digital infrastructure as central to national prosperity and security. Therefore, geopolitical responses to semiconductor dependence have reshaped investments and trade policy. China launched large-scale investment programs ($150 billion) to build domestic semiconductor capacity. The COVID-19 pandemic and resulting supply disruptions exposed vulnerabilities and underscored chips' strategic importance. The US enacted a Chips Act in 2022 with substantial funding ($52 billion) to reshore advanced chip manufacturing and announced export controls on high-end chips, and design and manufacturing tools to limit access by strategic competitors. In response to US export controls, China implemented export restrictions on certain critical raw materials in 2025, further intensifying competition. The semiconductor sector has thus become highly strategic and a central theatre in US-China technological and economic rivalry.

This book is an enlarged version of an international technical paper [1], which shortly described the technical and sustainable challenges for manufacturing nanoelectronic chips. Notably, it details the technical and energy-efficiency requests of AI chips, and provides more technical inputs about the performance of new AI computing paradigms such as in-memory and neuromorphic computing. It also describes the current vulnerabilities of the semiconductor supply chain, and the government's initiatives to reshoring critical parts of this supply chain.

This book also describes what a country or a region, being trapped in the China-USA technological rivalry and complex worldwide geopolitical shifts, must implement in order to keep its technological sovereignty and its prosperity. Europe has been used as a model since its GDP is comparable to that of the USA. Hence, it reviews the current involvement of the European semiconductor industry in the semiconductor supply chain, and proposes a long-term strategy for enhancing the EU supply chain resilience.

This book has been written to a very broad audience worldwide. Hence, it is devoted both to a technical audience– graduate students, PhDs, engineers, industry stakeholders– and to policymakers, providing simple insights into the ubiquity, status, and challenges of the semiconductor industry.

The first chapter maps semiconductor requirements across communications, mobility, energy, healthcare, agriculture, and defense, highlighting device, materials, and integration needs. 5G/6G drive RF and mmWave solutions using GaN, InP, III-V/Si heterogeneous integration, bipolar complementary MOSFET (BiCMOS), and silicon on insulator (SOI) for low-loss RF System on Chips (SoCs). Electrification and mobility require widebandgap (WBG) power devices (SiC and GaN), ultra-WBG (UWBG) research, high-performance sensors (LiDAR, radar), and edge AI for real-time control. Energy grid and conversion efficiency improvements rely on Si and SiC MOSFETs plus GaN high electron mobility transistors (HEMT) to raise power density, reduce losses, and enable smart bidirectional systems. In the agriculture, healthcare, and defense sectors, miniaturized sensors, implantable, advanced imaging, rad-hard and low size, weight, power, and cost (SWaP-C) components, and optimized on-edge deep neural network/spiking neural network (DNN/SNN) inference are cross-cutting priorities.

Chapter 2 quantifies the explosive growth of data (ZB scale) driven by sensors and Internet of Things (IoT), and argues that large-scale AI models are required to extract value from this data. It explains DNN fundamentals (layers, parameters, and backpropagation) and distinguishes training versus inference workloads, noting models can reach trillions of parameters. The von Neumann memory-compute bottleneck limits central processor unit (CPU)-based DNN processing, prompting adoption of massively parallel accelerators, i.e., graphic processor units (GPUs), tensor processor units (TPUs), which is really an application-specific integrated circuit (ASIC), and neural processor units (NPUs), which handle matrix and tensor operations used in neural networks, for training and inference. Training large AI models incurs very high power and infrastructure demands (multi-MW data centers), creating significant energy and scalability challenges. Promising mitigation strategies include energy-efficient devices, near-memory and in-memory computing, edge and neuromorphic architectures, photonic interconnects, and 3D integration to reduce data movement and power.

Since the GPU-based DNN training and inference processes consume multi-MW power, Chapter 3 presents some solutions for reducing the power consumption for processing AI models. Continued transistor scaling reduces its switching energy but exponentially raises fab, design, and R&D costs, concentrating leading-edge chip production. Near-memory and 3D/2.5D heterogeneous integration reduce data movement by placing computing units close to dynamic random-access memory/high bandwidth memory (DRAM/HBM), improving latency and energy efficiency. Analog and digital in-memory computing (IMC) using static RAM (SRAM), DRAM, resistive RAM (ReRAM), phase change RAM (PCRAM), ferroelectric RAM (FeRAM), and magnetic RAM (MRAM) promise large energy gains for multiply and accumulate (MAC)-dominated DNN workloads, with tradeoffs in speed, endurance, and maturity. Spiking neural networks (SNN) and neuromorphic hardware offer orders-of-magnitude energy and latency improvements for event-driven, noisy, and edge applications, but require new device and architecture codesign.

Chapter 4 presents the next challenges for the semiconductor industry. For logic devices progress beyond planar MOSFETs to fully depleted devices (FinFET, fully depleted SOI/FD-SOI) toward gate all around (GAA) and complementary FET (CFET) is required for sub-5 nm nodes, with new channel materials (transition metal dichalcogenide (TMD)). TMDs offer a very attractive potential but faces drastic issues such as wafer-scale synthesis, contact, doping, and integration challenges. Advanced node scaling raises fab, design, and mask costs, driving disaggregation into chiplets, 2.5D/3D packaging, and heterogeneous integration to sustain performance and reduce die cost and yield risk. Si photonics and 3D integration are positioned to replace copper for intra- and inter-chip links, addressing latency and bandwidth limits imposed by AI-driven data volumes. Main flash memory architecture (NAND) and DRAM have moved to 3D architectures; emerging NVMs (spin transfer torque-STT-MRAM, FeRAM and FeFET, ReRAM, PCRAM, and electrochemical RAM-ECRAM) approach DRAM/flash performance for embedded and IMC/neuromorphic roles but remain cost- and maturity-constrained. WBG/UWBG materials (SiC, GaN, β-Ga_2O_3, diamond, and AlN) and vertical device architectures promise higher Breakdown voltages (BV) and efficiency for power/RF, while sensors will trend toward higher resolution, broader spectra, quantum sensing, and 300 mm integration for edge-AI fusion.

Chapter 5 documents the high environmental footprint of semiconductor manufacturing, highlighting large electricity, ultra-pure water use, hazardous waste, and rising greenhouse gas (GHG) emissions as node complexity increases. GHG emissions are analyzed by scopes 1–3, with scope 2 (energy) dominant and scope 1 largely driven by fluorinated process gases (perfluorocarbons (PFCs)/hydrofluorocarbons (HFCs)) and per- and polyfluoroalkyl substances (PFAS)-containing chemistries that are hard to substitute. Advanced nodes raise energy and water intensity (more process steps, masks, and tools), motivating near-term mitigation via renewable power, process optimization, gas abatement, and higher water recycling rates. Supply-chain risks stem from critical raw material concentration (notably China for Ga, Ge, and rare earth metals), driving interest in urban mining and improved waste electrical and electronic equipment (WEEE) recycling to reduce upstream scope 3 emissions. The industry also faces a severe skilled-labor shortage, requiring expanded science, technology, engineering, and mathematics (STEM) pipelines and workforce development to sustain fab expansion and decarbonization efforts.

Chapter 6 maps the highly fragmented, global semiconductor value chain – raw materials, wafers, equipment, electronic design automation/intellectual property (EDA/IP), foundries, integrated device manufacturers (IDMs), and outsourced semiconductor assembly and test (OSATs) – and contrasts operating models (IDM, fabless-foundry, OSAT) with their differing R&D and capital expenditure (CapEx) profiles. It documents geographic specialization and concentration (Taiwan: advanced foundries, OSAT; US: EDA, IP, design, equipment; Japan/Europe: equipment, materials; China: materials and OSATs), creating single-point vulnerabilities. Key risks for breaking this supply chain are geopolitical tensions, export controls, natural disasters, critical raw material dependence and skilled-labor shortages, driving policy and industry responses such as diversification, reshoring, and strategic investment to improve resilience.

Governments worldwide have launched large-scale incentives and CHIPS-style programs (the USA, EU, Japan, Korea, China, Taiwan, and others) to onshore semiconductor R&D and manufacturing, driving unprecedented greenfield foreign direct investments (FDIs) aimed at diversifying leading-edge capacity and securing supply-chain sovereignty, that are described in Chapter 7. Despite substantial public funding, building competitive leading-edge fabs remains capital-intensive and ecosystem-dependent. Announced projects could materially rebalance global capacity but face cost, talent, infrastructure, and supplier-ecosystem constraints that limit guaranteed success.

Chapter 8 presents the European semiconductor ecosystem and its vulnerabilities. Europe excels in equipment and their subsystems, materials, power/analog, sensors and RTO-led R&D, providing a competitive base for embedded and industrial semiconductor markets. The region lacks scale in fabless design, leading-edge foundries, memory production, lithography resists/masks and back end of line (BEOL) capabilities, creating strategic dependencies. European demand is dominated by automotive, industrial and healthcare embedded systems (power, analog, optoelectronics), which favor mature to advanced nodes rather than ≤7 nm leading-edge technologies. Fragmented funding, weak deep-tech venture capitals (VCs), limited fabless unicorns, and high operational costs hinder scaling of design and manufacturing capabilities for AI and cloud-AI markets. To secure sovereignty and capture value, Europe must coordinate long-term industrial strategy: scale chip design, expand pilot lines, strengthen supply-chain resilience, and target AI and edge-AI chips, as well as power chips and photonic integrated circuits.

It is also worthwhile to notice that this book uses some economic data from semiconductor analysts. Figure 2 highlights that even expert firms that are specialized in semiconductor economics and build on years of experience cannot predict the next months of market development reliably. Although forecasts sometimes get close to the real market development, deviations of more than 10% are perfectly normal. In some years, analysts were not even right about the general trend of market development. This underscores how volatile and unpredictable the semiconductor market is [249].

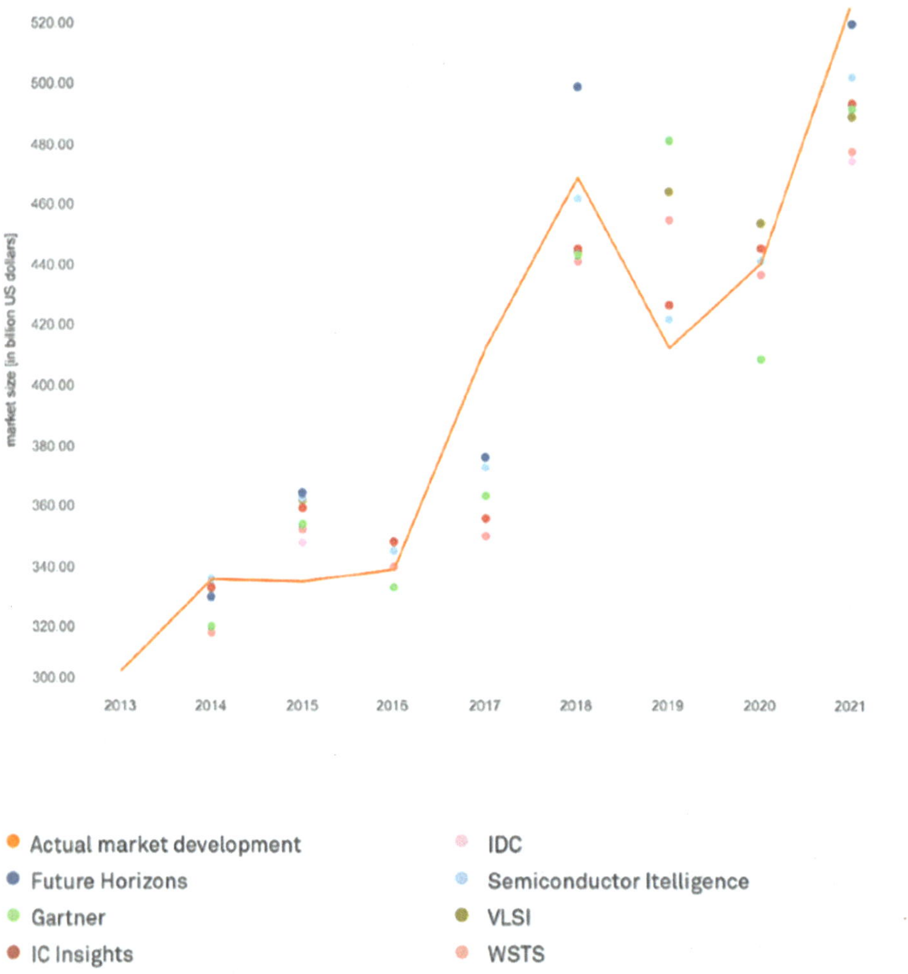

Figure 2: Comparison of 1-year market forecasts from multiple analysts and consulting firms (colored points) to the actual development of the semiconductor market (solid orange line between 2014 and 2021 [2]).

Glossary

Acronym	Definition
ADAS	Advanced driver-assistance system
ADC	Analog to digital converter
AI	Artificial intelligence
AIoT	Artificial intelligence IoT
ALU	Arithmetic logic unit
ANN	Artificial neural network
AR	Aspect ratio
AR/VR	Augmented and virtual reality
ASIC	Application-specific integrated circuit
ATP	Assembly, testing, and packaging
BEOL	Back end of line
BGA	Ball grid array
BiCMOS	Bipolar complementary MOSFET
BV	Breakdown voltage
CAGR	Compound annual growth rate
CapEx	Capital expenditures
CBA	Crossbar array
CBRAM	Conductive bridge RAM
CED	Cumulative energy demand
CFET	Complementary FET
CIS	CMOS image sensors
CNN	Convolutional neural network
COTS	Commercial off-the-shelf
CPP	Contact poly pitch
CPU	Central processor unit
CRM	Critical raw material
CSP	Chip scale package
CT	Charge trapping
CTE	Coefficient of thermal expansion
CU	Control unit
DAC	Digital to analog converter
DAO	Discrete, analog, and optoelectronics
DNN	Deep neural network
DRAM	Dynamic RAM
IDM	Integrated device manufacturer
IMC	In-memory computing
IoT	Internet of Things
IP	Intellectual property
IPCEI	Important projects on common economic interest
IPD	Interpoly dielectric
ISR	Intelligence, surveillance, and reconnaissance
LIF	Leaky integrate-and-fire
LLC	Last-level caches
LPDDR	Low-power double data rate
LRS	Low-resistance state

Acronym	Definition
ECRAM	Electrochemical RAM
EDA	Electronic design automation
EEPROM	Electrically erasable programmable read-only memory
EIC	European Innovation Council
EICF	Electronic IC
EUV	Extreme ultraviolet
EV	Electrical vehicle
FDI	Foreign direct investments
FD-SOI	Fully depleted SOI
FeCAP	Ferroelectric capacitor
FeFET	Ferroelectric FET
FEOL	Front end of line
FeRAM	Ferroelectric memories
FET	Field-effect transistor
FG	Floating gate
FLOP	Floating operation
FOAK	First-of-a-kind
FPGA	Field programmable gate arrays
FTJ	Ferroelectric tunnel junction
GAA	Gate all around
GDDR	Graphics double data rate
GDP	Gross domestic product
GHG	Greenhouse gas
GNSS	Global navigation satellite system
GPU	Graphic processor unit
GWP	Global warning potential
HBM	High bandwidth memory
HBT	Heterojunction bipolar transistor
HEMT	High electron mobility transistor
HFC	Hydrofluorocarbon
HPC	High-performance computing
HRS	High-resistance state
IBE	Ion-beam etching
IC	Integrated circuit
PCRAM	Phase-change RAM
PE	Processing element
PFAS	Per- and polyfluoroalkyl substances
PFC	Perfluorocarbon
PGM	Platinum Group Metals/Program
PIC	Photonic IC
QLC	Quadruple-level cell
RAM	Random access memory
RDL	Redistributed layers
REE	Rare earth element
ReRAM	Resistive RAM

 | https://doi.org/10.1515/9783112247952-206

MAC	Multiply and accumulate
MCM	Multichip module
MCU	Microcontroller
MEMS	Micro electromechanical systems
MFM	Metal-ferroelectric-metal stack
MIM	Metal-insulator-metal stack
MLC	Multilevel cell (2)
MMIC	Monolithic microwave integrated circuits
MOSFET	Metal oxide semiconductor FET
MRAM	Magnetic RAM
MS	EU member states
MTJ	Magnetic tunnel junction
MVM	Matrix-vector multiplication
NA	Numerical aperture
NAND	Main flash memory architecture
NMC	Near-memory computing
NPU	Neural processor unit
NVM	Nonvolatile memory
NVSRAM	Nonvolatile SRAM
OpEx	Operational expenditures
OSAT	Outsourced semiconductor assembly and test
PCB	Printed circuit board
RIE	Reactive ion etching
RTO	Research technology organization
SLC	Single-level cell
SM	Sensing margin
SME	Small and medium enterprise
SNN	Spiking neural network
SoC	System on chips
SOI	Silicon on insulator
SOT-MRAM	Spin-orbit torque MRAM
SRAM	Static RAM
STDP	Spike-timing-dependent plasticity
STEM	Science, technology, engineering, and mathematics
STI	Shallow trench isolation
STT-MRAM	Spin-transfer torque MRAM

RF	Radio frequency
RIE	Reactive ion etching
ROM	Read-only memory
RTO	Research technology organization
SNN	Spiking neural network
PCRAM	Phase-change RAM
PGM	Platinum group metals
PGM	Program
QLC	Quadruple-level cell
ReRAM	Resistive RAM
SWaP-C	Size, weight, power, and cost
TLC	Triple-level cell
TMD	Transition metal dichalcogenide
TPU	Tensor processor unit
TSV	Through-silicon vias
UPW	Ultrapure water
UWBG	Ultra-wide-bandgap materials
VC	Venture capital
WBG	Wide-bandgap materials
WEEE	Waste electrical and electronic equipment
WLCSP	Wafer-level chip-scale packaging
WSPM	Wafer start per month

1 Introduction

Nowadays, the term "semiconductors" or "chips" mainly refers to computing chips used in data centers, personal computers and cellphones, which form the basis of any artificial intelligence (AI) microprocessor and accelerator. Semiconductors are compulsory in many sectors, especially automotive [3], data and telecommunications [4–6], household appliances [7], health and wellbeing [8–11], crop and soil monitoring [12–14], electrical smart grid infrastructures [15, 16], and missile guidance systems [17]. Beyond computing chips, numerous electronic devices, such as memories, power and radiofrequency (RF) devices, sensors, and biosensors, are also based on semiconductors.

In 2025, the overall semiconductor industry revenue grew between $792 and $793 billion [18, 19], up from $630 billion in 2024, and drove around 12–15% [20] of the worldwide gross domestic product (GDP), which was valued at $117.2 trillion in 2025 [21]. Fueled by the amazing boom of AI infrastructures, it is expected to reach $975 billion or even more than $1 trillion in 2026. Annual semiconductor sales of $2 trillion seem likely by 2036 [22, 23]. In 2025, an estimated 1.05 trillion chips were sold at an average selling price (ASP) of $0.74 [22].

The semiconductor industry is notoriously cyclical and experiences brief periods of weakness (Figure 1.1). The great recession and sub-prime crisis in 2008–2009 as well as the recent COVID-19 pandemic (2019–2021) were major events that slowed down the growth of the semiconductor revenue. Taiwan – which accounts for more than half of the world's chip production – recently faced its worst drought in more than 50 years slowing down the semiconductor manufacturing, which is highly water-intensive. Taiwan was also hit in 1999 by a 7.3 magnitude earthquake that knocked out power to most of the country, including two nuclear power plants, which stopped the Taiwanese semiconductor production. Similarly, Russia's attack on Ukraine in 2022 disrupts the production of rare gases such as neon and krypton, which are essential for semiconductor production [24].

Since the COVID-19 pandemic and its associated chip shortage, the semiconductor industry revenue has increased at a very high growth rate (Figure 1.1).

The semiconductor industry has become the backbone of the modern economy and is highly strategic for national/regional security, competitiveness, and the green transition.

The chip or microelectronics industry really starts with the invention of the metal-oxide-semiconductor field-effect transistor (MOSFET) in 1933, and its first fabrication in 1960 [25]. Governed by the well-known Moore's law [26], which predicts that the number of transistors on an integrated circuit (IC) doubles approximately every 2 years, microelectronics gradually evolved into nanoelectronics by shrinking the dimensions and voltage of the transistors and electronic devices via the so-called Dennard's scaling [27]. Their miniaturization has been the key to improve their cost, en-

 | https://doi.org/10.1515/9783112247952-001

ergy consumption, and performance, and resulted in huge profit for the semiconductor industry.

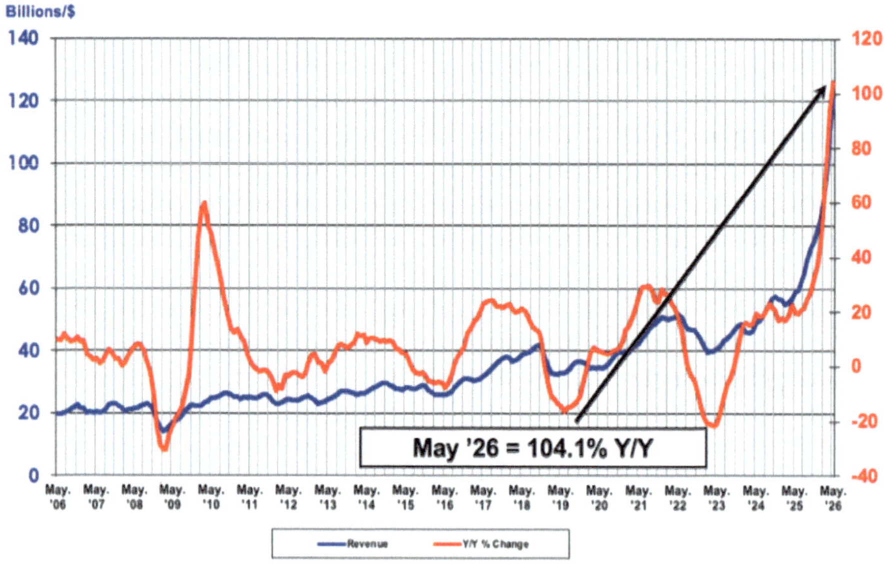

Source: WSTS

Figure 1.1: Evolution of the monthly revenue of the semiconductor industry (blue) and the year-to-year change (red) [18].

Dennard's rule is at the origin of the technology node concept. A technology node refers to a specific semiconductor manufacturing process and its design rules. Generally speaking, the smaller the technology node, the smaller the patterned feature size, the smaller the transistor, the faster the IC speed, and the more energy/power efficient it is. Historically, process node names referred to different characteristics of transistors, including gate length and first metal (M1) half-pitch. More recently, the number itself has lost its exact meaning due to various marketing moves and disagreements among foundries. The technology node concept will become less useful over time. While the metal pitch at the 5 nm node is likely to be about a quarter of 20 nm node metal pitch, the corresponding contacted polysilicon (gate) pitch (CPP) is closer to one-third of its 20 nm node value (Figure 1.2). Moreover, there is likely to be substantial divergence among manufacturers, with as much as 20 nm of variation in gate pitch within a single "node." Newer technology nodes refer only to specific generations of chips made with specific technologies. It does not correspond to gate length or M1 half-pitch. However, the naming convention is respected, which is what the semiconductor ecosystem calls technology nodes [28].

However, Dennard's law reached its fundamental physical limits at the 28–22 nm technology nodes, around 2015, because the simple scaling rule is no longer a sufficient strategy to meet future transistor density, performance, and power requirements [29], whereas the transistor density keeps following Moore's law at a slower pace [30].

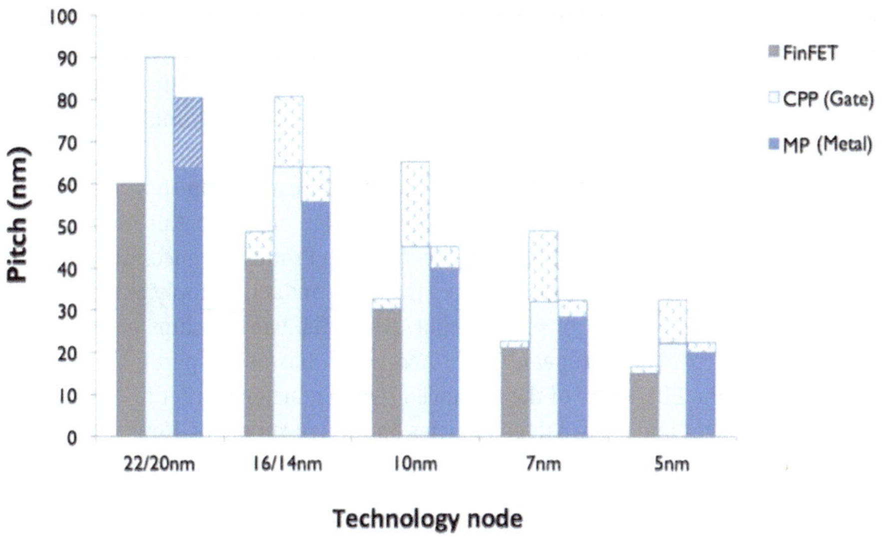

Figure 1.2: Correspondence between the "as-called" technology node and the dimensions of the transistor in case of FinFETs – Courtesy of imec.

Semiconductors strongly contribute to reduce the worldwide carbon footprint by allowing, for instance, the transition from thermal to electrical vehicles thanks to power electronic devices. However, the semiconductor industry may also have some negative impacts on the environment. Semiconductor manufacturing is resource-intensive. This includes significant consumption of electricity, water, and materials during the manufacturing process, which, as production scales up, can lead to global resource constraints and pose challenges to environmental sustainability efforts [25]. Some production processes use a wide range of various chemicals and contribute to greenhouse gas (GHG) emissions and long-term environmental contamination.

The semiconductor supply chain is a labyrinthine network of interconnected processes, each with its own set of vulnerabilities. It faces numerous risks that could significantly disrupt its operations. These risks include geopolitical tensions, weather events (drought, freeze, etc.), earthquakes, critical shortages and disruptions, and a shortage of specialized labor.

The COVID-19 pandemic was an unexpected risk for this supply chain. During this period, rapid changes in the semiconductor demand, led the shortages in microelectronic devices, especially for the automotive industry [32]. Those chip shortages cost the US economy $240 billion only in 2021 [33]. They induced an estimated loss of GDP in Germany alone of some €102 billion from 2021 to 2023 [34].

Due to the strategic importance of semiconductors, China was the first country to support its semiconductor industry with subsidies and equity investments. In 2014, the Chinese government called for about $150 billion investment from central, provincial, and municipal governments to facilitate the development of a closed semiconductor ecosystem in China [35]. This Chinese initiative results in geopolitical tensions between the USA and China, which still affect the export of semiconductors and of mineral resources today.

Since 2022, after the COVID-19 pandemic and its associated chip shortage, numerous countries or regions such as the USA, Europe, Japan, South Korea, and India launched various CHIPS (Creating Helpful Incentives to Produce Semiconductors) Acts in order to master at least numerous parts of the semiconductor supply chain [36].

In 2021 and 2022, Europe consumes about 20% of the world's chips but manufactures only 8–9% [37]. The European CHIPS Act, which has been operational in 2023, aimed to double the EU's share of the advanced semiconductor market to 20% by 2030 to mitigate shortages and reduce dependencies on foreign suppliers. However, achieving this goal faces significant challenges. Furthermore, Europe also suffers of several weaknesses in the semiconductor supply chain, which could also prevent to strengthen the overall position of its semiconductor industry.

2 Main semiconductor devices for the applicative sectors

Semiconductors have driven advances in communications, transportation, smart energy distribution, healthcare, agriculture, and military systems (Figure 2.1). Moreover, they are giving rise to new technologies that hold the promise to drastically transform the industry and society, including augmented and virtual reality, Internet of Things (IoT), automated devices, robotics, and, above all, AI.

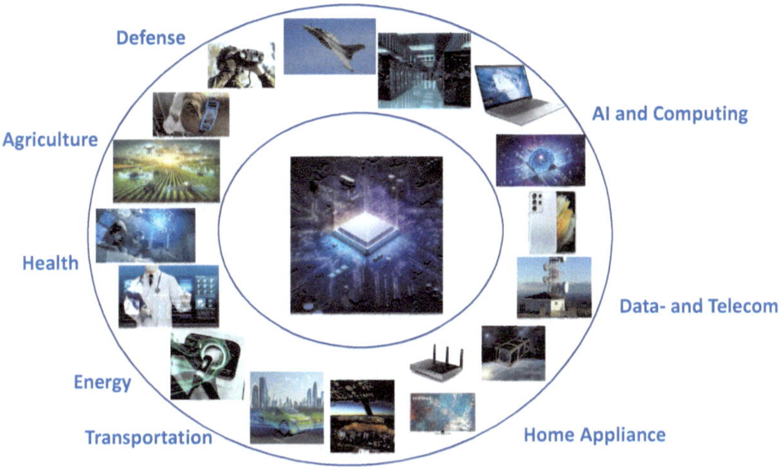

Figure 2.1: The ubiquitous semiconductor chip.

2.1 Communications/connectivity

Since 2019, 5G in the sub-6 GHz bands has been successfully deployed worldwide. One of the biggest differences between 5G and its predecessors is higher speed, reduced latency, and enhanced reliability. 5G offers theoretical speeds of up to 10 gigabits per second (Gbps), which is more than 30 times faster than the 300 Mbps theoretical maximum speed of 4G [38]. Millimeter-wave (mmWave) 5G in the frequency range 24–40 GHz has just started. These 5G technologies allow massively faster data transfer speeds thanks to greater bandwidth capacity, ultrahigh throughput, and ultralow latency [39]. Therefore, 5G technology opened up new possibilities, enabling applications such as autonomous vehicles, IoT, remote healthcare services, and smart cities.

Research activities moved toward the next generation of telecommunication systems, i.e., 6G, in order to be operational beyond 2030. One of the key aspects of 6G will be the shift toward higher spectrum portions within the 30 to 3,000 GHz range and latency

 | https://doi.org/10.1515/9783112247952-002

<1 ms, thus being at least 10 times better compared to 5G [40]. The most transformative aspect of 6G is also the fundamental integration of AI into the network architecture.

Semiconductors are applied to 5G technology and will be used for 6G technology as (Figure 2.2):

- **Base stations**: These chips manage increased data loads and handle complex signal processing, making them indispensable for efficient network operations. Due to their higher carrier mobility and breakdown voltage compared to Si-based devices, GaN on SiC, GaN on Si (for 5G), and InP (for 6G) based RF devices are or should be the most common semiconductors to be used as power amplifiers for those stations [41]. As GaN and InP-based semiconductors are more expensive than their Si counterparts, heterogeneous integration of III-V devices with Si CMOS and Si bipolar CMOS (BiCMOS) components will look compulsory [42].
- **Mobile devices**: Smartphones and other mobile devices rely on powerful semiconductors to support 5G and 6G connectivity. These chips ensure faster download and upload speeds, allowing users to experience seamless streaming and rapid data transfer. These chips include RF switches and low-noise amplifiers based on RF-SOI (silicon on insulator) and fully depleted FD-SOI, and RF filters based on surface and bulk acoustic waves [43].
- **IoT devices**: The IoT ecosystem significantly benefits from 5G technology, with semiconductors playing a key role. They enable IoT devices to communicate efficiently and reliably, supporting a network of interconnected smart devices. IoT devices require very low energy/power consumption for allowing their deployment in numerous applications.
- **AI**: One of the most important ways that AI is being used in the telecom industry is to improve network performance and customer support operations, enabling more flexible, agile, secure, and energy-efficient network management. AI can be used to analyze data from network sensors to identify potential problems before they occur. This allows telecom providers to take proactive steps to fix problems and prevent outages.

Figure 2.2 shows the basic semiconductor technologies that are mandatory for the telecommunications and data-communications sectors. Since those sectors use both digital and analog signals, their associated transistor architectures are radically different than the digital Si-based CMOS used for computing. III-V transistor architectures are designed for taking benefits of their excellent carrier mobility. BiCMOS is a combination of silicon CMOS and SiGe graded base bipolar transistors on a single substrate. BiCMOS technology uses Ge doping of the Si-heterojunction bipolar transistor base to increase switching speed and current gain with efficient power management, making it suitable for high-speed and analog applications like RF circuits.

As CMOS technology continues to scale down, allowing operation in the 10–100 GHz range, it provides the opportunity of low-cost integration of analog, digital, and RF functions on the same wafer for system-on-chip (SoC) applications. SoC circuits on

Si are prone to substrate losses and coupling, especially when RF, analog, and digital functions are integrated together into the same chip. The development of SOI technology presents the major advantage of providing high-resistivity silicon (HR-Si) substrate capabilities, which are mandatory for high-performance RF integrated circuits, leading to substantially reduced substrate RF losses and crosstalk [44].

The integration of both RF and digital circuits in the same SoC also allows getting access to AI-data processing and embedding some dedicated modules to protect the RF-SoC from cybersecurity attacks.

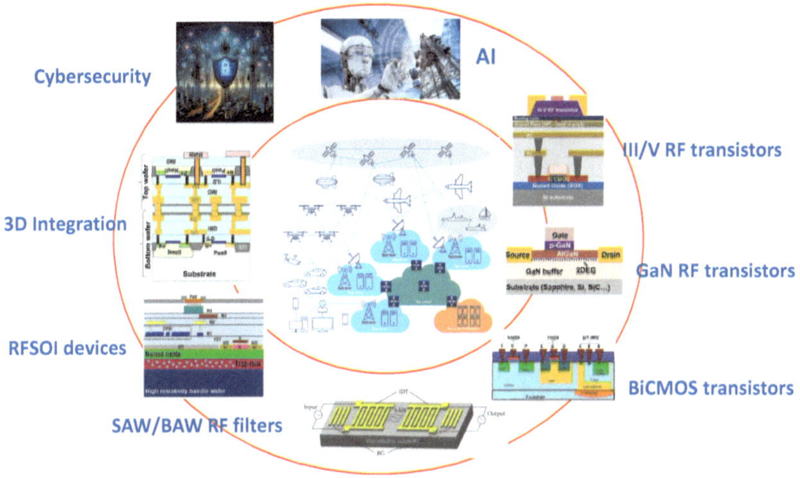

Figure 2.2: Basic semiconductor technologies in the communications domain.

2.2 Mobility

The future car will be fully connected, always online, carbon neutral, and autonomous [45]. Semiconductors are essential to mastering the challenges of decarbonization, as well as digitalization and digital transformation in the vehicle. The automotive sector is also the driver for innovation in many other transportation sectors, including aerospace, maritime, and rail, but also in other sectors, such as agriculture and industry (advanced robotics).

Therefore, semiconductors are mandatory for:

- **Electrification of vehicles** and development of powertrains for carbon-free energy carriers and powerful charging stations, including energy-efficient devices, power electronic components and systems, and energy (i.e., battery) management systems. To meet new and future power demands for EVs, electric infrastructure, and other novel electrical systems, wide-band gap (WBG) materials are being developed and introduced. Silicon carbide MOSFETs (SiC) are now available and being widely deployed, while gallium nitride (GaN) high electron mobility transistors (HEMTs) have been developed for power chargers. These WBG devices should gradually replace Si-based power devices (Figure 2.3). Since an increasing semiconductor bandgap increases the breakdown voltage of power devices, devices based on ultrawide band gap (UWBG) such as β-Ga_2O_3, AlN, and diamond are currently investigated for higher voltage applications like train and maritime [46].

- **Automated and autonomous vehicles** to make traffic more efficient and thus reduce pollution by new electronics architectures, smart and connected sensor systems, AI-based real-time software, higher performance in-vehicle controllers and networks, etc. Useful sensors are ultrasonic sensors, visible and IR sensors and cameras, LiDAR, radar, and global navigation satellite system [47].
- **Rapid advances in AI** and edge computing to operate in highly complex environments, for processing a large number of possible situations and real-time, safety-relevant decision-making.

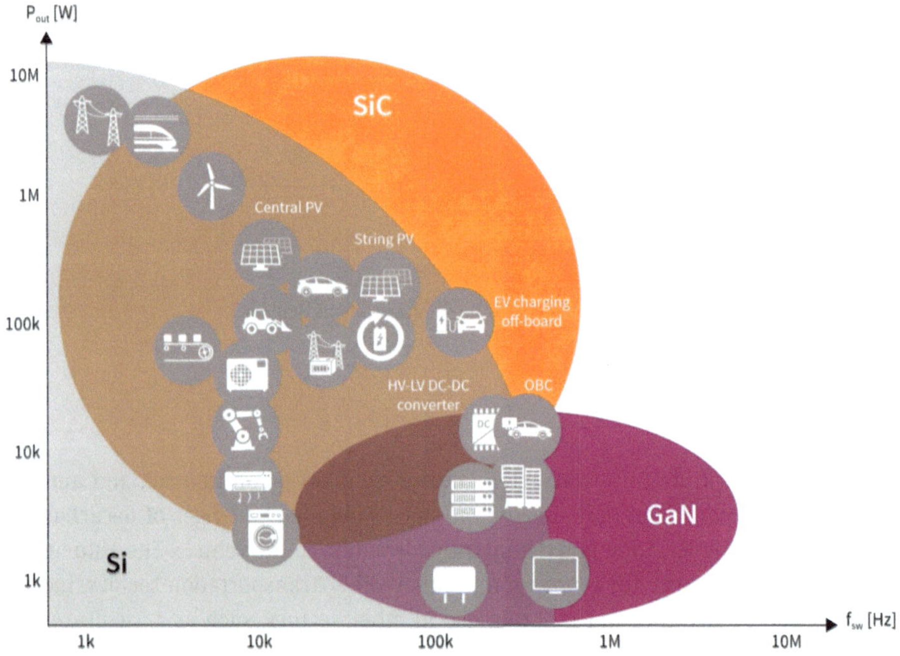

Figure 2.3: Evolution of power devices – courtesy of Infineon.

2.3 Energy

The development of energy systems is driven by action against climate change, mainly induced by greenhouse gas (GHG) emissions, booming decentralized renewable generation (solar, wind) as a substitute for fossil fuel. It is also compulsory for affording digitalization, and AI technologies, as well as cybersecurity issues (Figure 2.4). The power grid architecture developed for centralized, unidirectional, demand-driven power generation will be transformed into a smart, dynamic, and more reliable multimodal energy system architecture. It will include distributed renewable generation, energy conversion units for sector coupling, transmission and distribution grids allowing bi-directional power flow, and energy storage for all modes of energy (electric, thermal, and chemical).

Efficient energy storage and transfer can be achieved with Si power MOSFET technology coupled with the innovative digital control for power conversion. They have been improved with the adoption of new WBG devices such as SiC MOSFETs and GaN HEMTs (Figure 2.3). Indeed, thanks to advances in SiC and GaN power devices, the efficiency of system power conversion from the electrical grid to power computers can be improved by 1% to 3% from the silicon MOSFET-based solutions [48]. While it does not look like much in terms of absolute percentage, the power-loss reduction and cost savings due to the adoption of those semiconductor technologies are significant for large energy users such as large data-centers [48]. Increase in power density and reduction of losses (i.e., through exploitation of WBG and UWBG materials) and a decrease of system size by miniaturization and smart integration are forecasted in the near future [49].

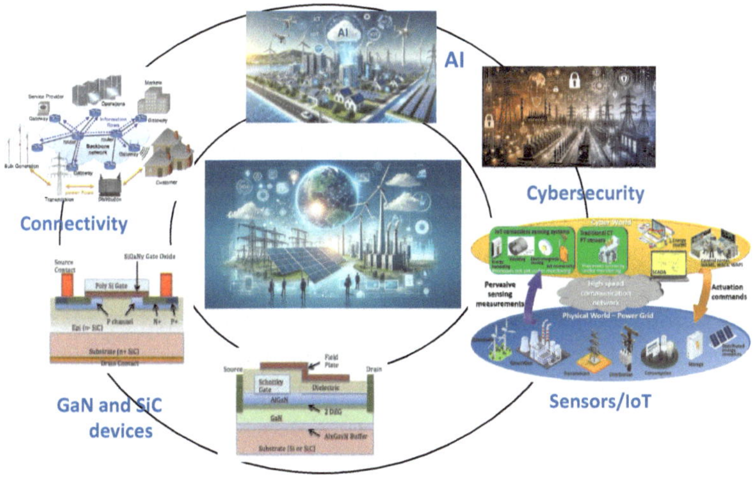

Figure 2.4: Basic technology needs in the energy domain.

Due to the complexity of the aforementioned energy supply chain, efficient electrical energy generation, transmission, and consumption require intelligent control and connection with digital technology. With the addition of microcontrollers (MCU) and wireless solutions, a high-efficiency, smart electrical grid can be implemented with advanced diagnosis and remote-control features. Sensor technologies such as radar sensing and temperature sensing can also help energy efficiency from a power consumption perspective. The sensors provide the data and information to the system that allows the MCUs to make decisions about the application, thanks to AI [50].

The use of AI in the creation and management of smart grids is helping to reshape the way energy is produced, distributed, and consumed.

2.4 Healthcare

The technologies provided by the semiconductor industry have the potential to significantly disrupt the healthcare industry. They will change the way healthcare is delivered and received, make healthcare more accessible, and improve patient outcomes. The semiconductor's ability to process information quickly, to store huge amounts of data, and to control complex systems allows improving the ways medical diagnosis, treatment, and patient care are performed (Figure 2.5).

Medical devices integrate a wide range of semiconductor-based sensor technologies, each tailored to measure specific physiological parameters. These sensors enable more accurate diagnostics, real-time monitoring, and advanced therapeutic interventions [51]. The miniaturization of semiconductor devices allowed the development of portable and wearable devices, which monitor the patient's health parameters (blood pressure, heart beat, glucose level, etc.) in real time. It also led to the deployment of implantable devices such as pacemakers, neurostimulators, and defibrillators, and of implantable pumps to deliver insulin or chemotherapy drugs with high accuracy, thus minimizing side effects [52].

Semiconductors are also at the heart of advanced diagnostic imaging technologies like magnetic resonance imaging, computed tomography scans, and ultrasound devices. Innovations in semiconductor detectors and processors have significantly improved the resolution, speed, and safety of these imaging techniques, leading to earlier and more accurate diagnoses [53].

The progress in connectivity and the pervasion of AI, both powered by advanced semiconductors, in medical devices, paves to telemedicine that allows patients to receive care remotely, reducing the need for hospital visits. Telemedicine is paramount for the ageing population and for people living in rural areas, improving the accessibility of healthcare for anybody [54].

In the era of digital healthcare, there has been an ever-lasting attempt to construct individual-centered healthcare systems for remote health monitoring, preventive medical care, long-term medical records, and user-friendly recovery support. To

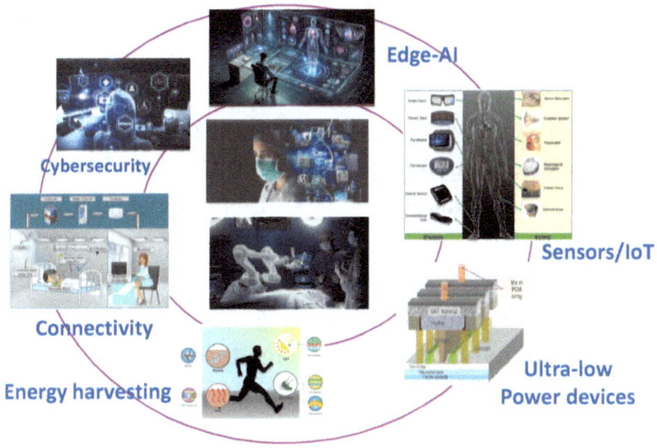

Figure 2.5: Basic technology needs in the healthcare domain.

realize this ambitious aim, the key step is to acquire enough data sheets and create digital representations of individuals' health. Therefore, powerful technologies that address personal health data sheets with a vast scale and high complexity are highly desirable.

Current research efforts focus mainly on integrating bioelectronics with specialized AI methods. For example, Machine Learning (ML) algorithms that are effective for extracting salient features, capturing spatial hierarchies, and/or analyzing time-series data have been widely integrated with healthcare bioelectronics such as point-of-care diagnostic devices, in vivo treatment arrays, on-patient recovery patches, and long-term health monitoring bioelectronics [55].

2.5 Agriculture

The fast increase in the Earth's population, reaching 9.7 billion by 2050 [56], may create a lot of issues, especially in the production of agricultural foods. Increasing food production is being hit hard by climate change in regions where food scarcity and inefficient resource management are most prevalent. The necessary acceleration in agricultural productivity growth is also being hampered by the degradation of natural resources (including soil), a reduction in biodiversity, and the spread of transboundary pests and diseases of plants and animals. Furthermore, agriculture is one of the main contributors of GHG emissions (CH_4 and NO_x, mainly). The semiconductor industry highly contributes to the current deployment of a digital agri-food system, often referred to as "Smart Agriculture," capable of having a more productive, decarbonized, and sustainable agriculture globally [57] (Figure 2.6).

Semiconductor-based IoT systems coupled to AI become very important for sustainable production and consumption of safe and healthy food, as well as for sustainable practices in agriculture, livestock, aquaculture, fisheries, and forestry. They can foster access to clean water, fertile soil, and healthy air in addition to helping fight against pests while preserving biodiversity, restoring the planet's ecosystems, and providing solutions for monitoring GHG emissions to ease decarbonization [13].

Smart agriculture includes advanced equipment, digital information technologies such as wireless connectivity, and high-performance sensors with higher spatial, spectral, and temporal resolutions. Location and position sensors, IR sensors installed in unmanned vehicles, monitor the organic matter and moisture content of soils. Electrochemical sensors trace the pH and soil nutrient levels, allowing the automated irrigation control for increasing the productivity yield, and reducing the use of pesticides and fertilizers for reducing GHG emission as well as surface and groundwater pollution [58].

Moreover, AI makes the sensors smarter and enables the sensor-based agriculture machines to be more efficient for smart agriculture practices [58].

For livestock monitoring, smart sensors can also be used to provide useful information for early detection of diseases and to control their spread, to quickly check for eventual bacterial infection, and to increase animal well-being [59].

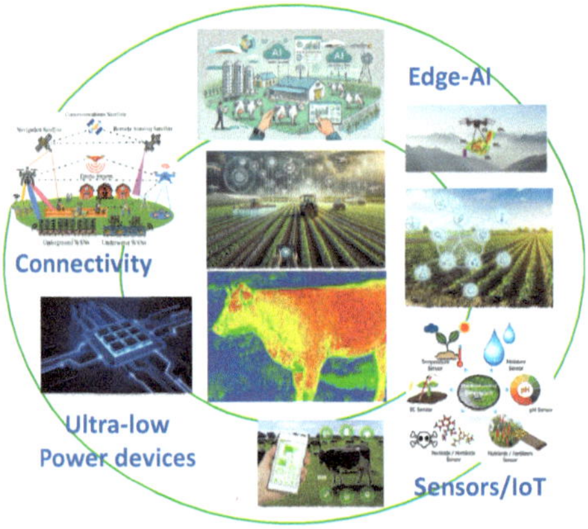

Figure 2.6: Basic technology needs in the "smart" agriculture sector.

2.6 Defense and aerospace

Fueled by the increasing complexity of contemporary warfare, where electronic systems are vital for communications, reconnaissance, and targeting, the demand for improved electronic systems is rapidly rising in military and aerospace applications. In military contexts, semiconductors are crucial because they can rapidly and efficiently process and transmit large volumes of data, which is vital for real-time data processing necessary for sound decision-making. They enable the creation of advanced radar, navigation, and communication systems with high performance, along with encryption technologies, missile guidance, weaponry, and electronic warfare. Semiconductors facilitate miniaturization, which is crucial as it allows the development of smaller, lighter military gear without sacrificing any capability (Figure 2.7).

The explosion of semiconductors has drastically changed the paradigm of the military industry. Whereas defense industries had traditionally functioned as a great engine of technological innovation and tended to transfer nonlethal technologies into the commercial realm, today the reverse is increasingly observed. Because the aerospace and defense market represents <1% of global semiconductor demand ($2 billion to $6.75 billion in 2024 [19, 60]), the defense industry thus increasingly relies on technologies originally conceived for civilian purposes, so-called commercial off-the-shelf (COTS). Civilian technologies are adapted into defense as COTS once their military applications have been highlighted. As an example, to illustrate this shift, the first drones were produced as children's toys. As the war in Ukraine has so clearly demonstrated, drones now play a vital role in battlefield management, intelligence gathering, and as offensive and defensive military platforms [61]. Therefore, nanoelectronic devices for defense are generally two or three technology generations (nodes) behind the civilian cutting-edge technologies.

Military systems require long lifetimes (20–40 years) and high reliability. Legacy microelectronics remain essential, despite obsolescence in commercial supply chains. Therefore, the military industry must have access to sources of "legacy microelectronics," or obsolete devices that are no longer made by commercial suppliers except through special arrangements.

Semiconductor uses in defense and aerospace sectors necessitate particular devices such as radiation-hardened ("rad-hard") components, RF chips, and optoelectronic technologies, along with elements that also allow reduced size, weight, power, and cost (SWaP-C) relative to civilian counterparts. SOI technology provides intrinsic radiation hardness for military and aerospace systems. SOI devices enhance electromagnetic immunity and prevent latch-up under extreme conditions [62, 63]. Moreover, SOI-based microprocessors and microcontrollers consume very low power, affording the use of batteries for numerous military devices.

GaN-HEMTs continue to outperform other devices in the field of high-power switching and RF electronics in terms of power and linearity, while InP HEMTs could be used for speed and low RF noise, and micro electro-mechanical systems (MEMS) for low RF loss switching. GaN monolithic microwave integrated circuits are firmly established as the leading solid-state technology for RF power amplifiers. GaN becomes a critical technology for defense applications, delivering a core capability for wireless communications, radar, missile seeker, and electronic warfare applications [64].

Advanced sensing technologies are compulsory in military applications in order to give enhanced situational awareness and fast decision-making in the battlespace. This superiority significantly improves defense personnel's survivability and mission success response. The modern military depends on reliable, advanced optical sensors from the IR to THz range with high sensitivity, high speed, and, sometimes, high oper-

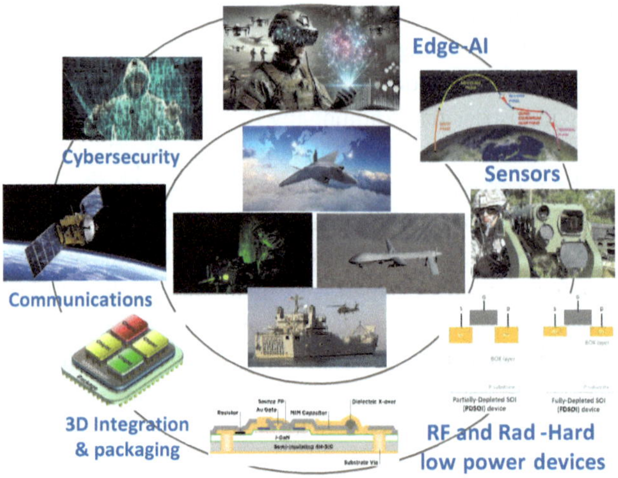

Figure 2.7: Basic technology needs in the defense sector.

ating temperature. They are, for instance, utilized in night vision systems to precision-guided munitions, intelligence, surveillance, and reconnaissance drones, command and control centers, and navigation systems for aircraft and ships. LiDAR for military and security applications includes target recognition, target location, aim point selection, tracking, and weapon guidance [65]. THz sensors are used to determine THz radiation in solid-state materials such as high explosives and biological tissues, etc., defining innovative sensing modalities (e.g., remote object and personnel identification) for defense and security [66].

In adversarial environments, intelligent sensing or IoT applications often cannot rely on connections to cloud infrastructures, as such environments frequently disrupt or block connectivity. As a result, these applications need to run their AI models locally on computing and energy resource-constrained edge devices. Given the limited resources available at the edge, it is crucial to optimize deep neural network inference performance to ensure robustness, predictability, and meet real-time requirements. Another approach is to use highly energy-efficient event-based or spiking neural networks to operate directly on the sensor output, leading to ultimately fast decision-making.

3 The AI revolution: advantages and shortcomings

All the aforementioned applicative sectors, as well as the industry, which was not presented in the previous section due to its diversity, claim access to AI models to efficiently process their own data. Indeed, data is becoming the "new oil" of the twenty-first century. In the last 15 years, the amount of data that have been created, captured, replicated, and consumed each year exponentially increases from about 2 Zeta-bytes (1 Zb = 10^{21} bytes) in 2010 to ~180 Zb in 2025 (Figure 3.1). They are expected to pursue such an impressive growth up to more than 2,000 Zb in 2035 [61]. Moreover, in 2025, 90% of these data did not arise from humans but from sensors, IoT, machines, etc., compared to 44% in 2018 [61]. For data to be processed effectively and to allow for conclusions, decision-making, and actions, it usually requires the application of large AI models. Conversely, the ongoing growth of AI, particularly with the development of generative AI models, will greatly accelerate a surge in demand for data.

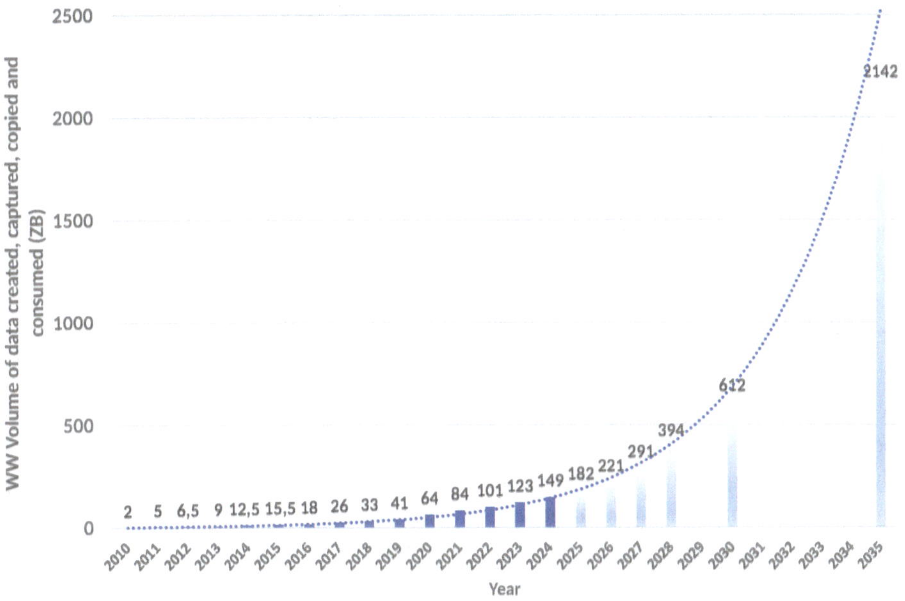

Figure 3.1: Worldwide volume of data created, captured, copied, and consumed per year from 2010 to 2035 – data from Mehonic and Kenyon [76].

AI is currently considered the best solution to efficiently and quickly analyze huge amounts of data in the era of Big Data and IoT, where digital data are perpetually generated (Figure 3.1). Therefore, AI developments, such as machine learning (ML), have been expanded widely across diverse applications in the last decade, like computer vision, healthcare, business, and finance, as well as for autonomous systems

 | https://doi.org/10.1515/9783112247952-003

(e.g., drones, autonomous cars, and robotics). Nowadays, the prominent ML algorithms employ brain-inspired computations: (i) artificial neural networks (ANNs), which include deep neural networks (DNNs); and (ii) spiking neural networks (SNNs), as shown in Figure 3.2 [68].

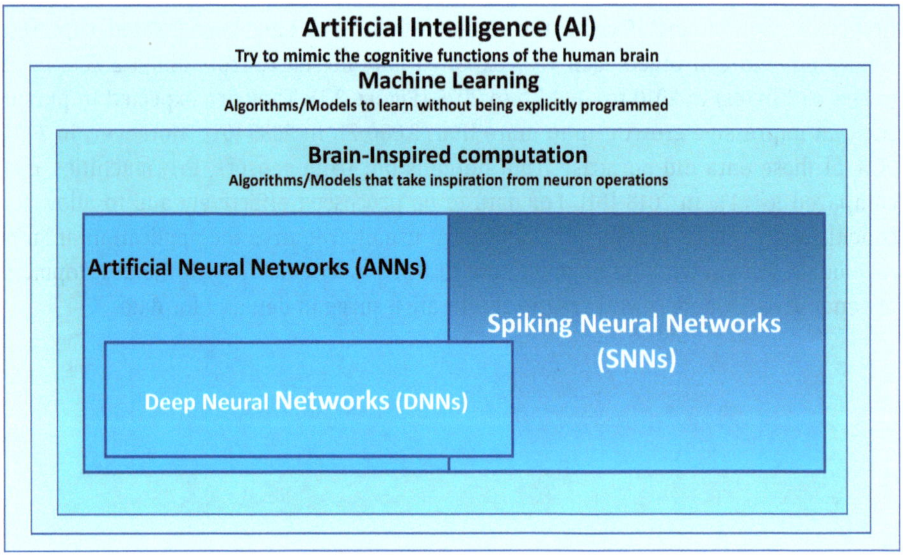

Figure 3.2: Subdivision of AI in ML, then ANNs and DNNs, and SNNs – inspired from Putra and Shafique [68].

3.1 ANN, DNN, and SNN models

A biological neuron can typically be categorized into three functionally unique components known as dendrites, soma, and axon (Figure 3.3, top). The dendrites serve as the "input component," gathering signals from neighboring neurons and sending them to the soma. The soma functions as the "central processing unit" (CPU), which executes a crucial nonlinear processing action. The axon acts as the "output component," transferring the output signal to additional neurons. ANN and DNN networks reduce complex and data-heavy problems to a set of tunable weights, typically arranged in multiple layers. ANNs and DNNs take their inspiration from the notion that a neuron's computation involves a weighted sum of the input values. These weighted sums correspond to the value scaling performed by the synapses and the combining of those values in the neuron. A nonlinear functional operation within the neuron, such as sigmoid or leaky rectified linear unit, that is performed on the combined inputs causes a neuron to generate an output only if the inputs reach some threshold:

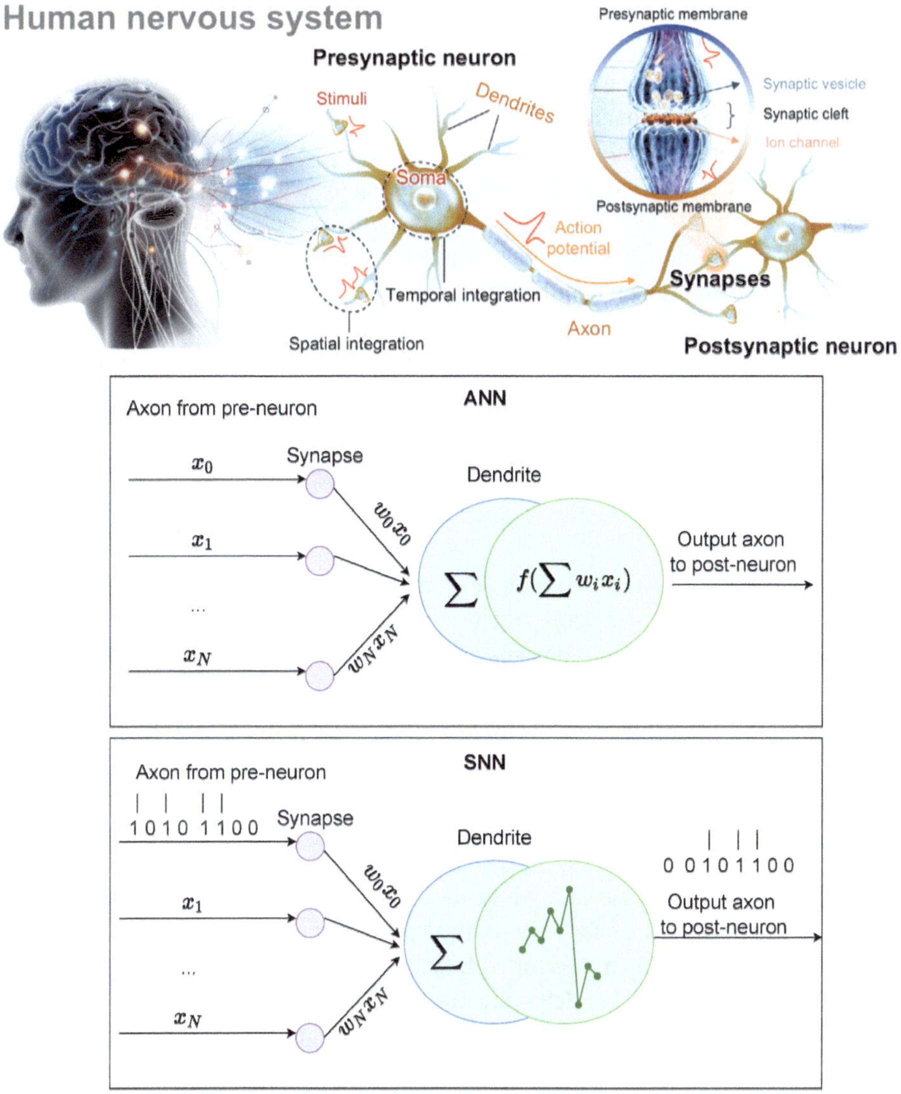

Figure 3.3: A comparison between the biological neuron [69], artificial neuron, and spiking neuron [70].

$$y = \phi\left(b + \Sigma_j x_j w_j\right) \tag{3.1}$$

where x, y, w, and b are input activation, output activation, synaptic weight, and bias, respectively, and j is the index of the input neuron. $\phi(-)$ is a nonlinear activation function, which can include rectified linear unit and various sigmoid functions [71, 72]. Neurons in ANNs communicate with each other using activations coded in high-precision and continuous values, and only propagate information in the spatial domain (i.e., layer by layer).

From eq. (3.1), it can be seen that the multiply-and-accumulate (MAC) of inputs and weights is the major operation in ANNs [70].

The biological neuronal signal is made up of brief voltage impulses known as action potentials or spikes. These impulses move along the axon and are sent to multiple postsynaptic neurons, where they generate postsynaptic potentials. When a postsynaptic neuron gets enough spikes from various presynaptic neurons within a brief time frame, its membrane potential can attain a threshold level, leading to the triggering of an action potential; this indicates that the neuron has "fired" a spike. This spike serves as the output signal of the neuron, which is conveyed to other neurons via the axons. A particularly simple model of a spiking neuron is the leaky integrate-and-fire (LIF) model, which is used in SNN artificial neurons (Figure 3.4) [70].

The LIF model can be simply described by Juarez-Lora et al. [73]:

$$\tau_m dv_m(t)/dt = -v_m + E_l + R_m I_{\text{syn}} \tag{3.2}$$

In eq. (3.2), $v_m(t)$ represents the neuron's membrane potential, E_l is the resting potential of the neuron, R_m membrane resistance, $\tau_m = R_m C_m$ is the decay time for $v_m(t)$, being C_m the neuron's membrane capacitance. I_{syn} stands for the injected current to the neuron. Each time a spike arrives at the neuron, I_{syn} increases. On the other hand, if no spikes arrive at the neuron, the current decays.

The evolution of I_{syn} can be expressed by the following equation:

$$\tau_m dI_{\text{syn}}/dt = -I_{\text{syn}} + C_m \Sigma_i^N \omega_{ij} \delta\left(t - t_i^f\right) \tag{3.3}$$

In eq. (3.3), ω_{ij} is the synapse strength value between a presynaptic, ith, neuron and a postsynaptic, jth, neuron. As for each postsynaptic neuron, there can be N presynaptic neurons connected, t_i^f is then a vector with firing times from each of the N presynaptic neurons. δ is the Kronedecker delta function, which $\delta(x) = 1$ for $x = 0$ and $\delta(x) = 0$ for $x \neq 0$. Once the neuron threshold voltage v_{th} is overpassed, the neuron spikes, emitting a pulse of magnitude v_{spike}, then the neuron resets to a reset potential $v = v_{\text{reset}}$, and it starts integrating again (Figure 3.4).

Biological neural networks in the human brain process information through distributed, parallel, and weighted update mechanisms, providing significant advantages such as high efficiency, fault tolerance, and low power consumption. However, compared with the human brain, artificial synaptic devices based on complementary metal-oxide (CMOS) semiconductors require complex interconnections of a substantial number of transistors to achieve fundamental synaptic functions. Moreover, the substantial amount of redundant data generated during computation exacerbates storage challenges and presents issues related to integration, power consumption, latency, and heat dissipation. Conversely, hierarchical biological neural networks exploit dynamic synaptic connections among neurons to facilitate efficient learning and memory. This process is governed by synaptic plasticity, which adjusts the excitatory and inhibitory chemical potentials (Figures 3.3 and 3.6). Therefore, simulating synaptic functions with a single device to achieve brain-like computation and learning capabilities is crucial for surmounting the pitfalls of von Neumann's architecture, which segregates the main memory from computing units (Figure 3.5).

Figure 3.4 shows a diagram of a computational ANN. The artificial neurons in the input layer receive some values and propagate them to the neurons in the middle layer of the network, which is called a "hidden layer." The weighted sums from one or more hidden layers are ultimately propagated to the output layer, which presents the final outputs of the network to the user. To align brain-inspired terminology with neural networks, the outputs of the neurons are often referred to as activations, and the synapses are often referred to as weights (Figure 3.4).

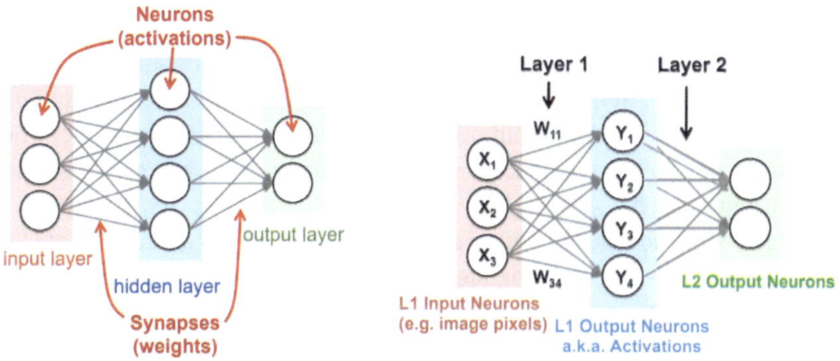

(a) Neurons and synapses (b) Compute weighted sum for each layer

Figure 3.4: Simple neural network example and terminology [74].

Furthermore, neuromorphic devices incorporating sensors or perceptual functions have garnered significant interest in various interactive domains, including machine vision, sound detection, and human-machine interactions. These interactive synaptic devices are poised to propel the advancement of next-generation neuromorphic electronics (Figure 3.5) [69].

Since they really mimic the communication and computation in the mammal brain, SNNs are designed to perfectly simulate the behavior of biological neurons. The primary difference between ANNs and SNNs is the idea that SNNs incorporate the notion of timing into their communication. Spiking neurons accumulate information across time from connected (presynaptic) neurons in the form of a membrane potential. Once a neuron's membrane potential surpasses a threshold value, it fires a binary spike to all of its outgoing (postsynaptic) connections. Spikes have been theoretically

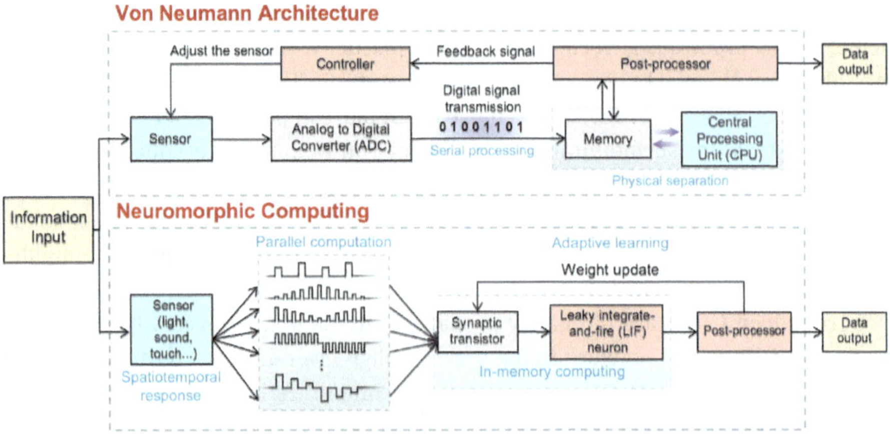

Figure 3.5: Information processing workflows: von Neumann architecture versus neuromorphic computing [69].

demonstrated to contain more information than rate-based representations of information (such as in ANNs), despite being both binary and sparse in time. In addition, compared to DNNs, SNNs should exhibit better energy efficiency, a lower latency, the ability to process noisy and dynamic data, and the potential for more robust and fault-tolerant computing [75].

Furthermore, SNNs possess the ability to effectively learn from unlabeled data by utilizing biologically plausible learning mechanisms such as spike-timing-dependent plasticity, thereby providing unsupervised learning features for systems based on these AI models, as shown in Figure 3.6. Labeled data is raw data that has been assigned labels to add context or meaning, which is used to train machine learning models in supervised learning.

Unsupervised learning, which involves using data that is not labeled, proves advantageous for numerous practical uses, such as autonomous agents. This method eliminates the need for expensive data labeling and allows for effective online learning, making systems based on SNNs adaptable to various working conditions. These benefits render SNNs ideal for tackling ML challenges on embedded platforms that have limitations in terms of resources and energy.

During the training process, SNNs execute a forward pass while applying a biologically plausible learning method, like spike-timing-dependent plasticity, to modify their weights. In the inference stage, the SNN carries out a forward pass with the adjusted weights, leading to significantly lower costs in memory access and computational energy compared to the training phase. The most advanced SNNs generally achieve better accuracy; however, this comes with an increase in model size, resulting in greater demands for memory access and computation, which leads to higher power and energy usage. Therefore, utilizing SNNs in embedded applications is highly

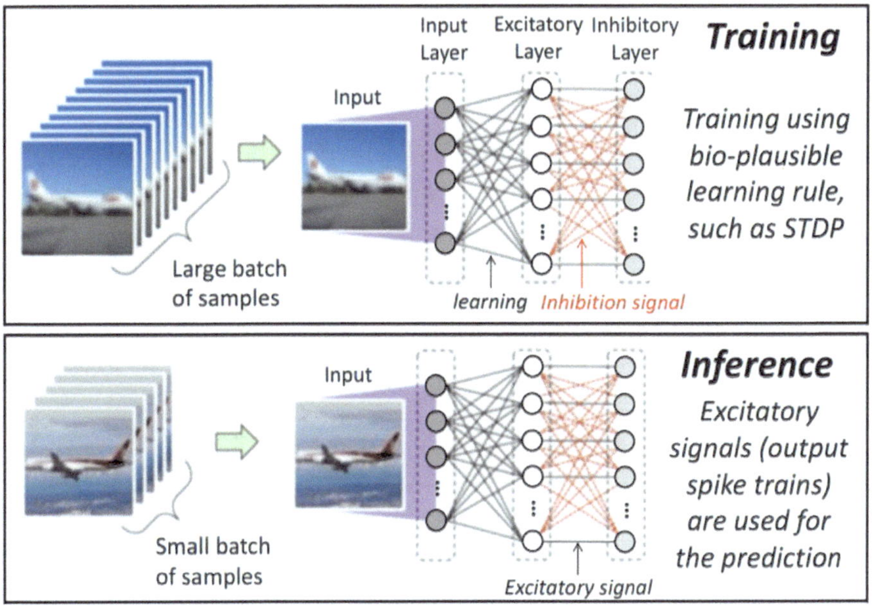

Figure 3.6: Training and inference phases in an SNN model [68].

sought after since they can offer improved services, such as real-time responsiveness, along with enhanced privacy and security. Nonetheless, embedded applications often have strict constraints concerning memory and energy, creating difficulties in efficiently executing SNN processing.

3.2 Computing with ANN and DNN models

ANNs played a vital role in AI deep learning over the past decades. The impressive AI revolution has been achieved through the use of large DNNs trained on tremendous amounts of accessible data, and innovations in computing hardware such as large-scale graphics processor unit (GPU) accelerators (Figure 3.7). Practical limitations of ANNs are that their low power/energy efficiency. Furthermore, they are not very good at processing dynamic and noisy data, which present drastic challenges toward their integration into edge computing systems.

A majority of AI systems utilize ANNs to enable computers to handle data similarly to the manner in which the human brain transmits signals and obtains information. Comparable to the human brain's complex web of neurons, ANNs function as simplified artificial neurons that act as computational units, linked through synapses that serve as memory components, collaborating to address a problem (Figure 3.4).

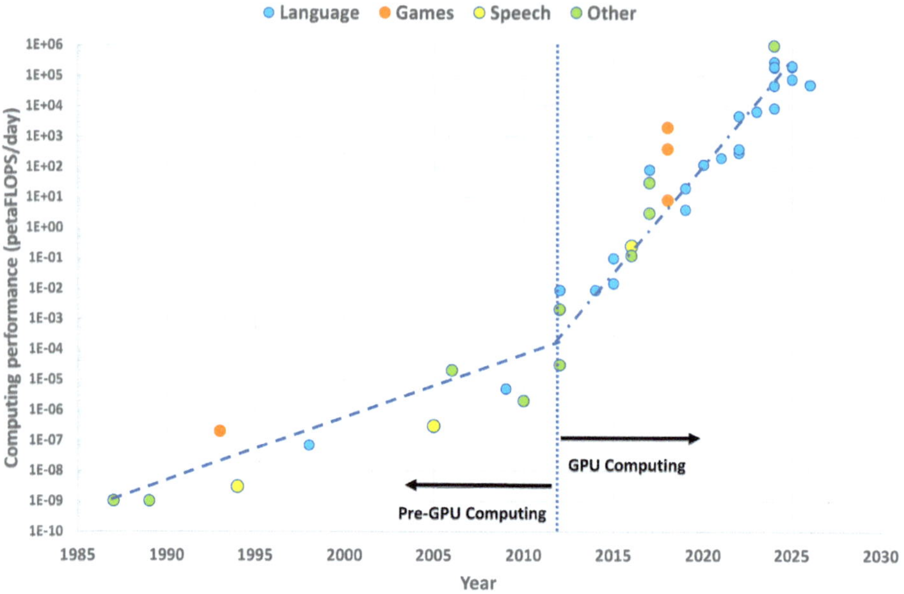

Figure 3.7: Increase in computing power demands over the past four decades expressed in petaFLOps (10^{15} floating point operation) per day – updated from Wyon and de Voorde [1] and Mehonic and Kenyon [76].

The AI revolution, which leads to natural language processing, computer vision, image recognition, healthcare, etc., has been enabled by the rise of DNNs. These models learn how to respond to new queries after being trained using very large datasets with the right answers to very similar queries. A DNN is an ANN with multiple hidden layers of units between the input and output layers. Each layer of neurons drives those in the next layer through their synaptic weight connections. DNNs can contain up to more

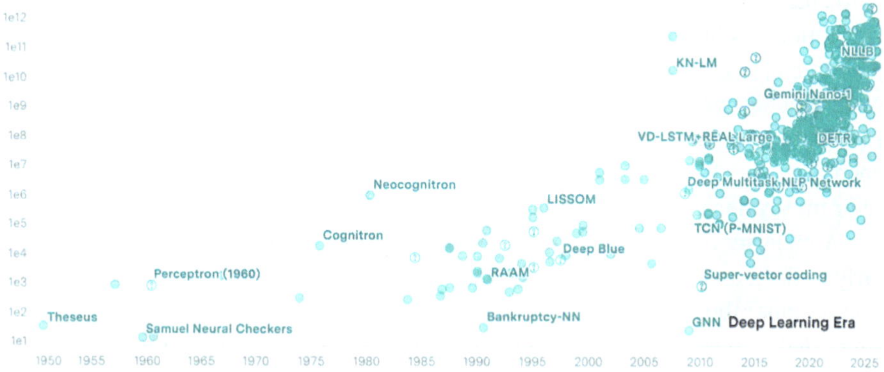

Figure 3.8: Evolution of the number of trainable DNN parameters with time [78].

than trillions of parameters (Figure 3.8). For instance, GPT-4 has 1.8 trillion parameters [77].

DNNs undergo training to establish the synaptic weights for every layer of neurons by utilizing backpropagation techniques that modify these weights with large datasets. After the training phase of DNNs is completed, they are capable of responding to inquiries in what is referred to as the inference phase. To enhance their precision, DNN models can be refined further by updating the synaptic weights with fresh data generated from various queries.

Traditional computing in laptops and desktops performs Boolean algebra and arithmetic under the Von Neumann architecture that separates the CPU and the main memory, consisting in dynamic random-access memory (DRAM) physically. During the computing operation, digital data (bits) are transferred back and forth from the DRAM to the CPU by metal interconnects. After retrieving data from the memory, the computing operations are performed in the CPU. To achieve high computing capability, an extremely large number of transistors/processors (cores) must constitute a single CPU (Figure 3.9).

The huge number of parameters to be trained and inferenced using large ANN and DNN models (Figure 3.8) pressures the computing performance of microprocessor CPUs. The requested computing performance, expressed in floating operations (FLOP), currently increases by 4.3× per year, to reach an outstanding computing performance of 1,926 FLOPs nowadays (Figure 3.10).

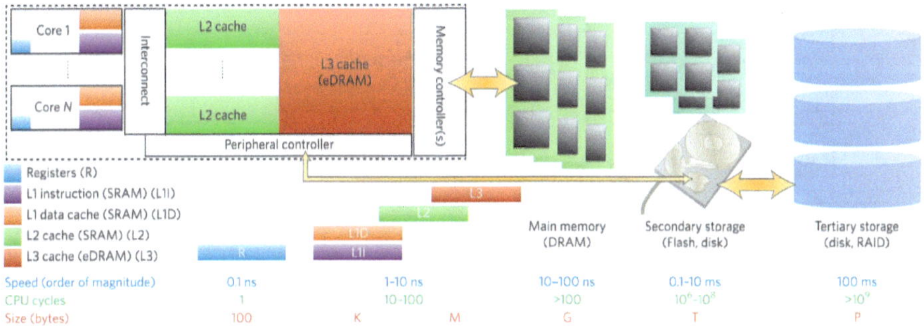

Figure 3.9: Conventional von Neumann architecture – physical distribution and hierarchy of memory in a computer. The lower panel shows speed (as order of magnitude), number of processor (CPU) cycles needed to access the memory, and size of the different memories [79].

To enhance the computational performance of a computer, two general approaches are considered: (i) increasing the transistor density (Moore's law) of the CPU, and (ii) increasing the data transfer rate between the CPU and the memory. However, boosting the speed of data exchange between the CPU and the DRAM array results in signif-

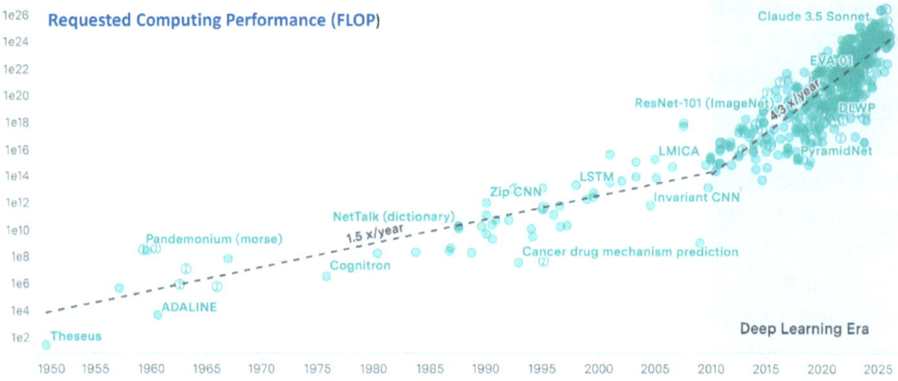

Figure 3.10: Evolution of requested computing performance for training DNN parameters with time [78].

icantly higher energy usage and extended processing times [80], which greatly restricts the applicability of CPUs for handling extremely large DNNs.

A CPU consists of three main blocks: (i) the arithmetic logic unit, which performs arithmetic and logical operations; (ii) the control unit, which extracts instructions from memory and decodes and executes them, calling on the arithmetic logic unit when necessary; and (iii) the memory management unit, which translates logical addresses into physical RAM addresses. Now, the CPU comes with single and multicore variants (Figures 3.9 and 3.11). Multicore means having more than one processor unit

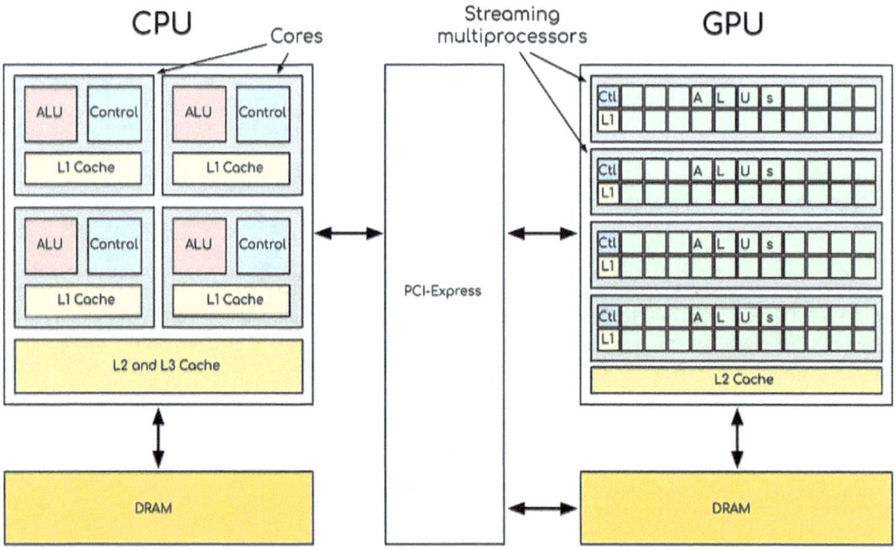

Figure 3.11: CPU and GPU architectures [82].

working side-by-side, which means that the CPU can manage more than one instruction simultaneously, drastically improving its performance.

A GPU is composed of hundreds of cores that can handle thousands of software threads simultaneously. The ability of a GPU with 100+ cores to process thousands of threads can accelerate some software by 100× over a CPU alone. Hence, GPUs are designed for maximizing parallel processing throughput and computational density. A tensor processing unit (TPU) is a custom-built integrated circuit developed specifically for ML [81].

Faster AI-related computing of DNNs was achieved with computing blocks that are designed for rapid parallel processing, such as GPUs, TPUs, neural processing units, and other AI-dedicated chips. GPUs, which were initially designed for gaming products, are now especially designed and produced for training DNNs in the cloud, thanks to dedicated data servers, in order to process massive datasets in parallel. Before the advent of GPUs for AI training, around 2012, computing power demand for training DNNs doubled every 24 months, following the miniaturization trend of transistors. Since 2012, doubling computing power has been shortened to approximately every 2 to 4 months due to the development of powerful GPUs, which allows strongly reducing the compute time for training DNNs compared to CPUs (Figure 3.7) [76].

To cope with the ever-increasing number of parameters to be trained in large DNN models (Figure 3.8) and for not spending too much time training and inferencing such models, the computing speed of AI accelerators has continuously increased during the last two decades (Figure 3.12). The computing speed only grew 1.6 times per year to currently reach 4 petaFLOPs/s. However, such an enhancement does not comply with the requested computing power that should increase by 4.3× per year (Figure 3.10).

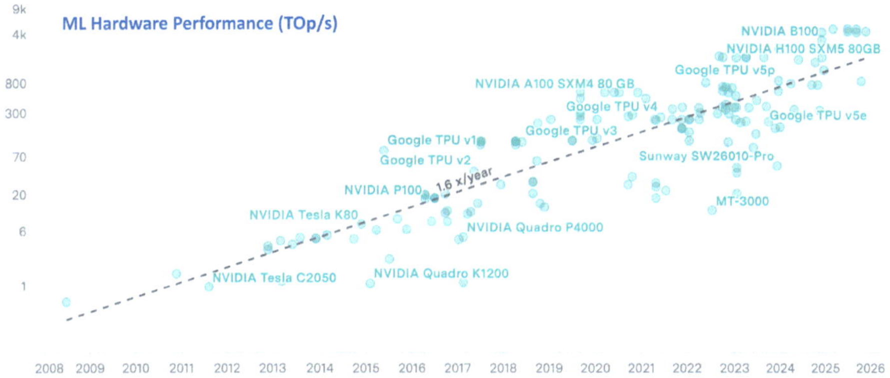

Figure 3.12: Evolution of requested computing speed for training DNN parameters with time [78].

3.2.1 The power wall

However, training large DNNs is time-consuming and energy-intensive. A single GPU can consume around 1,000 W in the case of the most advanced and most powerful chip. Large DNNs, such as GPT-4, are trained using 25,000 GPUs with a power consumption of around 10 MW. GPUs and high-bandwidth memories (HBMs) consume around 50% of an AI-based data center's power consumption, while CPUs consume roughly 10%. Additional power demand from equipment operating in the data server alongside the GPUs, HBMs, and CPUs used to train these models, such as networking equipment and switches, as well as the power requested for cooling the data server infrastructure, must also be added to the power consumption of GPUs. Adding all the power demands leads to a total rated power of the data server used to train GPT-4 of around 22 MW. Estimating that GPT-4 was trained for around 14 weeks, this results in a training energy demand of around 42.4 gigawatt hours (GWh), or around 0.43 GWh per day of training [83].

Higher operating temperatures pose multiple challenges for robust device operation: (i) high junction temperatures in Si will result in thermal runaway leading to catastrophic failure, even melting on occasions; and (ii) the device performance will continuously degrade with an increase in temperature [84].

The main parameter used for designing a component's cooling system is known as thermal design power (TDP). This parameter is used as a steady-state power target that the cooling solution must be able to cool. TDP is not the maximum instantaneous power that can be dissipated by the component, but is the "average" power that needs to be cooled while running a typical real-world stress application [85]. Since heat is a consequence of dissipated power, TDP indicates the power a chip consumes during a sustained, long operation. A chip may consume higher power than TDP for a short period of time, as long as it is still within its thermal and power envelopes. Figure 3.13 shows the evolution of the GPU computing speed, the number of constitutive transistors, and the corresponding TDP with time.

The so-called "power wall" for GPUs that are used for GenAI has been defined at a TDP of 1,000 W [507]. This limit has been reached by the GPU around 2024 (Figure 3.13). Moreover, the TDP of next-generation GPUs is expected to reach more than 4,000 W in 2027, which will induce more issues with the cooling of AI data servers. Figure 3.14 shows the very impressive cooling system required for cooling one GPU and its main memory.

Therefore, due to the thermal dissipation they have to endure, the expected lifetime of GPUs is no longer than 5 to 6 years. The GPU's short lifetime implies that data servers have to change their most important constitutive devices roughly every 5 years, which represents a quite complex task.

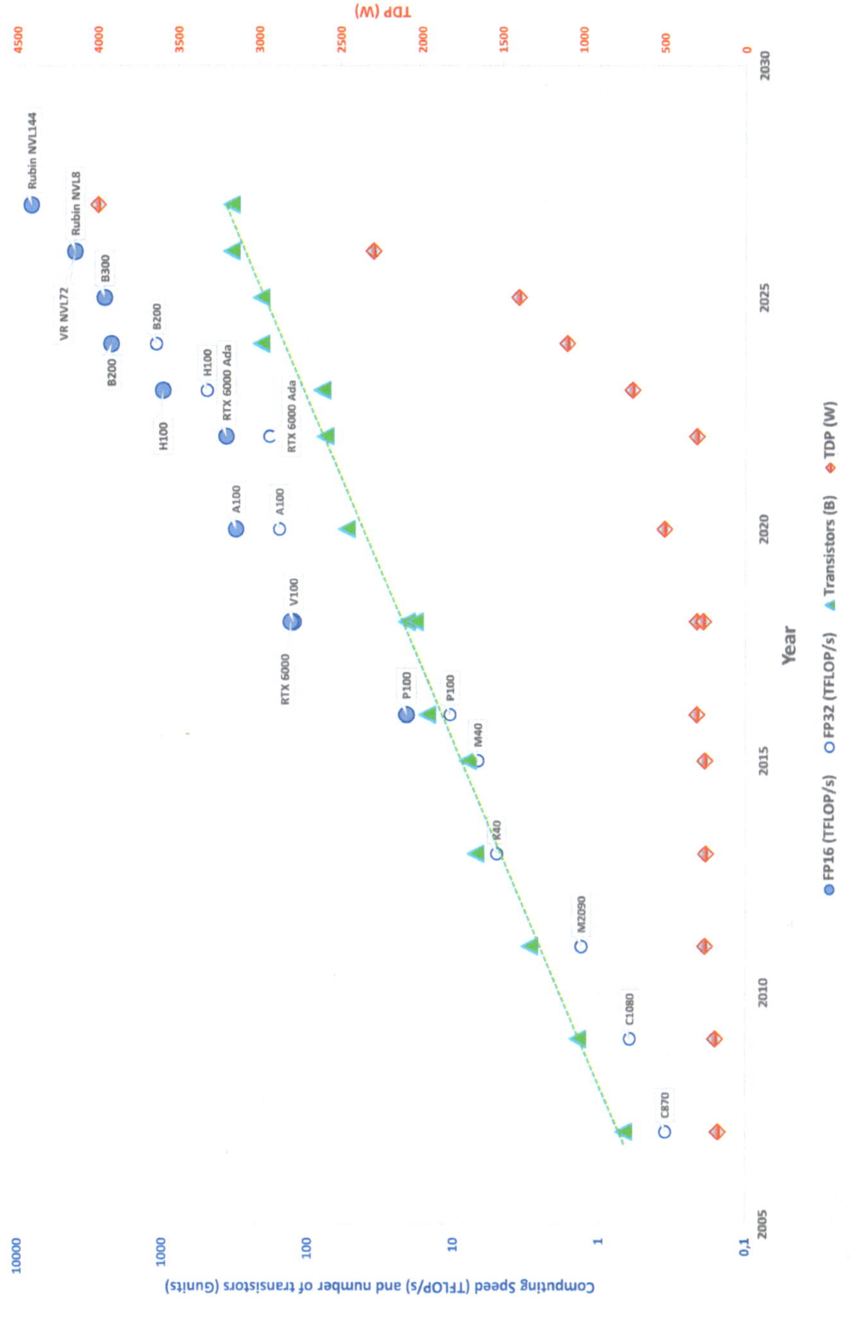

Figure 3.13: Evolution of NVIDIA GPU computing speed and their number of transistors (Log Scale) and the corresponding TDP (linear scale) – updated data from Hanindhito et al. [86].

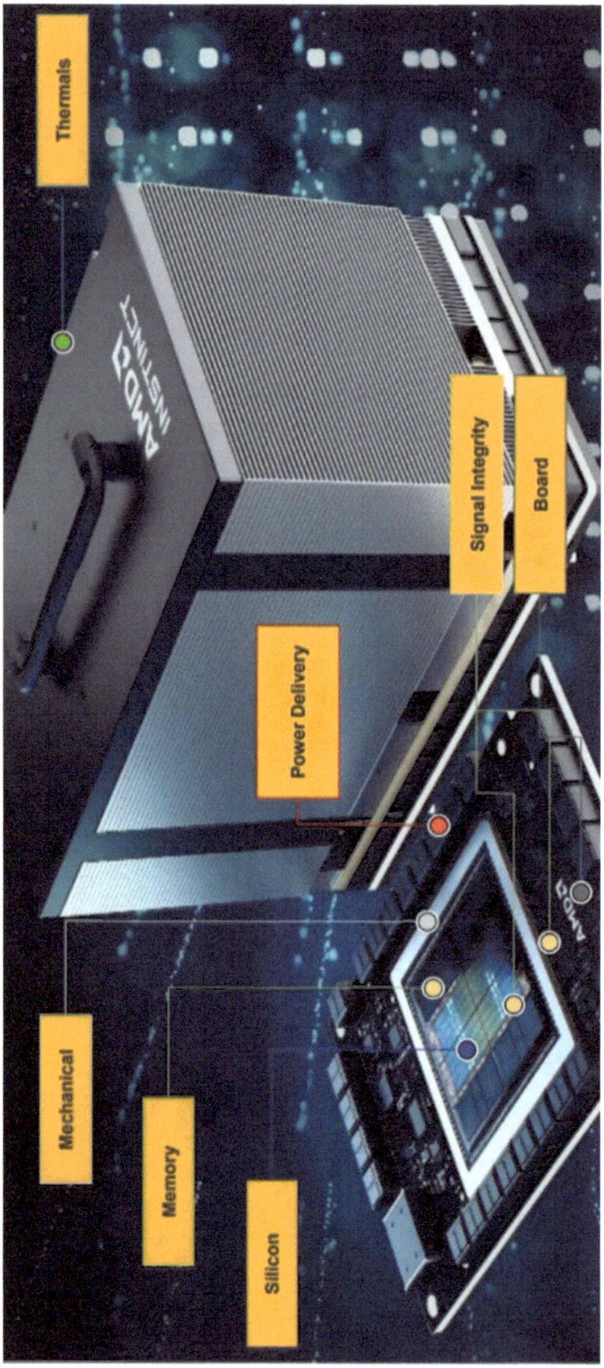

Figure 3.14: Hardware of a GPU and its main memory with the system for liquid cooling this computing device [87].

3.2.2 The memory wall or bottleneck

Currently, the processor (logic) side gets most of the attention from enterprise leaders, investors, and politicians for its contribution to AI. The rapid growth of AI workloads and other data-intensive applications has placed significant demand on data center infrastructures. The amount of compute power needed to train large language models has grown at a rate of 4.3× per year in the 2010–2025 timeframe. This exponential trend has been the main driver for AI accelerators that aim to increase the peak power of computing hardware, often at the expense of simplifying other parts, such as the memory hierarchy.

However, these trends miss an emerging challenge with training and serving these models: memory and communication bottlenecks. In fact, several AI applications are becoming bottlenecked by intra-chip, inter-chip, and communication across/to AI accelerators rather than compute [88].

Even though the compute/floating-point operations per second (FLOP/s) needed to train these recent models have only increased by a factor of 1.6× per year in the 2010–2025 timeframe (Figure 3.12), compute is not necessarily the bottleneck. First, the huge AI model (such as large language models) sizes have scaled at a rate of 410× per two years in that timeframe as well (Figure 3.8), exceeding the memory available on a single chip. Second, even when the model fits within a single chip, intra-chip memory still transfers from/to registers, Level 2 cache, global memory, etc. (Figure 3.9), and are increasingly becoming the bottleneck [88].

Moreover, the computing speed and performance are limited by the memory bandwidth. The memory bandwidth is the rate at which data can be read from or stored into a semiconductor memory by a processor. Memory bandwidth is usually expressed in units of bytes/second, in which 1 byte equals 8 bits. Increasing the computing speed is meaningless without increasing the memory bandwidth. Therefore, the overall computing speed of the system is limited by the memory bandwidth, which is often referred to as the memory bottleneck.

The memory bandwidth of AI accelerators for training AI models has increased in the last two decades but only at a rate of 1.1× per year (Figure 3.15), while the computing speed grows at a rate of 1.6× during the same period (Figure 3.12).

As shown in Figure 3.13, over the 2007–2027 timeframe, AI-server-dedicated NVIDIA GPUs computing speed expressed in FLOP/s has been scaled and is expected to scale at 2.45× per 2 years, outpacing the growth of DRAM storage capacity and memory bandwidth, which only scaled at 1.92 times every 2 years. This disparity has made memory, rather than compute, increasingly become a bottleneck, even for cases when the model can fit within a single chip.

Hence, in allowing the handling of extensive data collections and facilitating instantaneous processing, memory technology is fundamental to the implementation of AI models. For the foreseeable future, the role of memory in AI infrastructure – and

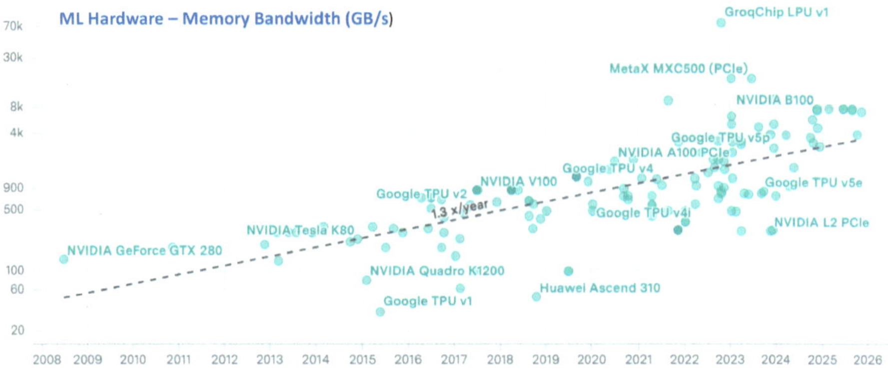

Figure 3.15: Evolution of the memory bandwidth of AI accelerators with time [78].

the technological innovations driving its advancement – and its impact on data center efficiency and sustainability will be paramount.

The memory wall refers to the increasing gap between the processor speed mandatory for processing large AI models and memory bandwidth, where the rate of improvement in processor performance outpaces the rate of improvement in memory performance due to limited inputs/outputs (I/O) and decreasing signal integrity. This disparity limits the overall system performance, as the processor spends more time waiting for data from memory, leading to a bottleneck. The memory wall or bottleneck is a significant challenge in performance-intensive computing, such as AI and HPC applications. As the number of parameters in AI models continuously increases with time (Figure 3.8), the memory bandwidth and storage capacity of AI chips become insufficient for fast processing of data using such huge AI models (Figure 3.16).

Memory plays an essential part in the efficiency and effectiveness of AI and data-heavy tasks, acting as a key enabler for advanced applications. AI models require large-scale data processing, and the performance of these workloads is directly linked to memory attributes such as bandwidth, capacity, and latency. HBM allows for faster data movement between memory and processing units, boosting computational speed, while higher memory capacity ensures that vast datasets can be stored and processed without CPU offloading and delays associated with CPU-to-GPU or GPU-to-GPU data transfers, which is critical for AI tasks. Low latency further enhances performance by reducing the time it takes to access data, speeding up real-time computations and decision-making, and improving the user experience in the case of inferencing.

In real-time applications, memory technologies are essential for handling and processing large datasets. These workloads demand high memory capacity and high bandwidth to ensure that vast amounts of data are efficiently processed and readily available. Efficient memory usage enables faster model training and inference, unlocking the full potential of AI models and allowing businesses to analyze data and

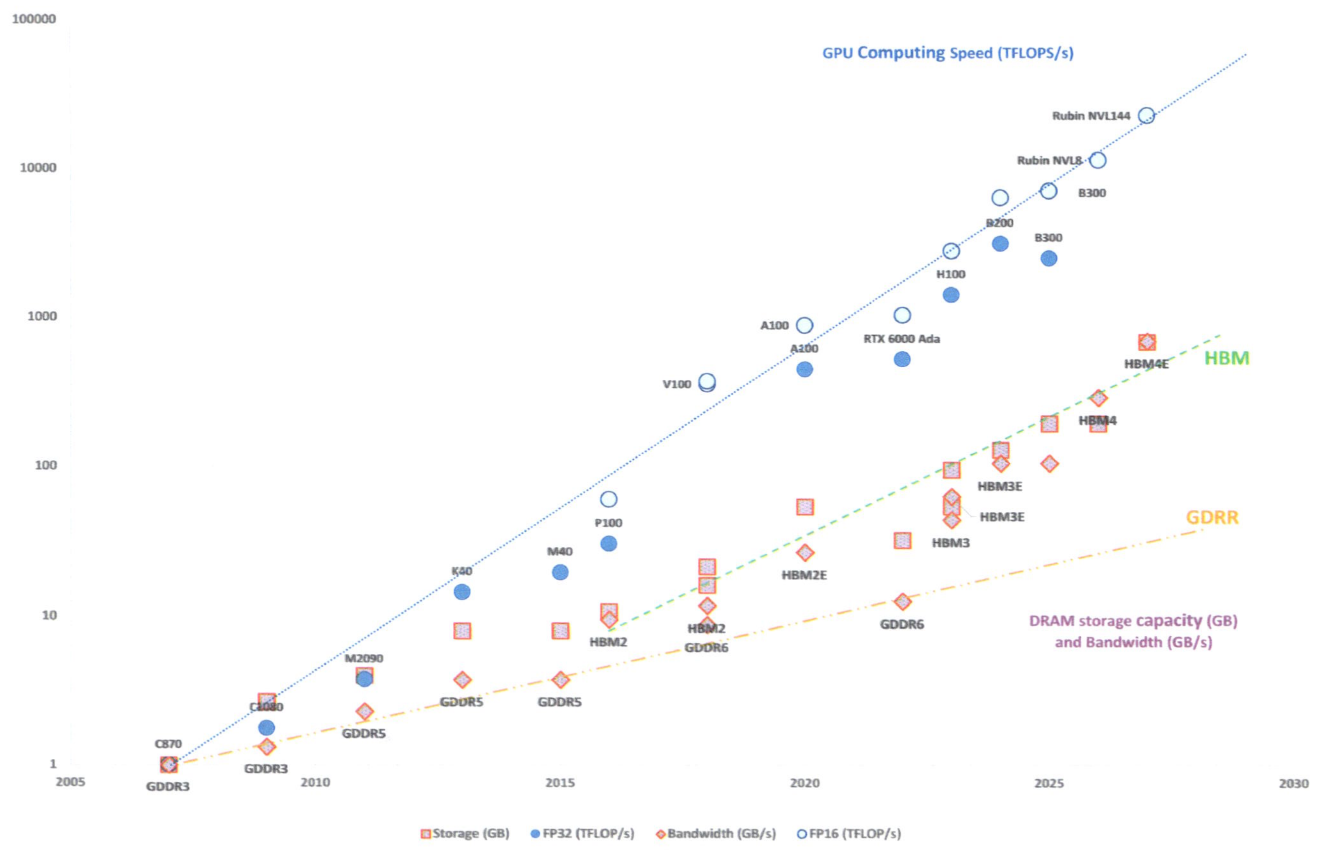

Figure 3.16: Evolution of the computing speed for training AI models using NVIDIA GPUs and the memory capacity and bandwidth for processing this data. The data has been normalized to 1 at the 90 nm node, i.e., in 2007 – data updated from Hanindhito et al. [86].

make real-time decisions. Ultimately, memory technologies are crucial in driving AI innovation and maximizing infrastructure efficiency [89].

As depicted in Figure 3.9, three main memory classes are compulsory for data processing and storing: SRAM, DRAM, and NAND flash memories.

3.2.3 SRAM performance

SRAM is widely used as the on-chip volatile cache for microprocessors, including CPU/GPU (Figure 3.9) and domain-specific accelerators such as TPUs.

SRAM is still irreplaceable owing to its sub-nanosecond access speed and unlimited endurance. Depending on the applications, SRAM's bit cell design features high-density or high-performance variants (mainly by sizing the number of fins in the FinFET era) [90]. Traditional SRAM is typically implemented as 6-transistor (6T) bitcells based on a cross-coupled inverter pair, which are heavily optimized for both power and performance for robust static data storage. However, this results in a substantial area footprint and suffers from relatively high static power consumption [91].

SRAM is made up exclusively of field-effect transistors (FETs) and lacks a specific storage component. Nevertheless, the electric charge, which is trapped within the confines of the energy barriers created by the FET channels and the gate dielectrics, can be viewed as the component for bit storage. This electric charge requires continuous replenishment from an outside power source, which means that SRAM is a highly volatile memory that must always be coupled to a power supply.

High-speed SRAMs create local cache memories inside the processing device. When the processor first requests data from the DRAM, a copy of that data is stored in the processor's cache. If the processor subsequently wishes to reaccess the same data, it uses its local copy, which can be accessed much faster. Memory cache plays a vital role in contemporary computing, serving as a link between the fast-processing power of CPUs and the more sluggish main memory. By keeping commonly accessed information, caches greatly enhance system performance, speed, and efficiency for various applications. With continual advancements in technology, creating new caching solutions will continue to be essential for fulfilling the requirements of future computing systems.

It's common to employ multiple levels of cache inside the computing device. These are called Level 1 (L1), Level 2 (L2), and Level 3 (L3) (Figure 3.9). The first cache level has the smallest storage capacity but the highest access speed, with each subsequent level having a higher capacity and a lower access speed. Assuming a 1-GHz system clock and DRAMs, it takes only 1.8 ns for the processor to access its L1 cache, 6.4 ns to access the L2 cache, and 26 ns to access the L3 cache. Accessing the first in a series of data words from the external DRAMs takes a whopping 70 ns [92].

As the process technologies used to develop integrated circuits have evolved to create smaller and smaller structures (Moore's law), most devices like SRAMs have become faster and faster. Their storage capacity also increased for technology nodes higher than 10–20 nm. However, the SRAM storage capacity is nearly constant for technology nodes lower than 10 nm (Figure 3.17). This slowing in SRAM storage density with the technology node does not impact the high-speed L1 cache since its needs in terms of storage capacity are rather limited to some hundreds of kilobytes (kB – 1 Byte = 8 bits) (Figure 3.17). But it could impact the SRAM L2 cache, which requires hundreds of MB (Figure 3.17).

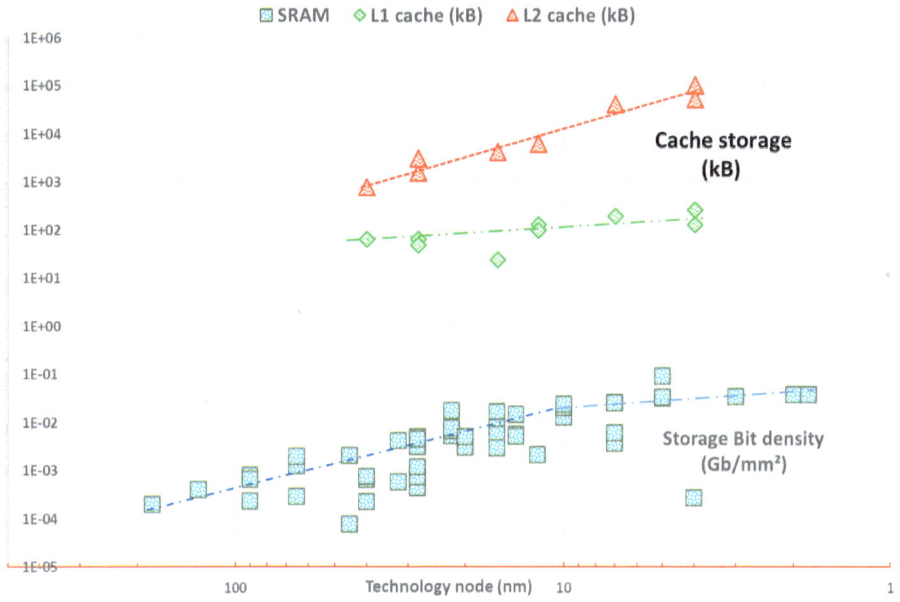

Figure 3.17: Evolution of the SRAM storage bit density, and the storage capacity of SRAM-based L1 and L2 caches in NVIDIA GPUs with the technology node [86, 93–115].

SRAM modules offer numerous benefits, such as high speed, low dynamic power, and reliability, and are compulsory for a wide range of applications. However, owing to their highly volatile nature, these memory modules face the challenge of static power loss during the idle and hold modes of operation.

One of the promising solutions might be a nonvolatile SRAM (NVSRAM) cell, which eliminates static power loss by storing data within the bit cell in a nonvolatile (NV) device. NVSRAMs store data present on the SRAM coupled inverter in the NV devices during standby/idle mode. The data stored in NV devices during standby/idle mode results in zero static power loss. When power is restored for the multiple read and write mode operations, the data stored in the NV devices are restored back to the SRAM coupled inverter. In addition to the store and restore operations, NVSRAM works as the SRAM cell during the conventional modes of operation and performs read and write operations. NVSRAM combines the nonvolatility of nonvolatile memory (NVM) with the high speed and durability of SRAM, offering a brand-new solution for building high-efficiency and large-capacity memory.

With the continuous breakthroughs in emerging NVM technologies (see hereafter), including FeRAM, MRAM, PCRAM, and ReRAM, NVSRAM architectures that deeply integrate volatile SRAM and NV storage at the cell level have demonstrated inherent structural advantages. Such hybrid designs not only satisfy the stringent requirements of high-speed computing in terms of low latency and high bandwidth but also enable flexible and efficient NV data management strategies. During periods of low computational workload or when SRAM cells are idle, data can be directly backed up to the embedded NVM within the same memory cell, thereby avoiding frequent refresh operations or external data migration and significantly reducing overall system power consumption [116].

3.2.4 DRAM performance

DRAM has been widely used as the main memory in most electronic devices due to its low-cost and high bit storage density. DRAM is often regarded as off-chip standalone memory with input/output (I/O) links communicating with microprocessors and accelerators. Depending on the applications, DRAM products have different I/O interface protocols such as double data rate (DDR), low-power DDR, graphic DDR (GDDR), and HBM (Figure 3.16) [90].

Existing computing systems have issues with the aforementioned memory bottleneck due to the discrepancy between the speed of processing elements and the DRAM access.

A DRAM cell generally consists of a capacitor placed in series with a FET: 1T1C DRAM configuration (Figure 3.18). The charge is confined within the capacitor insulator, which forms a fixed-height energy barrier, and the access FET. However, DRAM cells lose charge via the access transistors, and the stored charge lasts for only a brief moment (64 ms). As a result, it is necessary to periodically read the cells and refresh them to preserve the accuracy of the information. Thus, DRAM is classified as a volatile type of memory.

Moreover, DRAM suffers from charge destruction during the read operation that requires immediate, power-expensive refreshing for data restoration

DRAM cells are arranged in a row and have a bit line structure that connects to a memory address called a word line. The address provides a means of identifying a location for data storage, and the word line forms an electrical path allowing all the memory cells on that row to be activated at the same time for storage (Write) or retrieval (Read) (Figure 3.18).

DRAM's tiny 1T1C design makes it ideal for packing numerous memory cells into a small area to achieve high density and high storage capacity. In fact, billions of DRAM cells can be squeezed onto a single memory chip.

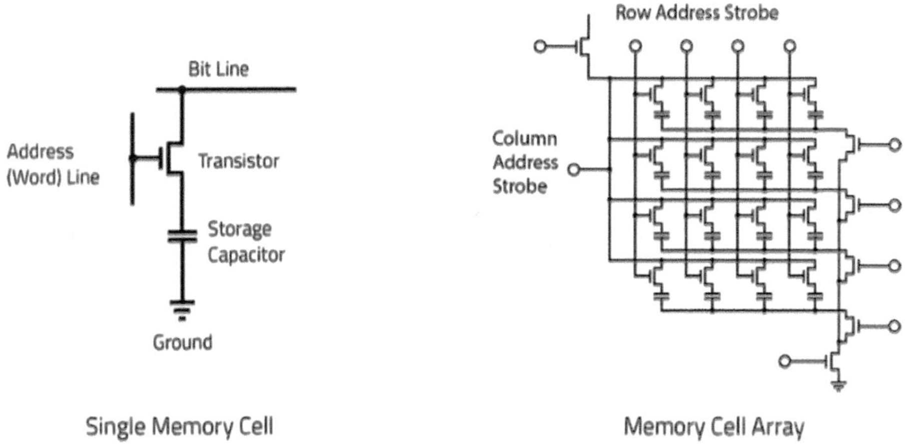

Figure 3.18: Single DRAM memory cell and array. Source: Lam Research.

However, it has become very difficult to scale or shrink the DRAM cell at the 20-nm node. The capacitor is a vertical, cylindrical-like structure. Inside the cylinder, the capacitor incorporates a metal-insulator-metal material stack. In the DRAM process flow, the access transistor is made first, followed by the capacitor. At each node, the goal is to maintain or increase the volume inside the cylindrical capacitor. But at each node, the capacitor shrinks, which may result in less volume inside the structure. That equates to less cell capacitance in the storage capacitor. At 20 nm, the industry hit a roadblock in capacitor scaling.

Starting in 2016, vendors moved to the 1xnm node regime, where suppliers had three DRAM products on the roadmap (1αnm, 1βnm, 1γnm, etc.). Originally, the 1αnm node was defined as a DRAM with 17 to 19 nm geometries, 1βnm was 14 to 16 nm, and 1βnm was 11 to 13 nm. Today, some vendors have relaxed the scaling specifications, creating some confusion in the market.

Due to the technological issue for scaling the DRAM cell, the DRAM storage capacity did not increase as the increase in computing performance of logic devices, further enhancing the memory wall.

Up to the 20 nm technology node, the DRAM storage capacity density increases at the same pace as the SRAM storage capacity (Figure 3.19). For technology nodes lower than 20 nm, due to the introduction of new DRAM technology nodes, the DRAM technology nodes cannot be compared to SRAM ones. The DRAM storage capacity density is typically 10 times higher than that of SRAM (Figure 3.19).

The memory bandwidth growth rate is the limiting factor in the AI ecosystem, thus creating bottlenecks for AI hardware system performance. DRAM bandwidth (GDDR) is growing at a rate of 2×/2 years, but it is still limited to 670 GB/s [86].

HBM with 1,024 or 2,048 data bus is the best of the available options for AI hardware memory subsystems (Figure 3.16). HBMs were originally targeted at the GPU. As

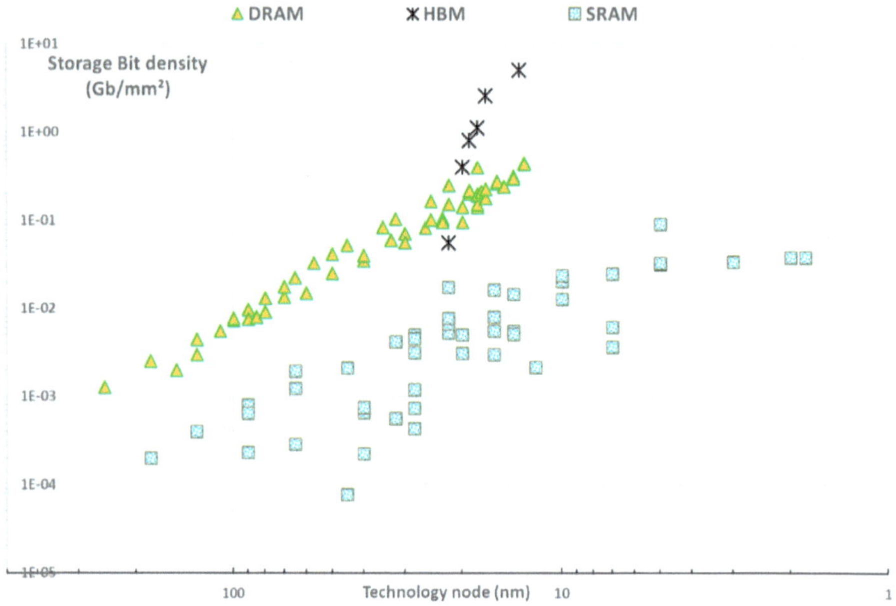

Figure 3.19: Evolution of the DRAM, HBM and SRAM storage capacity density with the technology node. Sources: Samsung, SK Hynix, Micron, and CXMT.

GPUs became popular for AI training applications, HBMs were the obvious choice for memory subsystems used for training state-of-the-art ANN and DNN models. The major advantages of HBMs over DDR or GDDR are the higher bandwidth, substantially lower power, and DRAM form factor.

HBM devices consist of a buffer (or base) die at the bottom and multiple 3D-stacked DRAM layers above it. The buffer die uses very-fine-pitch micro-bumps, allowing the memory stack to be co-packaged with an ASIC using advanced packaging technologies like silicon interposers or silicon bridges. Therefore, HBM uses 2.5 and 3D architectures (Figure 3.20). In 3D architecture, DRAM chips are stacked vertically and connected through TSVs (vertical electrical connections that pass through the silicon dies). In a 2.5D setup, multiple chips such as CPU, GPU, HBM device stacks are arranged adjacent to each other and/or on top of each other on a silicon interposer – a thin substrate of silicon that acts as a high-speed communication bridge. The interposer contains the fine-pitch wiring that enables fast, low-latency connections between the chips (Figure 3.20).

The data path between each HBM4 memory device and the processor requires 1,024 or 2,048 "wires" or traces. With the addition of command and address, clocks, etc., the number of traces necessary grows to about 3,000.

Thousands of traces are far more than can be supported on a standard printed circuit board. Therefore, a silicon interposer is used as an intermediary to connect

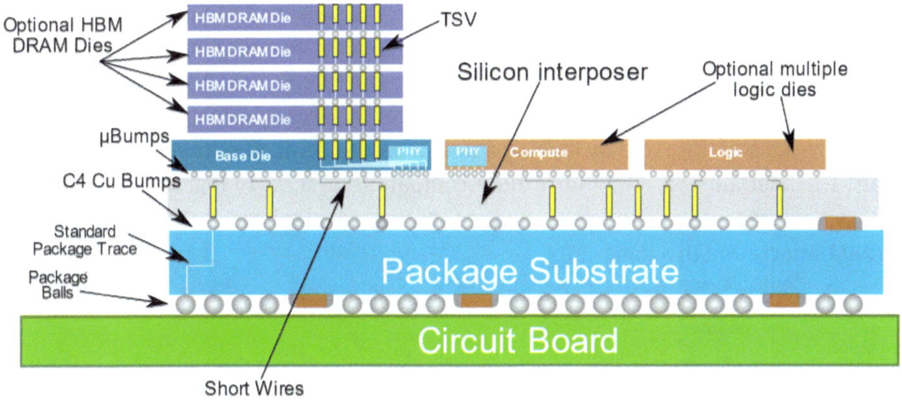

Figure 3.20: HBM cross section - Source: https://en.wikichip.org.

memory device(s) and the processor. As with an integrated circuit, finely spaced traces can be etched in the silicon interposer, enabling the desired number of wires needed for the HBM interface. The HBM device(s) and the processor are mounted atop the interposer in what is referred to as a 2.5D architecture.

The last HBM generation (HBM4, Figure 3.16) provides a bandwidth of around 2,048 Tb/s, which is more than three times that of the last generation GDDR one.

Furthermore, since HBM is constituted by a 3D DRAM stack, the storage capacity is also enhanced (Figure 3.19).

Compared to 6T SRAM-based caches, embedded 1T1C DRAM (eDRAM) offers higher memory density, which helps to reduce off-chip memory access and so lowers the power consumption compared to SRAM. This increased storage density enables a greater on-chip storage capacity within the same chip area. Moreover, eDRAM also offers much lower leakage power than SRAM. Hence, certain commercial devices have replaced SRAM with embedded DRAM (eDRAM) technologies for the last-level caches (LLC) in microprocessors and accelerators, such as the L3 cache in Figure 3.9.

Like DRAM, eDRAM has the disadvantage of charge loss on the capacitors. eDRAM also has a small retention time (the time interval over which the valid value of stored data is retained) in terms of microseconds and requires a periodic refresh before the charge leaks away after each interval, called the refresh period. Compared to normal DRAM, the refresh period of eDRAM is 10^3 times less. For instance, the refresh period of eDRAM is 40 μs, in comparison to the 64 ms retention time of the conventional DRAM. In other words, the eDRAM refresh occurs more often than the conventional DRAM. For the large LLC capacity, these frequent refresh operations consume a significant amount of energy and block the normal cache access requests when the cache banks are busy with the refresh process. In certain instances, these refresh activities can use as much as 70% of the overall energy usage and greatly impair the efficiency of the computing system [117].

Figure 3.21 shows the cross-sectional view of a stacked cell and trench eDRAM cell. Historically, a stacked cell approach was commonly used for commodity eDRAMs. However, it is necessary to build the capacitor over the device after the device has been built. This makes it difficult to keep the logic transistor performance. The trench capacitor can be fabricated by digging a hole in the silicon before the device has been built. This method will allow for a fully compatible process to logic technology and will not degrade transistor performance, giving an ideal technology solution for eDRAM integration in a logic chip.

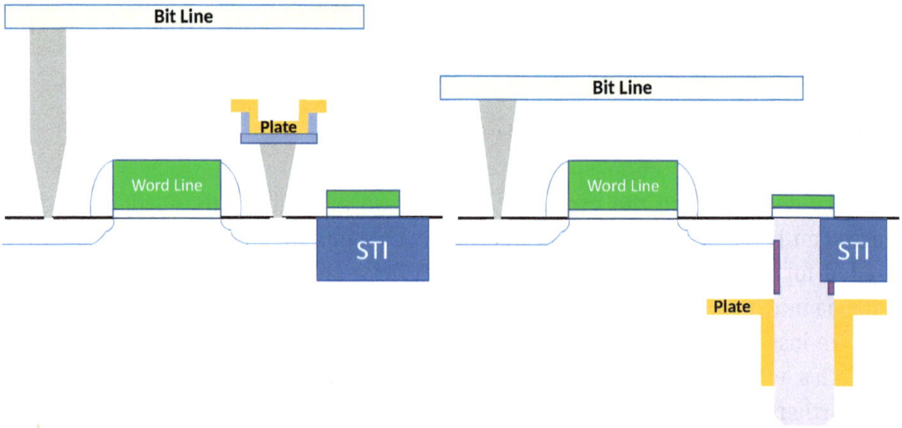

Figure 3.21: eDRAM cell (left) stacked capacitors, and (right) trench capacitors – inspired from Kirihata [118].

Embedded DRAMs are also used for in-memory computing (see hereafter). Performing in-memory computing using eDRAMs in last level caches also allows for drastically reducing the off-chip DRAM access, and, accordingly, the overall power consumption of the nano-electronic device [119].

3.2.5 NAND flash memory performance

As the preferred replacement for conventional mass-storage technologies, flash memories exhibit a potentially high capacity, fast access time, and stable operation. In flash memory, data is stored in the form of an array of floating gate (FG) or charge trapping (CT) transistors (cells).

The cell structure depicted in Figure 3.22 is similar to that of a standard FET, except that the transistor has two gates rather than one control gate. The additional gate is known as (i) a polysilicon FG, and (ii) a CT gate. The CT memory cell has been introduced in order to prevent any deleterious interference between the FG cells associated with the shrinking of memory cells. The CT memory cell consists of introducing embedded dielectrics (hypo-stoichiometric Si nitride) and the interfaces in the gate dielectrics of a MOS transistor for trapping the electronic charges (Figure 3.22). Both cell configurations are insulated all around by a dielectric oxide layer. Therefore, any electrons that are placed in the FG and CT gates can remain there under normal conditions (e.g., no applied erasing pulse), which leads to flash devices becoming an NVM.

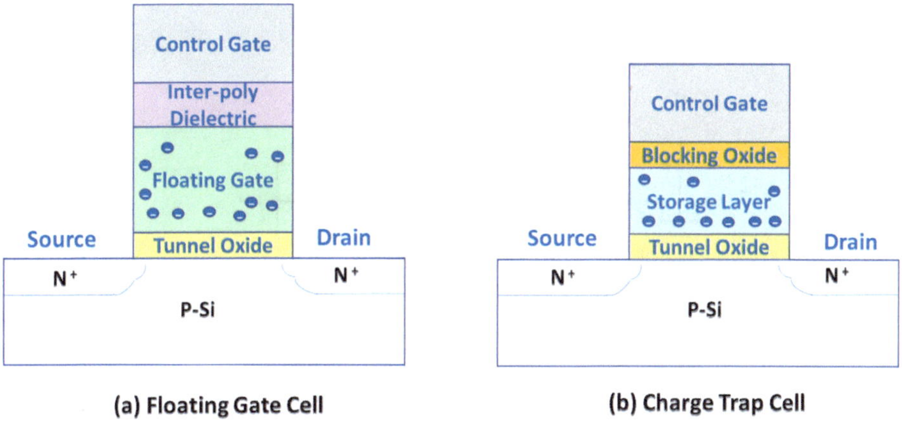

Figure 3.22: FG and CT flash memory cell - inspired from [120].

The FG and CT gates are sandwiched by (i) a tunnel-oxide layer and an inter-poly oxide (IPD) that usually consists of triple oxide-nitride-oxide layers, or (by) a blocking oxide layer (Figure 3.22). The tunnel layer isolates the FG and CT gates from the channel region inside the substrate, while the IPD and blocking layer can effectively prevent possible electron leakage between the control gate and FG or CT.

In flash memory, the programming/erasing operations are accomplished by injecting electrons into the FG or CT ("0" state) or by removing electrons from the FG or CT ("1" state) based on either Fowler-Nordheim tunneling or channel hot-electron injection [121].

Flash memory can be classified as either NAND flash or NOR flash in terms of its cell structure.

In NOR flash, cells are arranged in parallel such that one end of the cell is connected to ground and the other is connected directly to a common drain that is called a bit line. Owing to its architecture, NOR flash can allow for a fast read performance while yielding a high die cost and large power consumption. This primarily results from the increase in cell size.

NAND flash builds cells in series with adjacent cells that share the source and drain (Figure 3.23). This specific structure thus significantly reduces the cell size, enabling the NAND memory to have a much smaller cell size and lower die cost than the NOR memory. NAND flash gives a slow read performance, as the readout current is

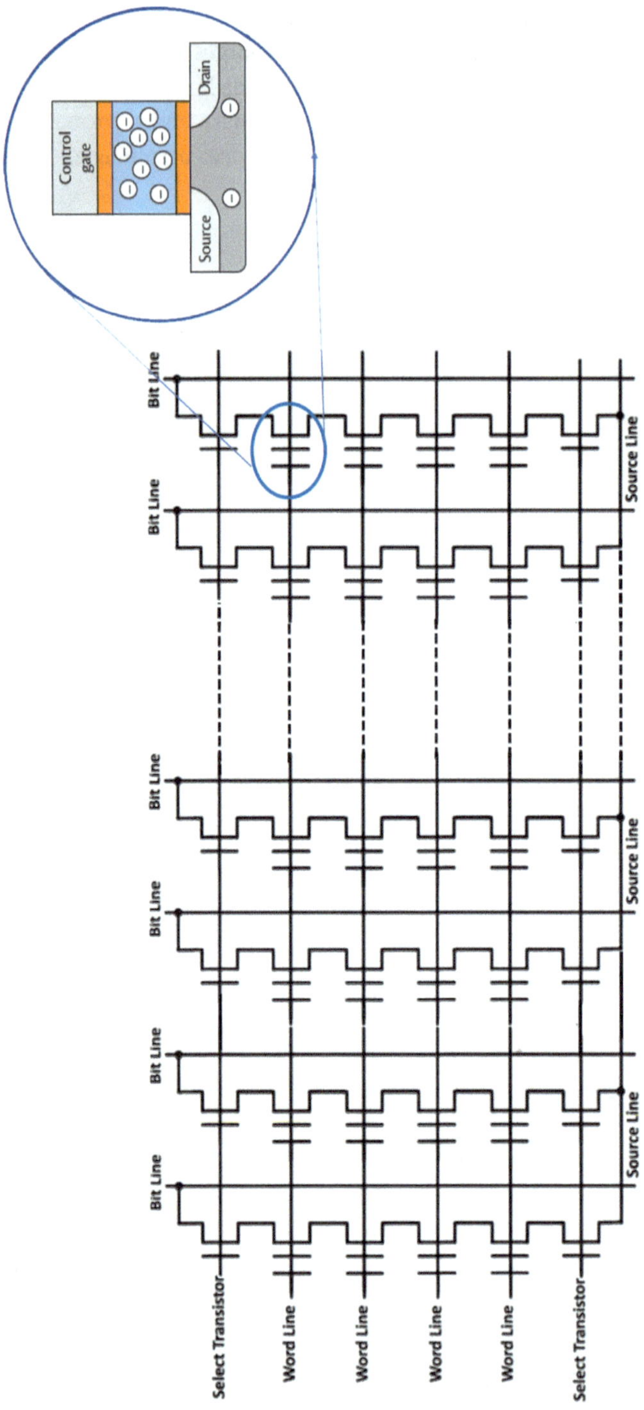

Figure 3.23: Typical layout of NAND flash memory cells.

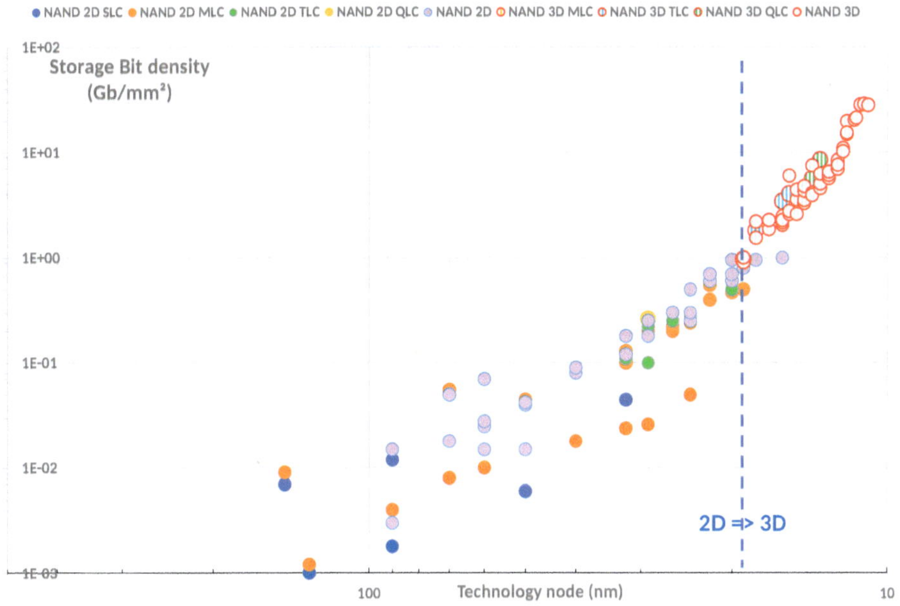

Figure 3.24: Evolution of 2D and 3D NAND flash storage density with the technology node – data from Samsung, SK Hynix, Kioxia, Micron, Solidigm, and YMTC and [122, 123].

lower when using a serial transistor [121]. The write operation of flash memories requires high voltages (typically >10 V) and entails significant latency (>10 µs).

Similar to their relatives, the logic MOS-based devices, planar Flash NAND memory cells have been miniaturized over many years by reducing their size, advancing fabrication technology, enhancing circuit architecture, and refining programming techniques. For instance, one way to increase the flash memory storage capacity without increasing the number of extra memory cells, of constitutive layers, and of process steps is the multilevel cell technology. In the past, one flash memory cell stored 1 bit of information (SLC) and was continuously increased to 2 bits (MLC), 3 bits (TLC), and now it stores up to 4 bits (QLC) of information (Figure 3.25). Extremely well-suited for typical complementary metal-oxide semiconductor manufacturing methods, their capability to demonstrate various distinct and stable storage states within each cell plays a vital role in minimizing energy use and memory size.

However, due to numerous parasitic effects within the memory array, shrinking the flash memory cell dimensions became impossible in the sub-20 nm technology range. To further enhance the storage capacity of flash memories, the semiconductor industry moved to 3D flash memory cell (Figure 3.24). 3D NAND flash arrays may utilize either CT or FG technologies. However, most 3D structures are constructed using CT cells, primarily due to the less complicated manufacturing procedure.

Major NAND chip manufacturers are racing to increase the number of vertical 3D NAND gates up to more than 400 layers at the time this book is written. A 321-layer quad-level cell (QLC) NAND with a data density of 28.8 Gb/mm^2 [124] and a 400+ layer triple-level cell (TLC) NAND with 28.2 Gb/mm^2 data density [125] have been demonstrated (Figure 3.25).

2D Flash NAND memories are still needed for some specific applications with higher reliability and low/high temperature operation. MCU, medical, robots, TVs/toys, game controllers, wearables, security cameras, smart speakers, IoT, AI, ML, printers, set-top boxes, and aerospace need 2D NAND products. Now, 3D NAND products are very popular for data centers, cloud, server, solid-state drives (SSD), PCs, mobile, and smartphones [126].

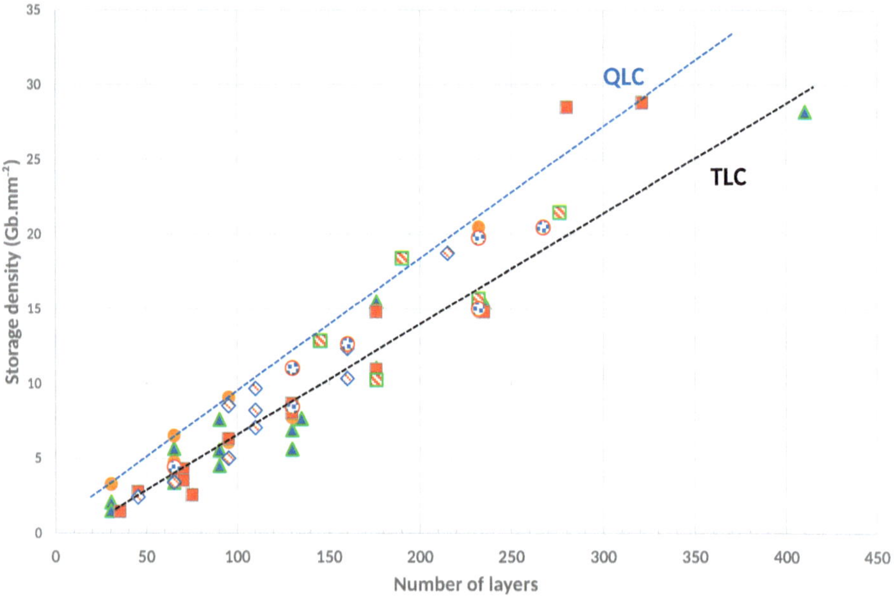

Figure 3.25: Evolution of the 3D CT NAND flash areal storage density with the number of layers – data from Samsung, SK Hynix, Kioxia, Micron, Solidigm, and YMTC and [122, 123].

3.2.6 AI hardware power consumption

The aspiration of mimicking the human brain's efficiency and smartness has brought notable progress in AI. While algorithms stemming from ANNs and DNNs have successfully demonstrated their high performance across diverse applications, their ever-growing size and complexity urge to develop highly energy-efficient hardware.

Despite the decrease in the energy required for switching a transistor with the technology node (Figure 4.2), the power consumption of AI accelerators, which are

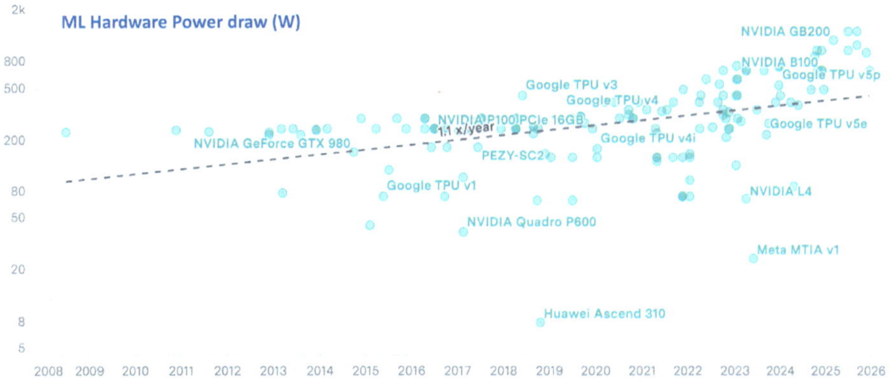

Figure 3.26: Evolution of the power consumption of AI accelerators [78].

manufactured using leading-edge technology nodes, is still growing to reach nearly 1,000 W (Figure 3.26).

Accordingly, power demand in the global data center market is forecast to grow substantially in the next half-decade or so. Analysts estimate current demand to be approximately 62 GW, comprised of cloud workloads (58%), traditional workloads (29%), and AI workloads (13%). AI is projected to grow to 28% of the overall market by 2027, while cloud drops to 50%, and traditional workloads fall to 21% (Figure 3.27a) [127].

Global electricity generation to supply AI data centers is projected to grow from 460 TWh in 2024 to over 1,000 TWh in 2030 and 1,300 TWh in 2035 (Figure 3.27b). This means that the AI energy demand increases from 1.25% and 1.5% of the worldwide energy production in 2022 and 2024, respectively, and should increase to more than 3% in 2035. Although such percentages could be considered as relatively weak, they can induce severe issues in regions where energy consumption is already strong or in regions where the energy resources are rather limited.

Large DNNs, such as GPT-4, require approximately 20 MW of power. This remarkable level of energy consumption is six orders of magnitude greater than that of the human brain, which is estimated to consume between 12 and 20 watts, utilizing around 10^{11} neurons and 10^{15} synapses, on average.

Managing power to lower the energy usage of AI data servers is considered extremely crucial. As a result, the semiconductor sector is exploring various methods to lower the energy demand associated with processing extensive AI models. Among the numerous thoroughly researched ideas, the creation of more energy-efficient devices and innovative computing paradigms, including near-memory, in-memory, edge, and neuromorphic computing, appears to be the most promising option.

Figure 3.28 highlights the main technologies needs for performing AI training and inference using AI data servers, i.e., cloud computing. They include high-performance processors, GPUs coupled to CPUs typically, massive data storage, high and quick con-

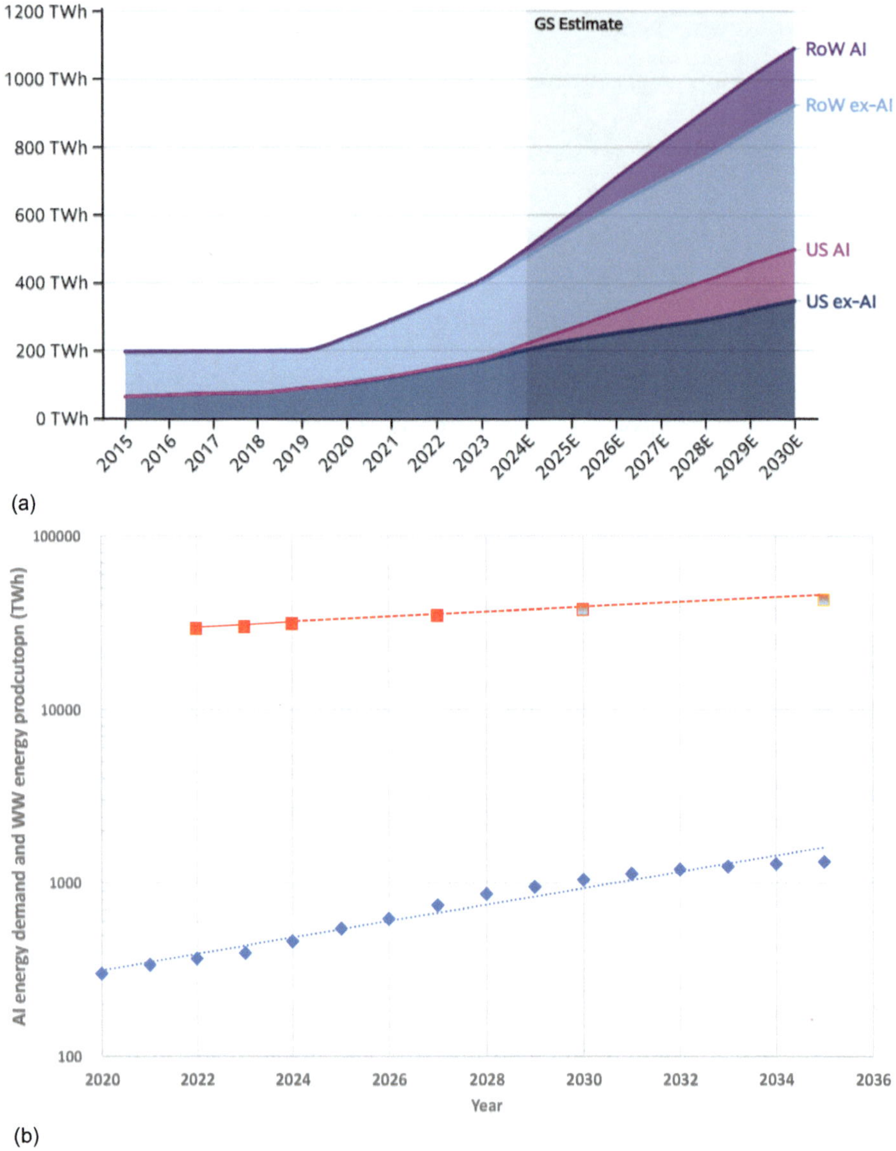

Figure 3.27: (a) Global data center electricity consumption [127]. (b) AI data center energy demand (blue) and worldwide energy production (red) per year – data from IEA [128, 129].

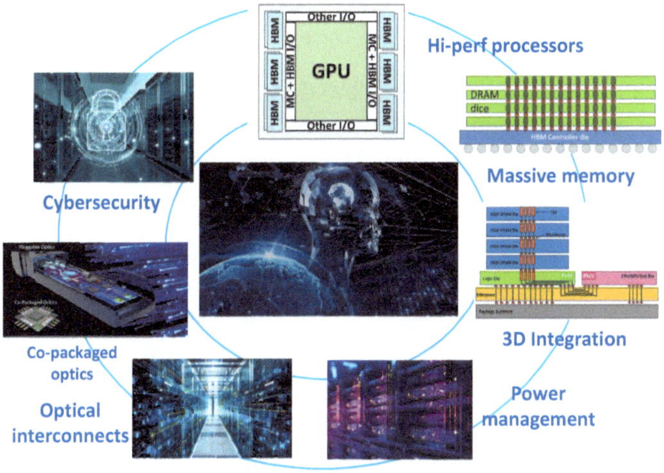

Figure 3.28: Basic technology needs for training and inference DNN models using data servers (cloud-AI).

nectivity between the modules of the data servers, using optical interconnects, and from the data sources to the AI servers, and all the cybersecurity blocks to protect data privacy.

Power electronics, photonic integrated circuits, and 3D integration are attractive approaches for reducing the power consumption and will be described later.

4 Approaches to reduce AI energy consumption

As depicted in Figure 3.7, the AI revolution was allowed by the use of GPU accelerators. However, since the power consumption of GPUs is very large, it could lead to some future energy shortages in certain regions. The semiconductor industry is researching for some GPU substitutes.

4.1 Logic devices

As aforementioned, the operating speed of CPU, expressed in giga-operations per second (GOps/s), does not comply for training large DNNs (Figure 4.1). GOps/s, which stands for Giga Operations per second, is the metric used to discuss the hardware effectiveness in processing AI workloads. Microcontrollers (MCUs) even exhibit a lower speed than CPU, but their power consumption is very low, which could comply for inferencing some lighter DNN models (Figure 4.1).

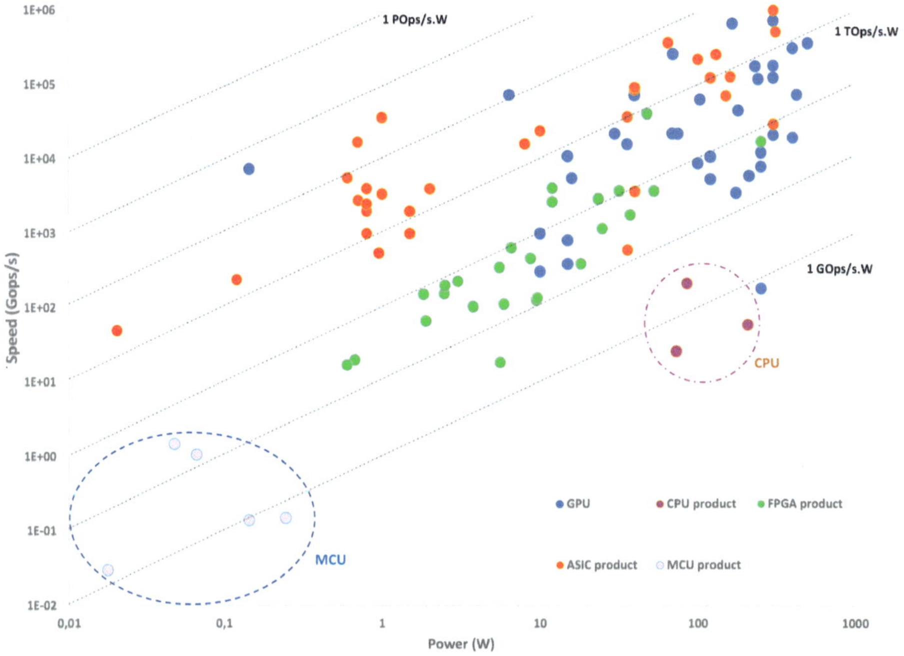

Figure 4.1: Speed in GOps/s versus the power consumption of AI digital commercial processors updated from [1] – data from [68, 131–137].

 | https://doi.org/10.1515/9783112247952-004

Field-programmable gate arrays (FPGAs) present computing speed and power consumption, which are lower than the GPU ones, but can comply with the training and inference requests of large DNN models. Custom AI application-specific integrated circuits (ASICs) look as the best candidates for replacing GPUs, as they present, on average, similar computing speed and slightly lower power consumption compared to GPUs (Figure 4.1).

For increasing the computing performance and the energy efficiency of various AI chips, a convenient way is to pursue Moore's law. This law claimed to double the transistor density in an IC every 2 years (Figure 4.2). The impressive progress induced by this law also allows for decreasing the energy for switching the transistor. This switching energy decreases from 1,700 aJ (1 aJ = 10^{-18} J) at the 180 nm technology node down to nearly 2 aJ at the 2 nm technology node (Figure 4.2).

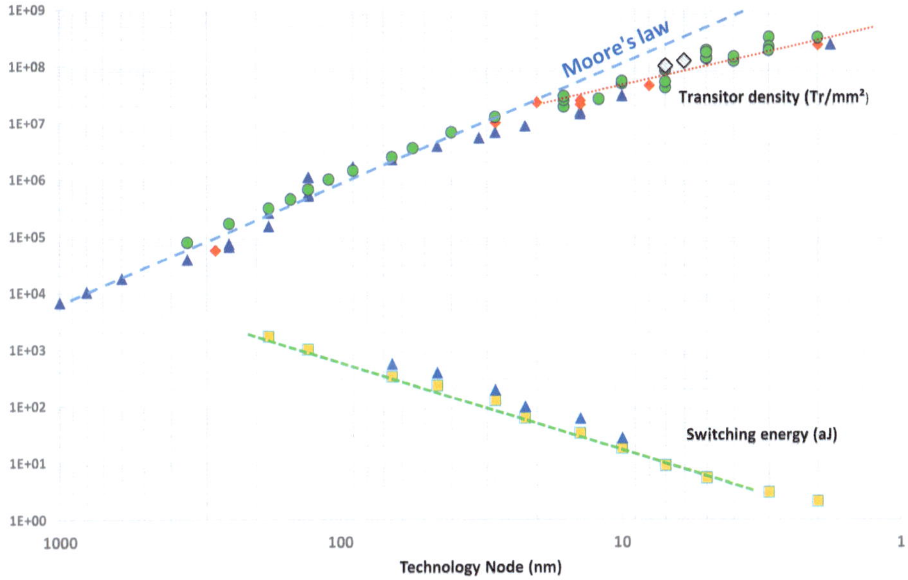

Figure 4.2: Evolution of the transistor density and the transistor switching energy with the technology node updated from [1]. Sources Intel, Samsung, TSMC, and [138].

Moving from a 55 to 65 nm technology node to more advanced 4 and 5 nm technology nodes allows increasing the computing speed by 4 to 6 orders of magnitude and, to a less extend, leads to an increase in the energy efficiency of AI digital chips (Figure 4.3). This scattered plot explains why AI chip designers always want the most advanced chips. AI chip designers must cope with the increasing demand for computing speed for training and inferencing ever more complex AI algorithms (Figure 3.10). Then, chip designers pressure the chipmakers to introduce as quickly as possible leading-edge technologies for chip manufacturing, at the expense of less advanced technologies.

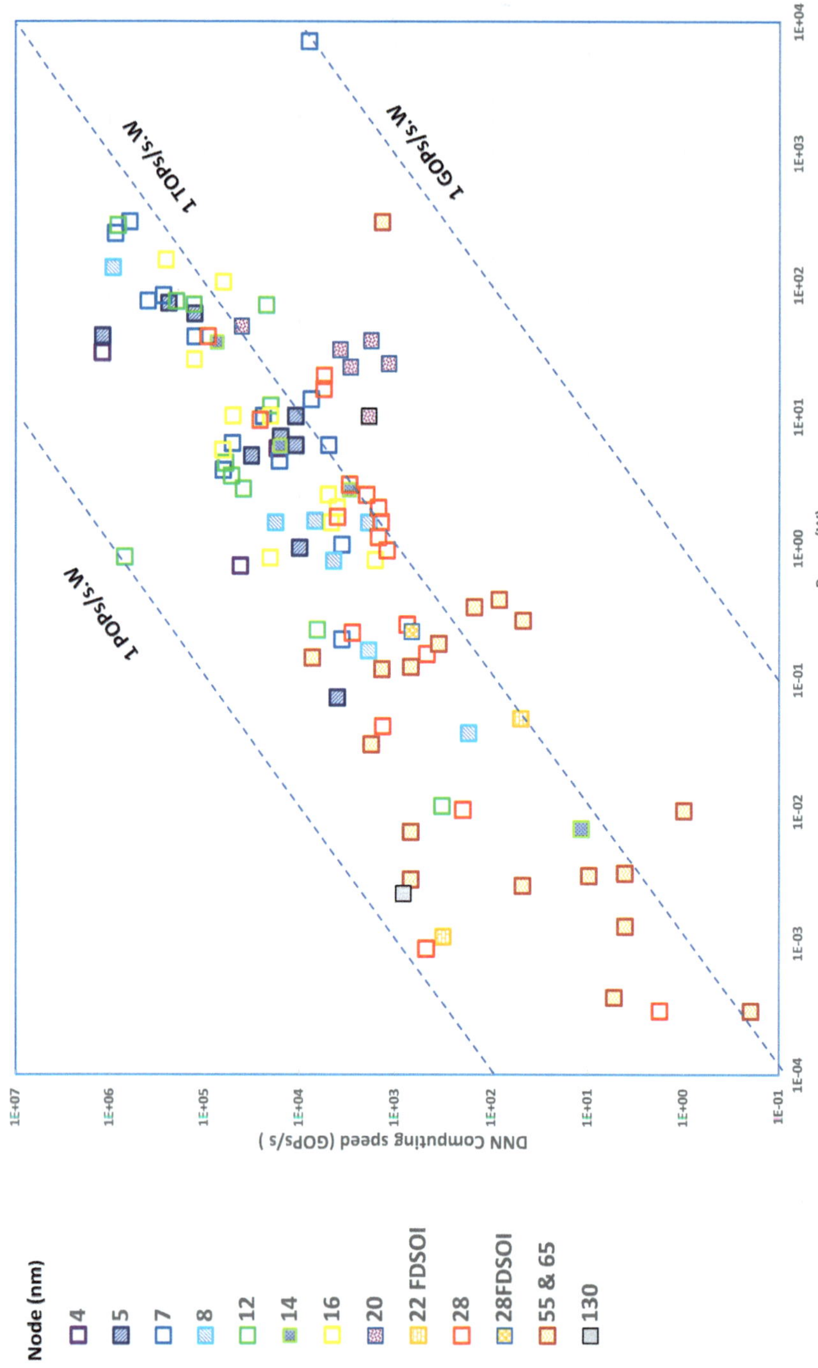

Figure 4.3: Speed in GOps/s versus the power consumption of AI digital processor chips as a function of their technology nodes [83, 139–153].

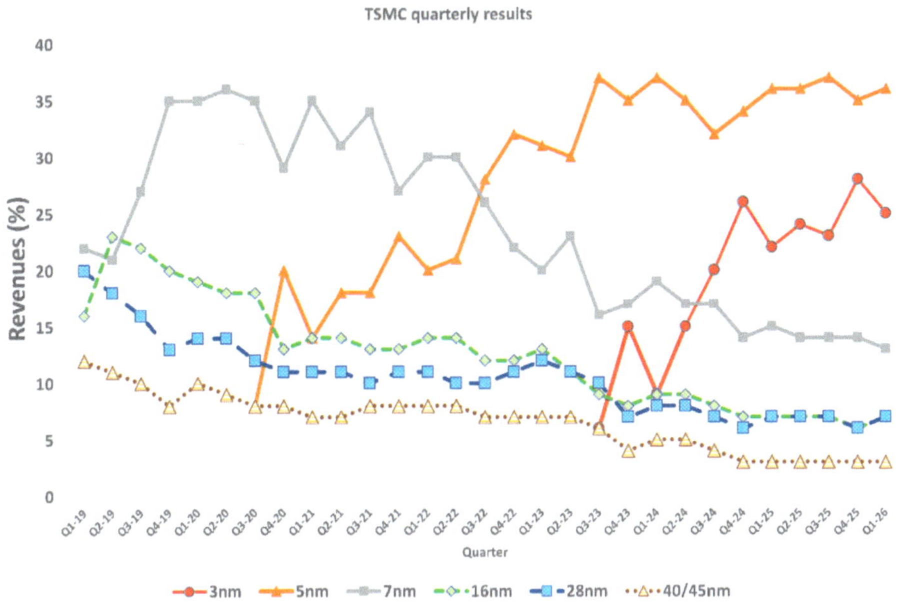

Figure 4.4: Evolution of the percentage of the TSMC quarterly revenues with the technology node – data from TSMC financial quarter reports.

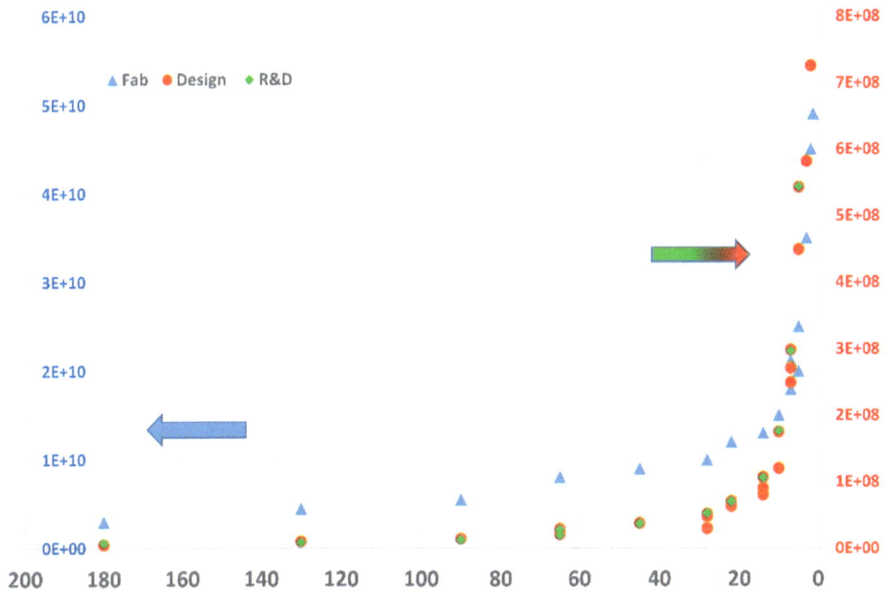

Figure 4.5: Evolution of the production line, chip design, and R&D costs with the technology node updated from [1] – data from [20, 158].

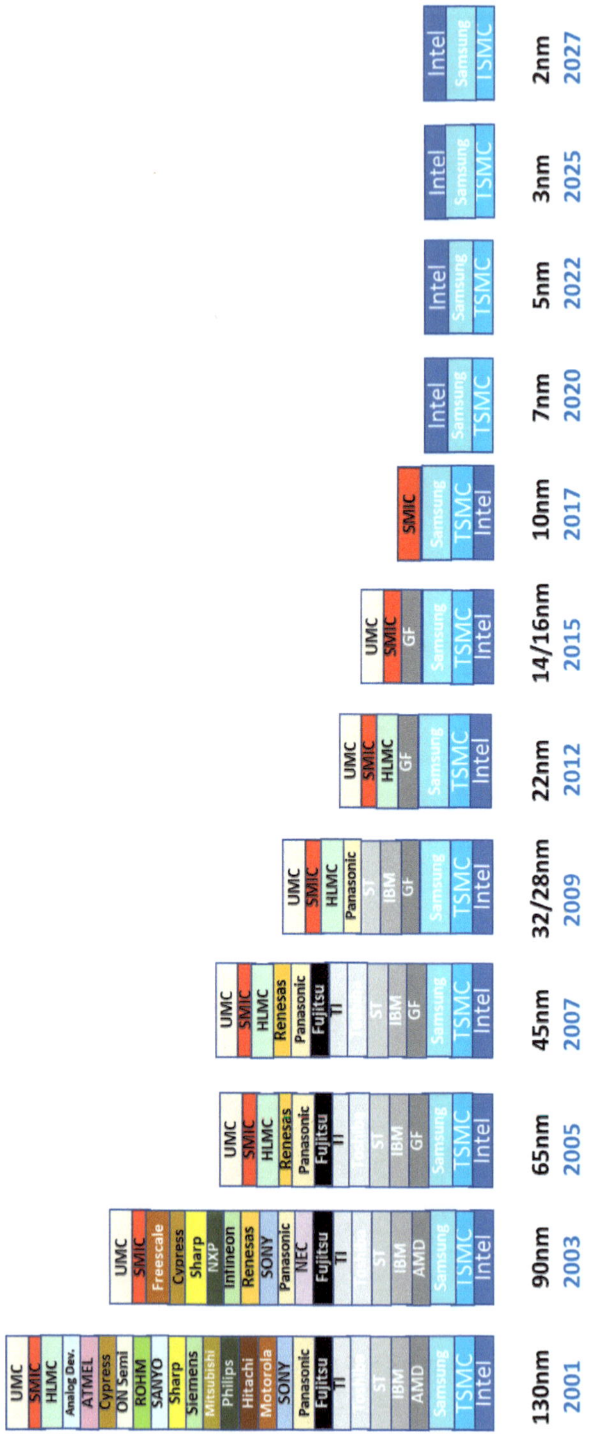

Figure 4.6: Semiconductor industry evolution at the leading edge – Number of players as a function of the technology node - Updated from Yole Intelligence.

This trend is highlighted in Figure 4.4, which shows the percentage of quarterly TSMC revenues from 2019 to 2025 as a function of the manufactured technology node. Driven by the huge AI and HPC demand, the 5 nm technology node, which dominated TSMC revenues in 2019–2020, has been replaced by the 3 nm technology node in 2022–2023. Then, the 3 nm technology node is currently being replaced by the 2 nm technology node.

Despite the decrease in the transistor switch energy with the most advanced technology nodes (Figure 4.2), and despite the progress accomplished this last decade for enhancing the energy efficiency of CPU, GPU, and neuronal processor unit (NPU) devices [154], the energy efficiency of AI digital chips stays mainly confined in the 1 to 100 TOps/s.W range (Figure 4.3).

If the transistor switching energy decreases with the technology node, the production line (fab), the design, and the R&D cost exponentially increase with the technology node (Figure 4.5). For instance, for the 3 nm technology node, the production line, design, and R&D costs are roughly $30 billion, $580 million, and higher than $600 million, respectively. Moreover, due to the increase (i) in the complexity and number of process steps from 250 at the 45 nm node to more than 1,300 at the 3 nm node, and, accordingly, (ii) in the number of expansive lithography masks that increase from 30 at the 90 nm node up to at the 3 nm node [155], the manufacturing cost of semiconductors follows similar trends.

The use of AI could reduce the overall cost for designing advanced chips as soon as AI-based electronic design automation tools are available and fully operational.

To highlight the increase in the overall chip production cost, the lithography tool cost of $450,000 per unit in 1979, while the most advanced high-numerical aperture extreme ultraviolet tool costs around $380 million for producing 2 and 3 nm logic chips and DRAMs, i.e., around 3 orders of magnitude increase in 45 years [156].

Such drastic increases in the overall cost of designing and producing semiconductors explain why the number of chip makers has decreased from 26 at the end of the last century (~130 nm technology node) down to 2 to 3 nowadays at the 3–5 nm technology nodes (Figure 4.6). While the number of lithography tool suppliers diminished from 4 at the 180 nm node down to only 1 at the sub-5-nm technology nodes.

4.2 Near-memory computing

Data movement between the memories and the processor(s) is the most critical resource to optimize computing architectures in terms of energy consumption. Nowadays, the energy cost of moving a 64-bit operand a mere 5 mm across the chip to the on-chip memory (static random-access memory – SRAM) exceeds that of a floating-point operation that uses that operand. Moreover, the energy cost of further moving this operand to the main off-chip memory (dynamic random-access memory, DRAM) is roughly two orders of magnitude higher than the floating-point operation [28]. Such a high energy consumption and the corresponding transmission latency often refer to the "memory wall" bottleneck of the von-Neumann architecture (see Section 3.2.2).

This constraint is further exacerbated by the main types of memory that are currently available, which include DRAM and SRAM. DRAM provides high memory capacity and is cost-effective, but it has limitations regarding access speeds (ranging from 50 to 100 ns) and must be housed on a separate chip from the processor chip due to the challenges associated with integrating DRAM with logic technology. Therefore, it functions as an off-chip memory option, linking to the processor chip through long metal interconnects.

The idea of processing data close to the memory is not really new, as it has been explored since the 1960s [157]. Coupled to the slowdown of the Dennard scaling law, the advent of data-centric applications such as AI, and the development of 2.5D and 3D integration/packaging, allowing blending memory and logic devices in the same package, near-memory computing (NMC) has regained interest within the semiconductor industry in terms of power consumption and compactness. The basic principle of NMC is to perform digital computations for AI training and inference, for instance, in the memory proximity by physically placing monolithic processing units (multicore CPU, GPU, FPGA, ASIC, MCU, etc.) near the DRAM memory chip to minimize data transfer cost (Figure 4.7).

NMC assigns a small amount of main DRAM memory directly to each processor located on the chip. Nevertheless, every processor within a CPU or GPU still must retrieve data from dedicated memory, move it to the computation unit, perform the necessary calculations, and subsequently store the results back into local SRAM and/or DRAM memory. As a result, even though placing DRAM units close to the processors can lessen the impact of the "memory wall" issue, this issue persists because of the existing performance disparity between processors and memory. This limitation of memory in current CMOS-based NMC systems must be acknowledged [159].

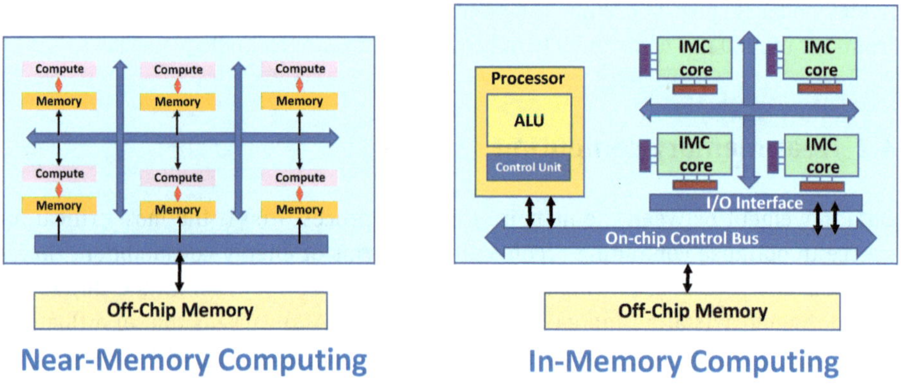

Figure 4.7: Basic architectures of near-memory and in-memory computing – inspired by Haensch et al. [159].

2.5D and 3D integration/packaging allow assembling smaller modular chips. This approach enables designing and manufacturing very complex chips. Such a heterogeneous integration enables a modular architecture and prevents the pitfalls of the production of large conventional monolithic systems on chip (SoC) devices. As a matter of fact, the production yield quickly decreases with an increase in the SoC area [138], leading to a further increase in the cost for large SoCs. 2.5D and 3D integration is a concept for increasing the functionality and performance of the chip while lowering its cost.

The vertical 3D integration exploits the third dimension by adopting the hybrid bonding technology or Cu micro-bumps for connecting the various chips or components by means of through-silicon vias. 3D integration can support a high interconnect density that can strongly reduce the processing latency and the energy consumption. Such a 3D stacking is touted as the true enabler of processing close to the memory.

3D-stacked DRAM architectures, such as high bandwidth memory (HBM), provide an opportunity to avoid unnecessary data movement between the main memory and the processor units. These multilayer architectures include a logic layer, where compute logic can be integrated underneath multiple (currently from 4 to 12) layers of DRAM cell arrays (i.e., the memory layers) within the same chip. The increase in the number of DRAM layers allows a strong improvement in the memory bandwidth from 128 to 896 Gb/s for the 4- to 12-layer DRAM, respectively [160]. Such a configuration allows reducing the memory access latency and the power consumption. Designers can take advantage of the logic layer to perform in-memory computing (IMC), or NMC, where some of the computation is moved from the CPU/GPU to the logic layer underneath the memory layer.

Compared to an 8-core CPU, near-memory acceleration using DRAM arrays leads to a 4.9×–5.4× speed up with an energy reduction ranging from 5.9× to 12.1× [161]. Compared to GPU, NMC with DRAM arrays can offer a 4× energy efficiency for the inference of a large language model and improve the inference performance [162]. The energy efficiency of NMC with DRAM arrays outperforms FPGAs by 33.1× on average for both compute-intensive and data-intensive applications [163]. Such trends about NMC by coupling FPGA with DRAM are not clearly highlighted in Figure 4.16, in which their reported speeds and energy-efficiency are closer to the CPU ones and much lower than the best GPU ones (Figure 4.1).

NMC using SRAM and spin transfer torque – magnetic random-access memory (STT-MRAM) coupled to CPU leads to lower computing speeds than GPU and AI custom ASIC, but can lead to higher energy-efficiency AI computing (Figure 4.16). Nowadays, NMC is relatively mature with significant commercial success, such as HBM.

4.3 In-memory computing

Conventional computing processors are based on the von Neumann architecture (Figures 3.5, 3.9, and 4.8a), with physically separated memory and processing units. The two units are linked through a data bus, which eases the transfer of information from the memory unit to the processing unit and then back again to save the processed information. This arrangement enables the use of computing systems with varying specifications and needs; however, it is not designed for workloads that require the processing of massive datasets. While computational delays can be reduced by utilizing HBMs or substantial on-chip SRAM, both methods lead to increased energy use during persistent data access and transfer.

IMC has the potential to resolve the limitations associated with the von Neumann architecture. In IMC, calculations are conducted directly within the memory elements, which removes the necessity for data movement (Figure 4.7). This approach can utilize different memory technologies, including volatile SRAMs, DRAMs, and nonvolatile memories (NVMs). Both types of volatile and NVMs can support various IMC designs: digital IMC, which employs logic gates for executing multiply-accumulate (MAC) functions, and analog IMC, which leverages physical representations of mathematical concepts, such as Ohm's law and Kirchhoff's law, for performing multiplication and accumulation, to facilitate efficient, low-energy computation at high speeds (Figure 3.26) [164].

An analog IMC core generally consists of an array of NVM units and specific peripheral components designed for programming and addressing the memory array (Figure 4.8b). These IMC cores are often regarded as fundamental functional elements linked together by specialized on-chip networks in more intricate IMC processors (Figure 4.8c).

IMC is especially beneficial for enhancing performance in data-heavy matrix-related tasks by utilizing the compact, highly parallel crossbar array (CBA) format, where memory components are positioned at the junctions of row and column conductors in the BEOL of logic devices. Figure 4.8d illustrates a typical CBA-driven IMC accelerator used for matrix-vector multiplication (MVM), a highly resource-demanding process in standard computing systems that finds applications in areas such as neural networks, image analysis, and optimization problems. In an open-loop IMC MVM setup, elements Y_{ij} of the target matrix Y are initially stored in the CBA's memory units as either electrical conductances G_{ij} or resistances R_{ij}. A voltage vector V, with elements V_j that denote the components x_j of the input vector x, is subsequently applied to the CBA's columns while the rows are grounded. Based on Ohm's law, each element produces a current $I_{ij} = G_{ij}V_j = V_j/R_{ij}$, which accumulates on the CBA rows in accordance with Kirchhoff's law [164]:

$$I_i = \Sigma_{j=1}^{N} G_{ij} V_j = \Sigma_{j=1}^{N} R_{ij}^{-1} V_j \tag{4.1}$$

which is the same as the MVM function $i = GV = R^{-1}\,V$, that is represented by the mathematical formula $b = Yx$. Due to the inherent spatial parallelism present in the CBA configuration, the MVM function can be executed in a single computational step. This results in a significant advantage in terms of the complexity of MVM when compared to traditional digital computers [164].

The computational potential of IMC can be enhanced by placing the CBA within a feedback loop that utilizes analog operational amplifiers. In this type of closed-loop IMC, the rows of the CBA act as virtual ground nodes for the circuit due to the feedback effect (Figure 4.8e). By introducing an external current vector i at the CBA rows, the resulting output voltage vector V measured at the outputs of the operational amplifiers will meet the requirement:

$$V = -G^{-1}i = -Ri \tag{4.2}$$

which offers a physical means to speed up the resolution of a linear equation system represented as $x = Y^{-1}b$, where Y, x, and b signify the matrix of the linear system, the unknown term vector, and the known term vector, respectively [164].

It is important to recognize that performing an inverse MVM poses significant challenges for traditional computers, as it demands high computational intensity. However, in IMC-based inverse MVM, the time complexity remains unaffected by the size of the matrix and is instead determined by the smallest eigenvalue of the matrix, potentially leading to energy-efficient acceleration of inverse matrices across various domains such as telecommunications, sensor integration, ranking systems, machine learning, combinatorial optimization, and general linear algebra (Figure 4.8).

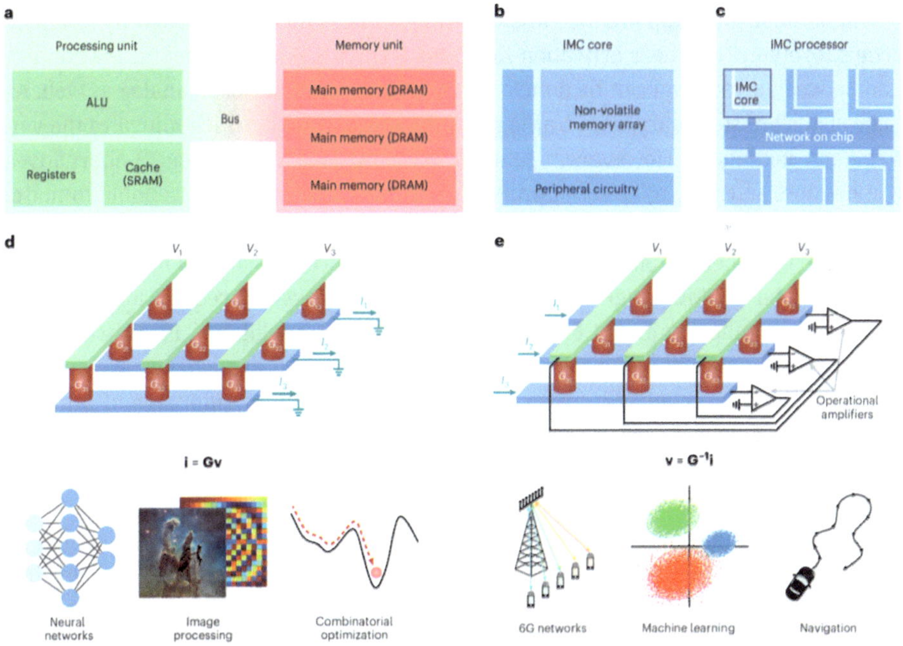

Figure 4.8: (a) Prototypical von Neumann architecture, (b) IMC core, (c) IMC processor, (d) open-loop IMC-based MVM circuit and selected applications, and (e) closed-loop IMC-based IMVM circuit selected applications [164].

The definition of IMC is not universal, and sometimes, confused with the NMC concept. In this chapter, IMC is defined in two ways [159]:

– Digital memory and analog compute (inference): 2D and 3D memory arrays store the synaptic weights as electric resistances or capacitances in one bit or in multi-bit capacity, at the memory cells of a CBA connected with metal interconnects. This approach increases the parallelism, like in GPU, and reduces the data transport between the processor and the off-chip memory (memory wall bottleneck). The MVM, coupled to MAC operations, that constitutes the basic ingredient of DNNs, is carried out in an analog fashion using Kirchhoff's and Ohm's laws.

– Analog memory and analog computation (training and inference): The weights located at the cross-bar junctions are represented in memory components that exhibit analog values across a continuous range from high to low conductance/resistance levels. The MAC function resembles the previously mentioned procedure. Both the MAC function and the update process gain advantages from the highly parallel array operation that maintains static weights.

Analog accelerators avoid the von Neumann bottleneck by performing the computation in place. The analog approach uses arrays of nonvolatile, programmable memories that encode the neural network weights. Computations like vector matrix multiplication or element-wise matrix updates can be done in parallel and in constant time, with no movement of the weights.

In an analog AI chip architecture, the AI network is then represented by stringing arrays together. Nonlinear activation functions (Figure 3.3) are inserted in the connection between arrays and can be done either in the digital space or in analog as well. As the weights are stationary, data traffic is significantly reduced, which mitigates the von Neumann bottleneck. However, in contrast to digital solutions, analog AI alternatives will be more sensitive to material properties and are intrinsically susceptible to noise, drift, mismatch, and variability. Algorithmic innovations can also be employed to mitigate material imperfections and improve the performance of analog AI hardware [146].

Analog-based IMC designs are frequently used for tasks that require low precision in ML and provide significant energy savings. However, analog IMC faces issues such as nonlinearity, variations in manufacturing processes, and mismatches, which complexify the achievement of high energy efficiency, accurate readouts, and minimal computation delays for precise MAC operations simultaneously. In contrast, digital IMC is resilient against variations in processes, accommodates any level of precision, and is anticipated to adapt easily to more advanced technology nodes [146].

Due to the inherently parallel characteristics of DNN computations, a majority of IMC frameworks are structured around a tiled architecture. The upper-level illustration of a typical tiled IMC framework is depicted in Figure 4.9a, which mainly consists of tiles (marked as T), crucial operational components for ANN calculations, centralized control, and storage linked through on-chip connections. These crucial operational components typically comprise accumulation units, pooling layers, and activation units, among others. Accumulation units are necessary for aggregating the

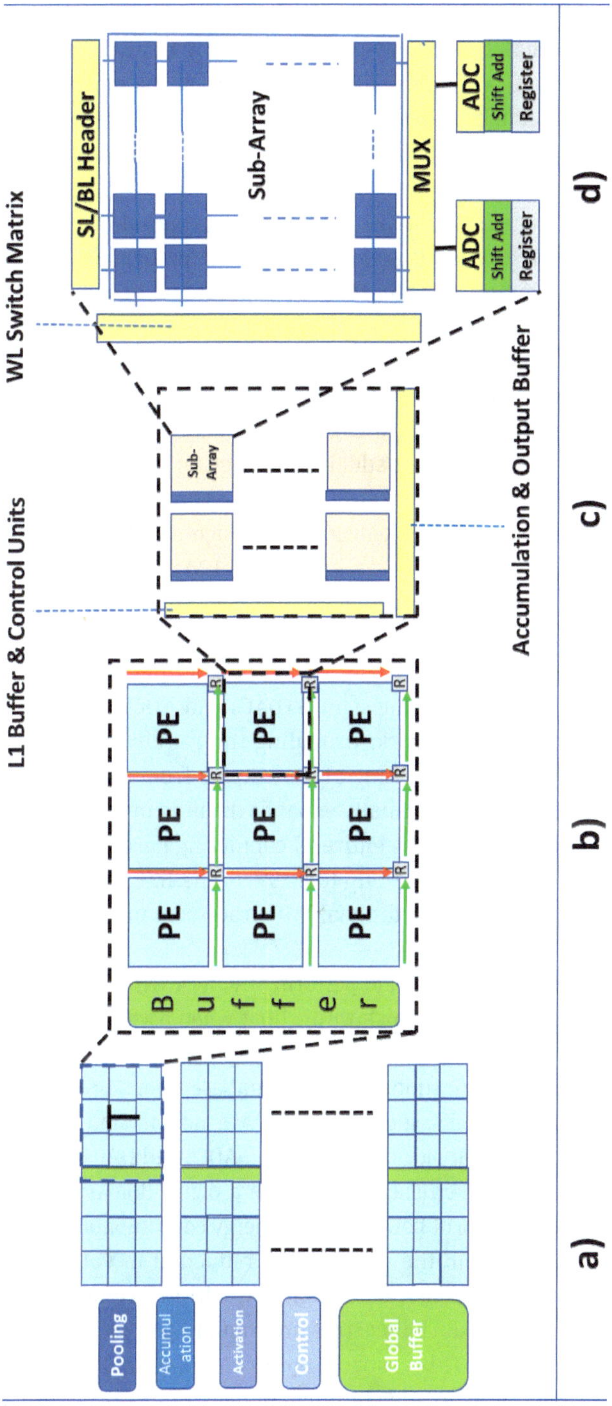

Figure 4.9: Hierarchical architecture of IMC: (a) Chip-level structure composed of tiles, (b) tile-level structure composed of PE, (c) PE-level structure composed of sub-arrays, and (d) sub-array structure – inspired from [166].

partial sums from various tiles. The results of these accumulations are subsequently forwarded to activation units. Ultimately, the activations will return to the centralized buffer. The global control component will manage the dispatching of the new feature maps to additional tiles for subsequent computations [**166**].

As illustrated in Figure 4.9b, a tile typically comprises several in situ processing elements (PEs) along with input/output buffers that are linked via routers. The routers facilitate communication between the PEs and assist in transferring partial sums from the PEs to the upper levels. Figure 4.9c presents a PE configuration that includes one or more IMC subarrays, a localized input/output buffer, and potentially accumulation units. Within the PE, the intra-accumulation units are generally used for aggregating the partial sums obtained from the subarrays. Figure 4.9d demonstrates a standard structure of an IMC subarray, which incorporates a crossbar memory array along with computing peripherals such as input encoders (DACs), a switch matrix for word lines (WLs), analog-to-digital converters (ADCs), shift-adders, and registers. Notably, DACs and ADCs enhance the scalability and versatility of mixed-signal communications between the subarrays and the upper levels. In conclusion, the design of IMC frequently relies on a multicore hierarchical framework where fundamental MAC operations are conducted in the analog domain at the array level, while additional processes such as activation functions and accumulation are carried out digitally.

Numerous digital and analog IMC methodologies necessitate peripheral electronics for the conversion between digital and analog signals (DACs and ADCs) to interact with the primary digital computing framework, providing the required input for memory arrays and allowing for further processing of the outputs from the IMC. The incorporation of ADC and DAC converters marginally expands the area of the memory array and slightly boosts the energy usage of the entire AI computing hardware setup. As depicted in Figure 4.10, the ADC can represent up to 80.9% of the IMC area and up to 57.4% of the overall computing energy in traditional IMC processors utilizing resistive random-access memory (ReRAM) cells [167].

Some eDRAM macros have also been designed to perform MAC in ANNs, either in the analog or digital domain. These macros achieve high throughput when accelerating neural networks; however, they are either incapable or inefficient in performing logic operations, which are critical for some AI applications, such as database query processing. Moreover, the overhead of additional circuits, such as ADCs, DACs, adders, and shifters, remains nonnegligible. For instance, in an analog-domain macro, ADCs and DACs contribute approximately 37% of the total area overhead. Similarly, in a digital-domain macro, adders and shifters account for around 40% of both area and energy consumption [168].

Moreover, the resistance in wires and the IR drop (the reduction in voltage between both ends of a conductive wire linking each memory cell) in extensive CBA leads to a decline in voltage and could induce incorrect programming or an incorrect selection of the targeted memory cells. This restricts both computational accuracy and the ability to scale effectively. Addressing these challenges requires advancements at the circuit level in addition to optimizing devices, ensuring reliable integra-

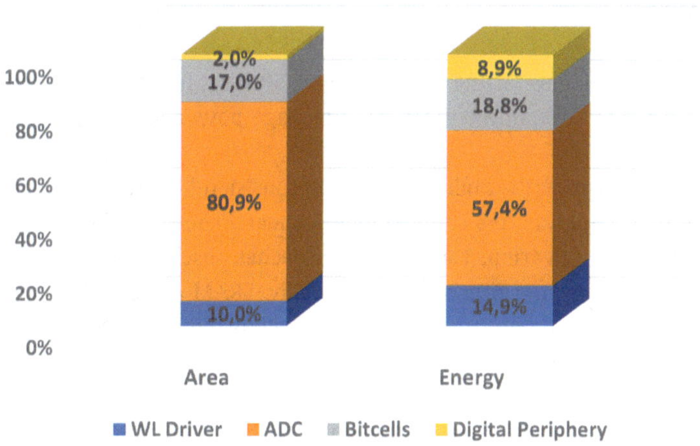

Figure 4.10: Area and energy overhead for analog-to-digital domain conversion – inspired by Kim et al. [167].

tion at high memory density. This entails the development of architectural approaches that are customized to meet particular application requirements [169].

IMC-based AI hardware system architectures are specifically crafted to address the unique needs related to training and inference phases. Systems focused on training must ensure robust endurance, maintain linearity, and allow for adaptable programming methodologies for continual weight adjustments. This necessitates that the memory cell features exceptional endurance, notable linearity and symmetry concerning voltage or current pulses/inputs, the capability to store multiple levels, and minimal variability.

On the other hand, systems that emphasize inference are engineered for the reliable long-term retention of pretrained models and the ability to execute rapid read processes with minimal energy consumption. Essential characteristics of the memory cell needed for training include an extended retention time, reduced power consumption during reading and writing, a significant ON/OFF distinction between various storage levels, and a high degree of integration density achievable through CMOS technology [169].

Analog IMC generally presents lower computing performance than digital processors and accelerators, but a higher energy efficiency than digital IMC. As an example, using SRAM arrays, the measured maximum energy-efficiency and area efficiency are 42.6 TOps/W and 4.4 TOps/mm^2, and 65.9 TOps/W and 4.9 TOps/mm^2 for digital and analog IMC, respectively [170].

Figure 4.11 compares the performance of IMC based on SRAM and embedded DRAM and NVM with that of GPU and ASIC. It unveils that IMC can provide computing speeds similar to or close to those of digital CMOS devices, but with a quite higher energy-efficiency than the digital accelerators.

4.3.1 SRAM-based IMC

Regarding the memory technologies for the IMC paradigm, SRAM is highly mature but is volatile, consumes large Si area, and suffers from leakage power in the constitutive FETs.

SRAM-based IMC is founded on the ability of SRAM to consistently keep data stable through static storage that employs cross-coupled inverter logic. The reliability of this storage enables the application of various computing approaches, resulting in continual enhancements in their efficiency and precision. Additionally, SRAM has benefits such as low energy usage during read and write processes, as well as the presence of high-density cells that are optimized across different manufacturing techniques.

However, SRAM-based IMC also comes with certain pitfalls. To begin with, the SRAM cell alone requires six transistors, and additional transistors are needed for the SRAM-based IMC. Consequently, the total number of transistors needed for a single MAC unit ranges from 6 to 18, leading to increased consumption of silicon area by 1.4 to 2 times. This fundamental constraint on the on-chip capacity of SRAM means that when processing large DNNs or comparable tasks in SRAM-based IMC, there is a frequent need to access external DRAM. This situation leads to a significant decline in the overall performance of IMC, as the high ratio of weight reloading can negatively impact task-level energy efficiency. Furthermore, SRAM-based IMC is limited to holding just 1 bit per cell,

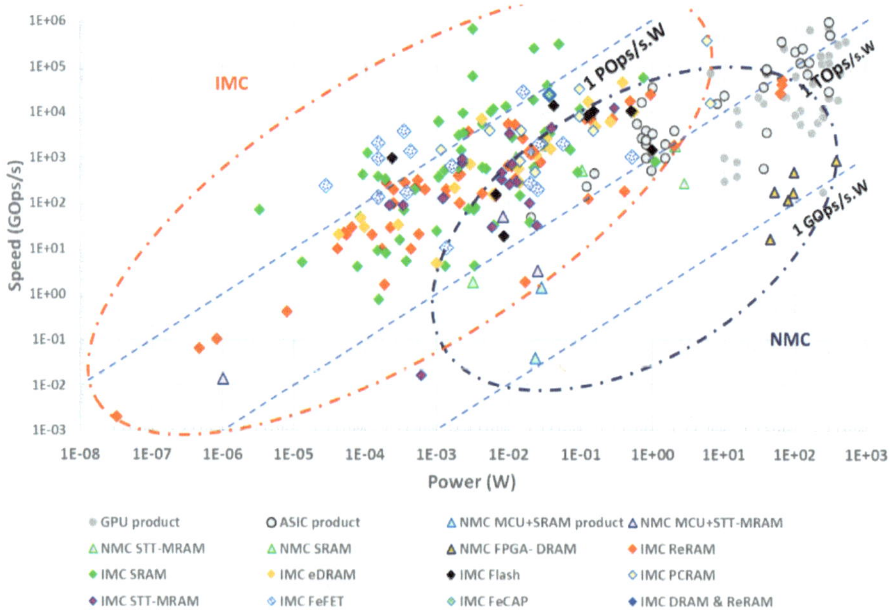

Figure 4.11: Speed in GOps/s versus the power consumption of NMC and IMC accelerators compared to some digital AI devices (GPU and ASIC) updated from [1] – data from [83, 132, 167, 170, 174–208].

which restricts weight density. Lastly, SRAM is not ideal for uses that need long-term data retention since it loses information when the power is off. This characteristic of volatility makes it less suitable for applications driven by events [171].

SRAM-based IMC presents the highest computing speed and energy efficiency among all the IMC approaches (Figure 4.11). SRAM-based IMC is interesting because of its high design flexibility and mature fabrication support. Unfortunately, SRAM and SRAM-based IMC face the challenge of low density: it is challenging to store all weights inside one chip for large neural networks. Consequently, for such large DNNs, the data in the SRAM needs to be dumped and reloaded from the DRAM, which results in undesired energy and latency costs, and does not solve the "memory-wall" bottleneck issue. To allow SRAM-based IMC to perform complex operations involved in DNNs, the basic 6T cells in the circuit may be modified, and peripheral circuits must be added. 7T cells have been investigated for splitting the read and write paths. 8T cells have been studied to perform logic IMC Boolean operations. 8T1C configuration which utilizes analog-mixed-signal capacitive-coupling computing to perform Boolean-and-accumulate operations for binary DNNs. 10T SRAM bit cells were studied for supporting binary MAC operation to achieve a high throughput and energy efficiency, and for supporting multibit MAC operations to increase the inference accuracy [172].

Experimental results highlight that the 7T SRAM-based IMC can deliver energy efficiency that is 3.38 times greater while increasing by 50% the chip size (approximately 81.80 mm^2) when contrasted with the standard architecture utilizing a traditional 6T SRAM array, which only enables row-by-row reading. Utilizing a more sophisticated 8T bit cell, the 8T SRAM-based IMC can achieve even greater energy savings with an acceptable increase in area (121.51 mm^2) in comparison to the 7T SRAM-based IMC [173].

While the energy-efficiency and computing speed of digital DNN accelerators and processors tend to increase with a decrease in the device dimensions (Figure 4.3), the computing speed and the corresponding energy-efficiency of SRAM-based IMC accelerators do not look to be strictly correlated to the technology node (Figure 4.12). This behavior could result from the number of FETs constituting the SRAM cell, and/or the memory bandwidth between the SRAM cells and the off-line DRAM cells. Such a behavior has already been unveiled in Shanbhag and Roy [181], which highlights that 18T SRAMs produced using the 22 nm technology node show the highest computing speed, compared to 8T and 12T SRAMs manufactured using the 5 and 7 nm technology nodes. Moreover, the SRAM-based IMC energy efficiency tends to decrease with the computing density, expressed in GOps/mm^2, whatever the technology node used to produce the SRAMs [181].

The energy efficiency of SRAM-based IMC accelerators mainly ranges from 10 to 1,000 TOps/s.W (Figure 4.12), which is roughly one order of magnitude higher than the average DNN digital accelerators or processors (Figure 4.3).

Accommodating all the weights on-chip for large-scale DNNs remains a great challenge for SRAM-based IMC with limited on-chip capacity, and requires a large amount of off-chip DRAM access. The foundry-mature SRAM-based IMC array possesses high computation speed and energy efficiency, but at the cost of a large area with quite lim-

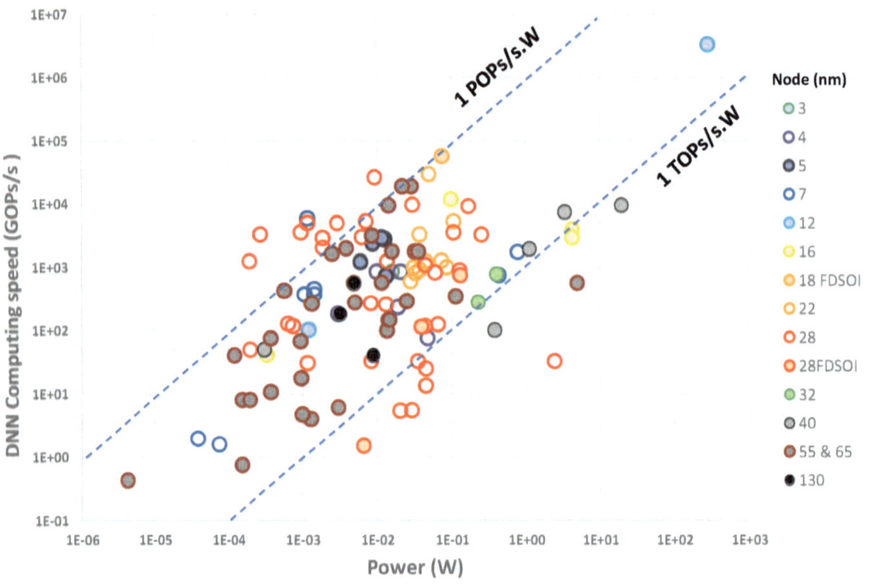

Figure 4.12: Speed in GOps/s versus the power consumption of SRAM-based IMC DNN AI accelerators as a function of the technology node – data from [146, 147, 149, 152, 201, 209–245].

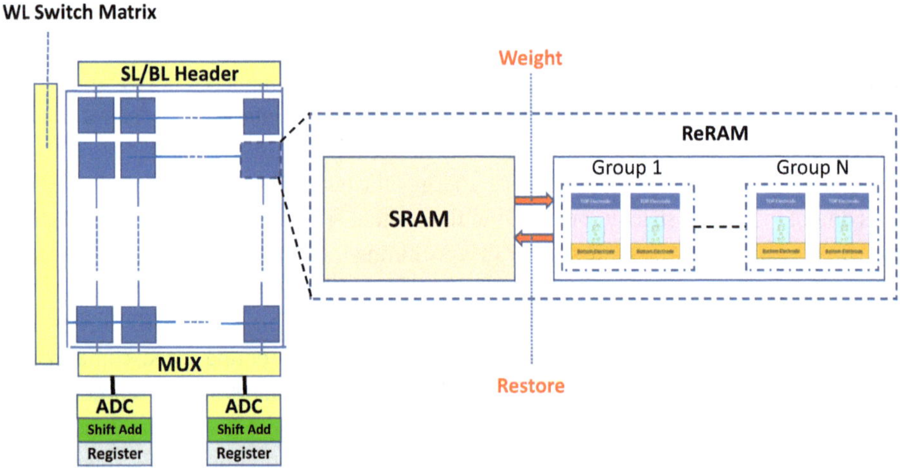

Figure 4.13: NVSRAM sub-array and bit cell for IMC.

ited capacity of on-chip volatile SRAMs. Therefore, IMC has been performed by combining fast, robust, and energy-frugal SRAM with NVMs such as ReRAMs (Figure 4.13). The nonvolatile ReRAM-based IMC with single-level cells or multilevel cells enables higher weight-storage density with a smaller device footprint than SRAM-based IMC. However,

ReRAM-IMC is generally sensitive to stochastic resistance variations, which result in error-prone MAC results and accuracy loss.

Combining SRAM with ReRAM to form an NVSRAM allows to circumvent these issues. Each NVSRAM-IMC cell contains one SRAM-IMC cell for high-performance MAC operations and multiple grouped ReRAMs for high-dense weights storage (Figure 4.13). The capacity of weights stored in ReRAMs is magnified compared to the previous SRAM-based IMC counterpart. The capacity of on-chip ReRAMs can accommodate the weights of large-scale DNNs, thereby eliminating the off-chip DRAM access. In addition, by keeping weights in ReRAM, the system can be turned off to conserve energy when the accelerator is not in use.

The bits of each quantized weight are stored in the selected ReRAMs, which share the same source line from a certain group of NVSRAM-CIM cells. Before inference, each bit of weights is restored from one ReRAM to an SRAM cell in array-parallelism. The MAC operations are then performed parallelly for the signed weight based on the SRAM-IMC mechanism. The output on each column is sensed by ADC, and the result of the MAC operation is generated by the shift-and-add unit (Figure 4.13). Weights in other ReRAMs are restored to SRAMs for completing the DNN. Considering the limited SRAM-based IMC resources, multiple cycles may be needed for MAC operations of

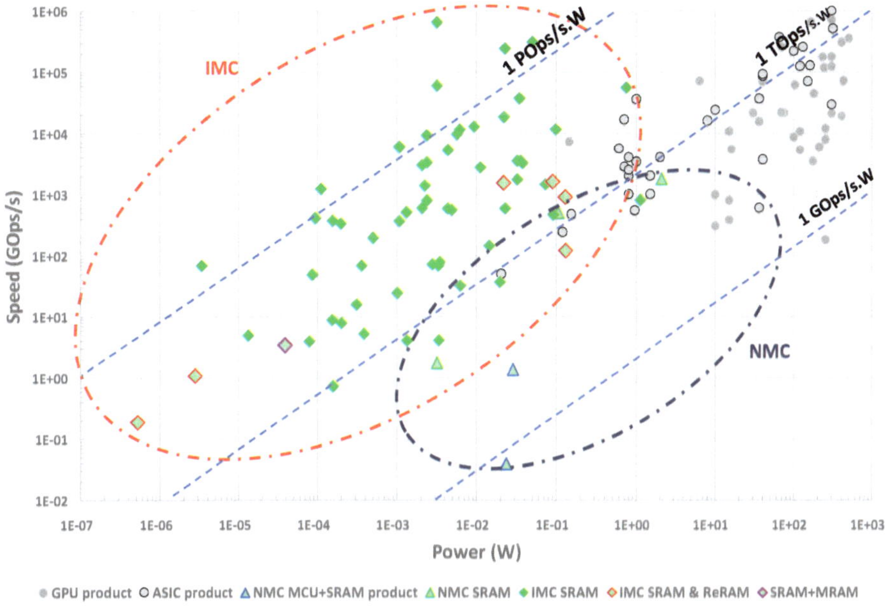

Figure 4.14: Speed in GOps/s versus the power consumption of SRAM-based IMC, SRAM-based NMC, and NVSRAM-based DNN – data from [189–191, 193, 195, 203–205, 207, 229, 232, 247–251].

large DNN layers. The overhead of energy and latency induced by restore operations can be omitted in comparison to SRAM-based IMC [246].

Using similar technology nodes, NVSRAM-based IMC can show up to 2.9× and 2.0× enhanced energy efficiency, respectively, compared to the baseline designs of SRAM-IMC and ReRAM-IMC, respectively [246]. Figure 4.14 compares the SRAM- and NVSRAM-based computing speeds and their energy efficiency. From this data, computing performance and the energy efficiency of NVSRAM-based IMC are lower than those of SRAM-based IMC, while their energy efficiency is higher or similar to that of digital-based IMC. NVSRAM-based is more dedicated for inferencing DNN models and for edge-AI applications, which require low-power consumption, and to mitigate any power disruption could impact the inferencing process.

Various NVSRAM cell architectures have been recently proposed by combining SRAM with eDRAM and FeFET. However, at the time this book was written, only simulation results on those architectures have been published.

4.3.2 Embedded DRAM-based IMC

Embedded memories have become a viable substitute for traditional SRAM because of their potential impressive storage abilities (Figure 4.15) and minimal dynamic

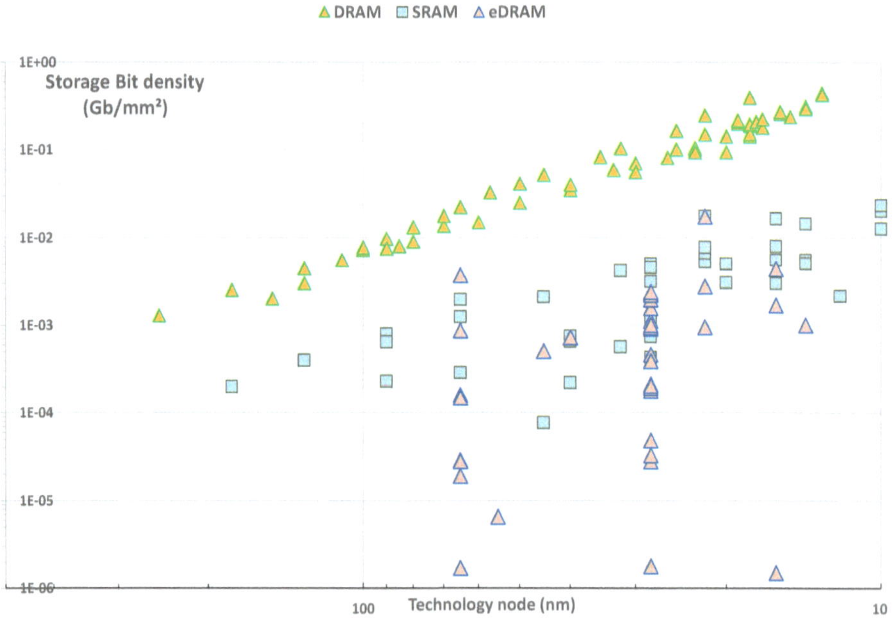

Figure 4.15: Comparison of the storage capacity of DRAM, SRAM, and eDRAM with the technology node – data from [119, 196, 198–202, 252–263].

power consumption. Although embedded memories can face issues with data retention, they present considerable advantages for uses that do not necessitate extended data storage. In this regard, embedded memories have been investigated for IMC with a focus on machine learning applications [165].

Due to their high technology maturity and despite the fact that eDRAM is a volatile memory and needs to be always connected to a power source, eDRAM-based IMC exhibits quite attractive results. Embedded DRAM-based IMC can perform both digital, analog, and analog-digital mixed computing. They lead to AI computing speeds similar to GPU and ASIC but with a far better energy efficiency (Figure 4.11). Therefore, they represent reasonable alternatives to digital devices as accelerators for training and inference of DNN models.

In their IMC configuration, the storage capacity of eDRAM is lower than that of stand-alone DRAM, but similar to that of SRAM (Figure 4.15). Therefore, eDRAM-based IMC represents a very interesting approach as digital devices for energy-efficient training and inference of DNN models.

4.3.3 Embedded NVM-based IMC

Since the connection of volatile memories to an external source is not often possible for numerous AI applications, NVMs represent an attractive solution for efficient IMC-based DNN acceleration.

FG- and CT-type NAND flash memory represent an exciting development for inference-driven computing platforms, allowing for nonvolatile retention of synaptic weights by holding charge within a trapping layer to accurately control the conductance of the transistor's channel. Recent progress has concentrated on optimizing the charge trapping process to improve the efficacy of IMC uses [169]. NAND flash-based IMC shows comparable effectiveness than e-DRAM-based IMC regarding processing speed and energy efficiency (Figure 4.11).

Another promising solution for the IMC architecture is using the emerging NVMs, including ReRAM, phase-change random-access memory (PCRAM), magnetic memory (MRAM), ferroelectric random-access memories (FeRAMs), and electrochemical RAM (ECRAM).

ReRAMs exhibit a structure composed of metal-insulator-metal similar to DRAMs. The core functionality of ReRAMs revolves around the creation of a conductive filament within a dielectric layer located between two metal electrodes. The formation or disruption of this filament allows for the detection of at least two different resistance levels: high resistance state and low resistance state. Devices based on ReRAM can perform computations at speeds comparable to GPUs and ASICs while also being more energy-efficient. ReRAMs can be categorized into three main types according to the method used to create the conductive filament.

In conductive bridge random access memory (CBRAM, also referred to as ECM for electrochemical metallization memory), the conductive filament is formed due to the migration of metal cations. Improving the confinement of filament formation by reducing the interfacial Schottky barrier height and doping metal clusters results in better variability and more consistent conduction traits. These features make CBRAMs suitable for high-performance immediate memory computing applications that emphasize training [169].

The oxygen-ion-based ReRAM acts as a two-terminal memristor, allowing the drift and redistribution of oxygen anions within the oxide dielectric, which aids in smooth and reversible changes in electrical conductance or resistance. This device has been extensively studied as a promising solution for supporting multiple storage states while demonstrating reduced variability. Compact 1R CBA setups containing interface-type oxygen-ion-based ReRAM devices can be effectively used in training-focused projects or IoT applications. An approach such as establishing an oxygen concentration gradient layer greatly improves durability and the ability to achieve high-resolution multi-level storage states, which are critical for creating IMC hardware suitable for demanding computational tasks [169]. Within this book, these ReRAMs are further integrated in the OxRAMs category.

OxRAM, also known as valence change memory, is a type of filamentary memristor where conductive filaments form from either oxygen vacancies or both metal cation and anion vacancies. The rapid operational capabilities, strong retention, and compatibility with CMOS integration offered by the switching layer made from amorphous oxide position OxRAM as a substantial candidate for IMC applications. Current investigations into OxRAMs focus on enhancing uniformity in switching and retention to boost overall device reliability. Novel filament-guiding architectures, employing confined structures, vacancy-profile adjustments, and carefully chosen materials, have made significant strides in reducing random filament formation while ensuring stable operation across device arrays. These advancements position OxRAMs as appealing choices for applications in IMC that require high precision in inference, thanks to their low-variation memory characteristics [169].

PCRAM or PCM (phase change memory) cells are essentially thin-film resistors, where the electrical resistance significantly varies under low-field conditions depending on the phase state of the active area. The main materials used for these cells are chalcogenides, chosen for their distinctive physical properties. Chalcogenides can be quickly and reversibly changed between amorphous and crystalline phases when an electric current is applied, generally resulting in a marked difference in resistivity [116]. However, the creation of materials for these technologies encounters major challenges, including the need to balance crystallization speed with the stability of the amorphous state and to control resistance drift, complicating consistent multilevel storage. Recent advancements in PCRAMs highlight broadly applicable methods, such as enhancing thermal confinement, minimizing the size of the active phase change, and managing resistance drift to boost the dependability of multilevel storage. While these strategies generally assist in optimizing devices, specific applications like superlattice designs, heterogeneous nanocomposites, projection liners, and nano-filament structures offer distinct avenues for achieving ultralow power efficiency. These cumulative innovations position PCRAM as an attractive choice for energy-conserving, inference-oriented IMC systems [169]. Furthermore, PCRAM-based IMC demonstrates particularly fascinating features when placed alongside GPU and ASIC (Figure 4.11).

Nowadays, ferroelectric nonvolatile memories (FeRAM) mainly use the polarization properties of ferroelectric materials like doped-HfO_2 films. There are three types of ferroelectric (FE) devices being investigated for memory use: the one transistor one capacitor (1T1C) (or 2T1C) FeCAP, 1T FeFET, and the ferroelectric tunnel junction (FTJ). The write process for these ferroelectric memories involves applying a specific electrical field, making it a process that is inherently efficient in terms of power consumption.

Much like a DRAM cell, a FeCAP memory cell is composed of a metal-ferroelectric-metal film stack. In the case of FeCAP, data reading is destructive, as it requires switching the ferroelectric layer in a particular direction and observing the difference between a switching event and a non-switching one. Therefore, reprogramming is needed after every read operation, which does not conform to the IMC standards. Significant research and development are ongoing to realize nondestructive reading capabilities for FeCAP cells, aimed at potentially replacing DRAM in the near future [265, 266] and enabling IMC functionalities with FeCAPs [267].

In the FeFET structure, the ferroelectric material is integrated within the gate stack of the FET. The gate stack is made up of a metal/ferroelectric(/insulator)/semiconductor (MF(I)S) configuration. Data in FEFETs are held through bound ferroelec-

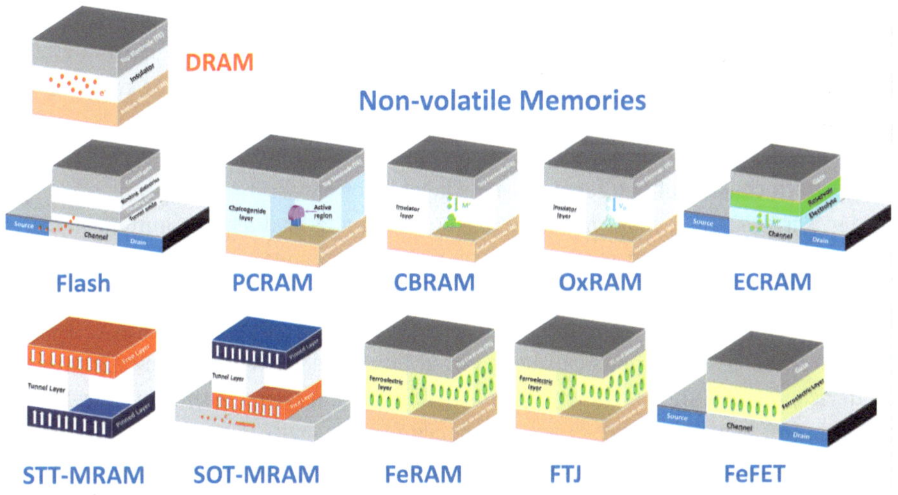

Figure 4.16: Schematic operations of the volatile DRAM cell and of some nonvolatile memories. Red circles: electrons, green circles: metal ions, and light blue circles: oxygen vacancies.

tric polarization, with operational states adjusted by the gate voltage: a positive pulse aligns the polarization toward the channel, thereby decreasing the threshold voltage V_{th} to a "program" (PGM) state, whereas a negative pulse reverses the polarization, increasing V_{th} to an "erase" (ERS) state. This polarization-driven modulation of V_{th}, which is similar to that of NAND flash memories, reduces reliance on trapped charges, thus enhancing retention and durability at smaller dimensions, which in turn improves scalability. Consequently, by leveraging the internal gain of transistors, a nondestructive read operation can be accomplished by detecting the drain-source current. The data can be kept in a nonvolatile state. The FeFET approach provides ultralow power consumption and, since the FeFET cell operates as a three-terminal device (1T), it should scale effectively at higher logic levels. The FeFET memory cell facilitates simultaneous computing capabilities and data storage, which is advantageous for deep neural network training and inference, while also providing multilevel storage, a high ON/OFF ratio, and minimal power usage. Additionally, given that FeFET-based IMC achieves processing speeds comparable to ASICs but with considerably better energy efficiency, FeFET memory presents itself as a compelling option for NVM in IMC, both for maintaining synaptic weights and functioning as neurons.

The FTJ functions as a memory apparatus with two terminals (Figure 4.16). This device utilizes a very thin layer of ferroelectric material that facilitates electron tunneling. The current that tunnels is influenced by the polarization states of the ferroelectric, which are altered through switching of the polarization. When an electric field is applied, the polarization state of the FTJ can be observed by measuring the flow of current. The ferroelectric polarization can be adjusted partially, allowing for continuous and analog tuning of the tunneling conductance, which goes beyond mere binary configurations. This capability for multilevel and nearly linear conductance variation is particularly beneficial for training-dependent IMC applications, where precise and incremental adjustments to weights are crucial. Like the FeFET, the FTJ supports a nondestructive method for reading. However, there is a potential issue with the read current being very low, posing challenges for future industrial usage of FTJs.

The primary component of MRAM technology is the magnetic tunnel junction (MTJ), which consists of two ferromagnetic layers that encase a nonmagnetic tunneling barrier. Ferromagnetic materials have magnetic domains that align in a uniform direction, leading to a net positive magnetism in these compounds. The magnetization's orientation in one of the ferromagnetic layers can move freely, allowing for both parallel and antiparallel configurations between these layers. The varying electrical resistance between these configurations can be detected by an electric current flowing through the device, representing the binary or analog information stored in the MRAM cell.

The most widely researched and developed MRAM device is the STT-MRAM, which features a perpendicular MTJ and is a well-established, reliable NVM option for industrial MCU and IoT applications. The concept of STT deals with adjusting the magnetic moment of a ferromagnetic layer without any applied magnetic field, solely through the transfer of spin angular momentum via a spin-polarized current. In MRAM cells, STT results from the journey of spin angular momentum between two magnetized layers across an ultrathin tunnel barrier [269]. The significant benefit of STT-MRAM lies in its rapid switching speed when compared to other emerging NVM technologies. However, a primary challenge for MRAM is its scalability, which is difficult due to the complexities involved in etching multiple ultra-thin films with good conformity. Additionally, STT-MRAM exhibits a small ON/OFF current ratio in comparison to other newer memory types, making multilevel storage more challenging to achieve. Furthermore, STT-MRAM encounters intrinsic trade-offs concerning speed and endurance because of a shared circuitry for writing and reading. STT-MRAM-based IMC provides exceptional energy efficiency for AI computing compared to GPU and ASIC, although its computation speed does not match that of its digital counterparts (Figure 4.11).

While STT is produced when a spin-polarized current flows through an MTJ, spin-orbit torque (SOT) occurs when a current is applied laterally through a bilayer of magnetic and nonmagnetic materials. Both mechanisms can be utilized for changing magnetization and storing data. Typically, a film made from heavy metals like platinum (Pt), tantalum (Ta), or tungsten (W) acts as the nonmagnetic layer. Due to significant spin-orbit interaction present in the heavy metal and/or at the junction with the ferromagnetic part, the in-plane current gives rise to a pure transverse spin current, which can impart angular momentum to nearby magnetic moments. In the SOT setup, the current flows sideways within the heavy metal or non-magnetic layer rather than moving through an MTJ as it does in the STT method. The SOT configuration has the potential to create SOT-MRAM featuring quicker spin response, enhanced durability, lower power consumption, and operation without a magnetic field. Because of these benefits, SOT-MRAM is viewed as essential for future MRAM technologies.

SOT-MRAM with a three-terminal arrangement allows separating read and write paths. SOT devices have matured in the past few years with demonstrated CMOS array functionality, BEOL 400 °C process compatibility, and improved memory performance in single devices, showing their overall potential for higher technology readiness. However, a significant open challenge in SOT-MRAM development is the need for symmetry breaking to enable deterministic write operation of the memory states, which is conventionally achieved with an external applied field. This field requirement imposes strong technology design constraints on achieving a dense device integration in an array, and therefore innovations in field-free switching solutions to the write operation are currently investigated [270].

An electrochemical RAM, abbreviated as ECRAM, is an innovative memory component that utilizes the movement of ions for data storage and features three terminals. It is made up of layers that indicate electrical conductivity, electrolytes that facilitate the movement of ions, and reservoirs that manage the number of ions present (see Figure 4.16). The gate electrodes are situated on top of the reservoirs, with the source and drain electrodes positioned at opposite ends of the channel. To write data into an ECRAM cell, a voltage or current is applied to the gate electrode, which controls the conductivity of the channel. To read information from the device, the conductivity of the channel is assessed at both the source and drain ends. ECRAM has been extensively studied for its applicability in compute-heavy IMC hardware due to

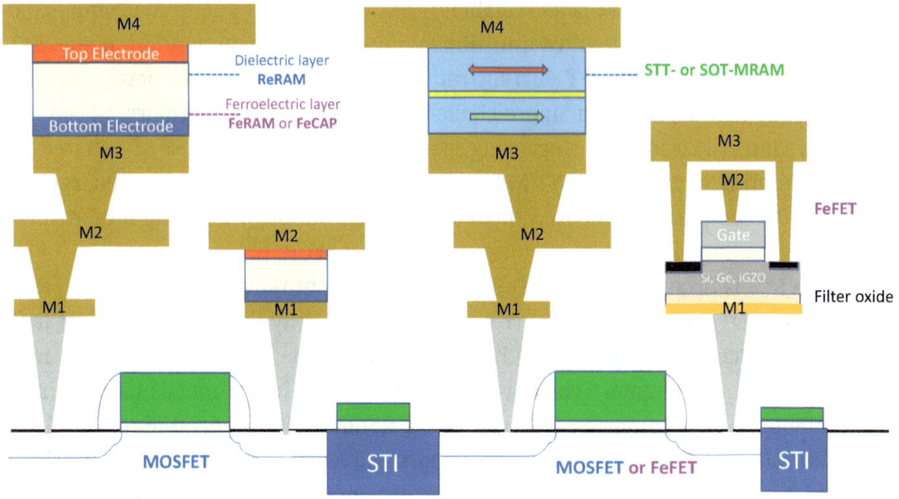

Figure 4.17: Schematic integration of emerging NVMs.

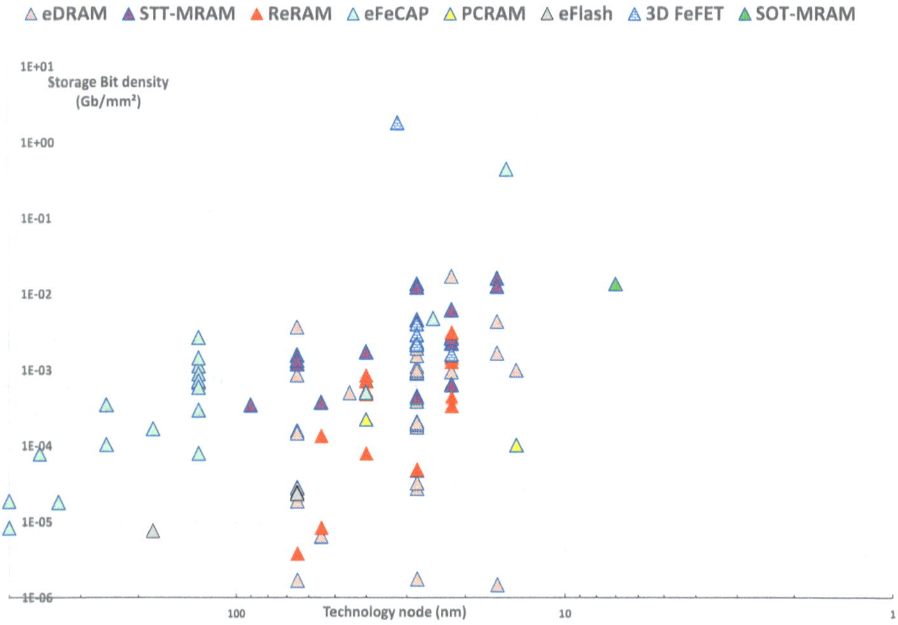

Figure 4.18: Comparison of the storage capacity of eDRAM and eNVM with the technology node – data from [119, 189, 192, 196, 198–202, 212, 248, 252–263, 271–298].

its adjustable analog states, ability to update conductance in a linear fashion, and its feature of non-destructive reading.

While FeFET cells can be integrated close to the addressing FET (Figure 4.17), emerging NVM cells such as ReRAMs, FeRAMs, STT-, and SOT-MRAMs are generally embedded in the metal lines of logic devices. They can be embedded close to their addressing transistor between the first and second metal layers, which are made of Cu since the 90 nm technology node, or in the upper metal lines, like between the third and fourth Cu layers, as shown in Figure 4.17. Therefore, embedded non-volatile memories need to be manufactured at a temperature lower or equal to 400°C for complying with CMOS thermal budget. FeFETs can also be integrated in the first metal layers of logic devices as soon as the FeFET channel, which can be constituted of Si, Ge, or IZGO ($In_2Ga_2ZnO_7$), an oxide semiconductor, can be easily integrated in those metal layers (Figure 4.17).

The storage bit density of the eNVMs and eDRAM is quite similar (Figure 4.18). The development of ferroelectric-doped HfO_2 films, which perfectly comply with CMOS, as substitutes for $Pb_{1-x}Zr_xTiO_3$ layers, has allowed the development of 3D-stacked FeFET and FeRAM memory arrays, which increases the storage bit capacity. ReRAMs can also be stacked for enhancing their storage capacity.

Figure 4.20 shows the IMC computing speed versus the consumed power for embedded mature memories-based IMC eDRAM, ePCRAM, and embedded NVM and 3D Flash.

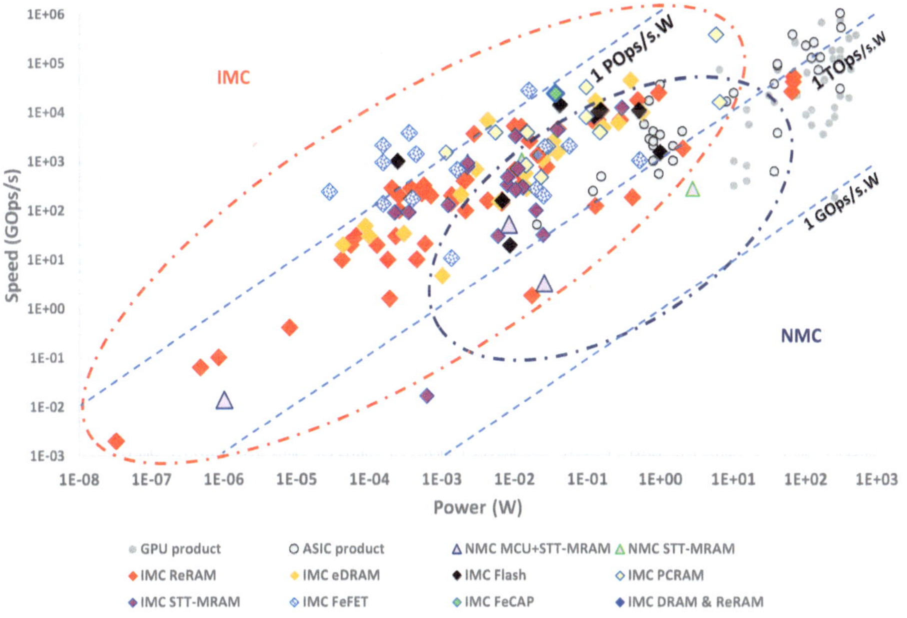

Figure 4.19: Speed in GOps/s versus the power consumption of eDRAM- and eNVM-based NMC and IMC accelerators compared to some digital AI devices (GPU and ASIC) – data from [75, 83, 167, 170, 187–208, 245, 299–304].

The IMC speed using those memories is lower than that of digital accelerators (Figure 4.1). It is currently limited to around 10^5 GOps/s, while digital accelerators reach 10^6 GOps/s. However, when the energy efficiency of digital accelerators ranges from 1 to 100 TOps/s.W, the energy efficiency of eNVM-based IMC can be higher than 1 POps/s.W for FeFET, and close to 1 POps/s.W for 3D NAND flash, eDRAM, FeCAP, and ReRAM (Figure 4.19). Their energy efficiency is similar to that of SRAM-based IMC, even if SRAM-based IMC can achieve a higher computing speed than eNVM-based IMC (Figure 4.11).

The computing speed for eDRAM-, eFlash-, and ePCRAM-based IMC tends to grow for advanced technology nodes, highlighting the utmost importance of bit density for performing the IMC (Figure 4.20). Such a trend is difficult to unveil for Flash memories since experimental IMC values are available for various types of flash memories: eFlash, 3D Flash, and flash coupled with SRAM. The computing performance and the energy efficiency are lower than those of SRAM-based IMC. Their energy ranges from 10 to 500 TOps/s.W, which is comparable to the best digital accelerators or processors (Figure 4.3).

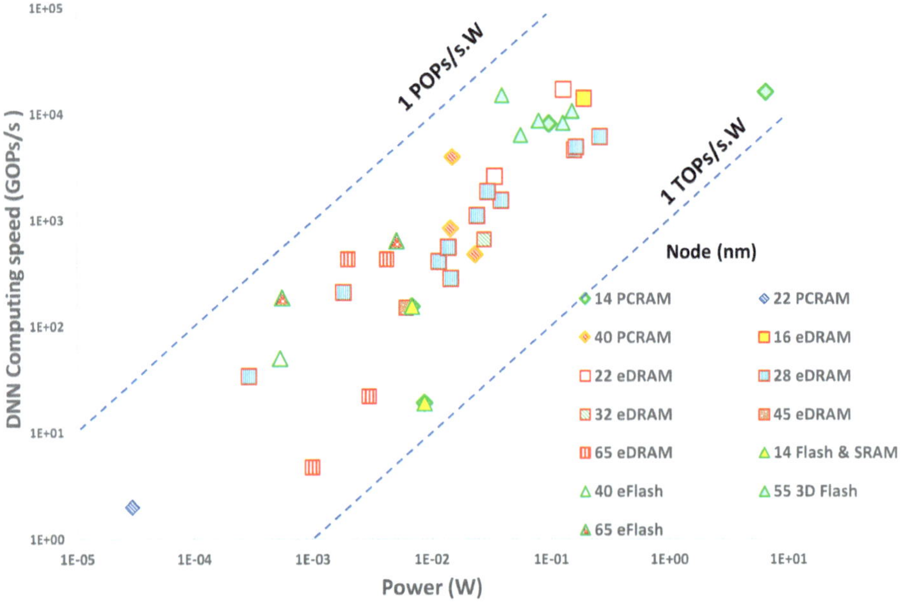

Figure 4.20: Speed in GOps/s versus the power consumption of eDRAM, ePCRAM, and flash-based IMC DNN accelerators as a function of the technology node – data from [119, 147, 177, 192, 196–201, 209, 232, 253, 260, 305, 306].

Similarly, Figure 4.21 shows the IMC speed as a function of the consumed power for embedded emerging memories as a function of the technology node. Globally, the IMC computing speed is also lower than that of SRAM-based IMC and digital CMOS

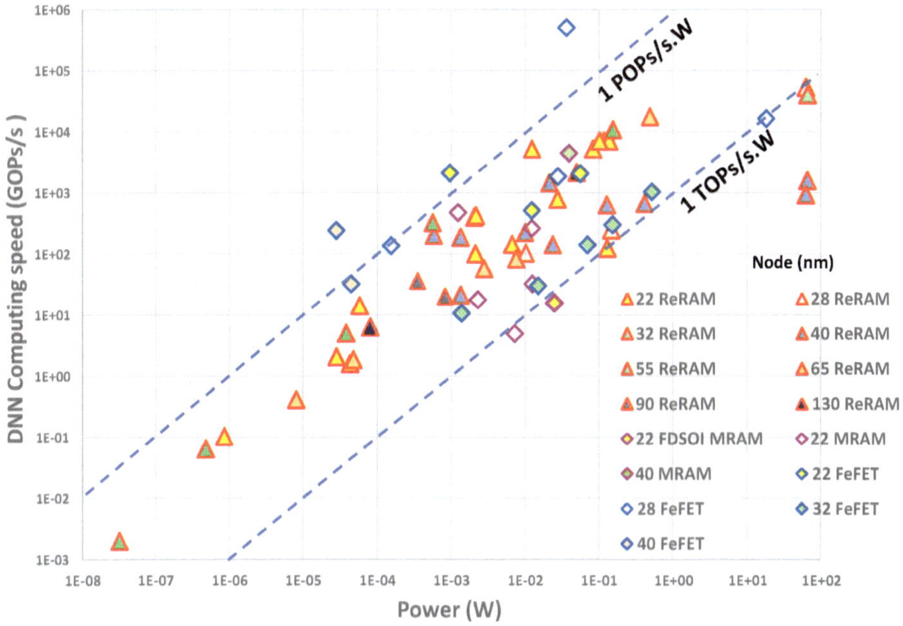

Figure 4.21: Speed in GOps/s versus the power consumption of embedded emerging memories-based IMC DNN AI accelerators as a function of the technology node – data from [83, 115, 119, 147, 167, 177, 178, 183, 184, 186–195, 204–208, 229, 232, 243, 244, 247, 248, 278, 279, 304, 306, 307].

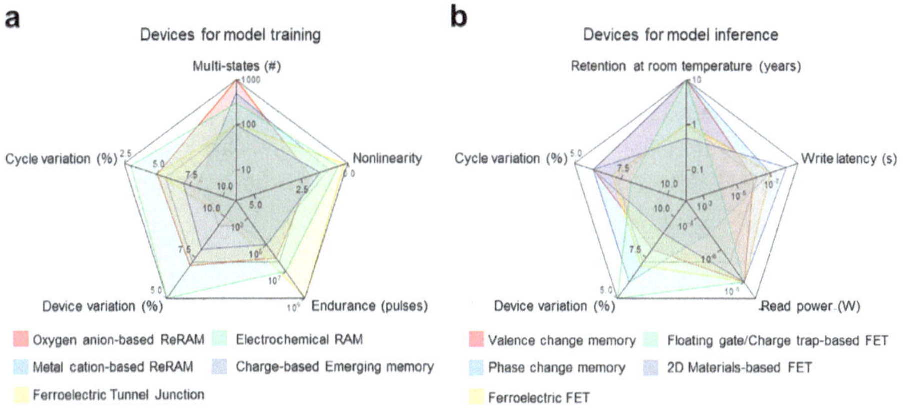

Figure 4.22: Selected device properties illustrating distinct performance trade-offs across device classes for (a) train-centric IMC applications and (b) inference IMC applications [169].

(Figures 4.3 and 4.12), except for 3D FeFET, which also exhibits the highest bit density. Their power efficiency, ranging from 10 to 1,000 TOps/s.W, is higher than that of digital accelerators and processors (Figure 4.3), and comparable to that of SRAM-based IMC (Figure 4.12).

The eNVM-based IMC concept is still at the R&D level, especially with emerging memories such as FeFETs, ReRAMs, and STT-MRAMs, and deserves deeper investigation before its industrial deployment for the inference and, potentially, the training of DNNs.

To elucidate the assessment of performance among different new memory categories, crucial device characteristics have been integrated – such as nonlinearity, durability, variability, data retention, response time, and energy usage – into a consolidated evaluation based on the distinctive levels of each category (Figure 4.22). These radar charts illustrate the comparative connections between the memory categories, facilitating the analysis of their potential appropriateness for IMC applications focused on training and inference. Figure 4.22 excludes the parameters for STT and SOT MRAM, in addition to those for embedded DRAM devices.

Table 4.1 gathers the main properties of the emerging NVMs, which are compared with those of DRAM and NAND flash. Among the aforementioned emerging NVMs, FeFETs and ECRAMs could eventually be useful for DNN training, and their current storage and computing properties can comply for the inference of DNNs. PCRAM and ReRAM, as well as flash memories, can also be used for the inference of DNNs (Table 4.1).

The markets for MRAM, ReRAM, and PCRAM are growing slowly, especially for embedded applications and for some niche applications as stand-alone memories. FeRAMs are very attractive for replacing DRAMs, and could be produced in high volume in the forthcoming years. New emerging NVM technologies such as ECRAM, FTJ, etc., are still in their development phase, while their volume production should not occur before 8 years [308]. Furthermore, none of those NVMs are currently close to DRAM and SRAM in terms of performance, and to NAND flash memories in terms of cost.

4.4 Neuromorphic computing

In this book, the term neuromorphic computing concerns the use of SNNs for various AI applications. SNNs utilize sequences of action potentials or spikes, commonly referred to as spike trains, for data representation. An SNN is made up of various elements, including the network's layout, neuron models, synapse models, spike coding, and learning mechanisms [68].

The architecture of the network determines how neurons interconnect. Looking at the network structure in Figure 3.6, this setup includes input, excitatory, and inhibitory layers. The input image corresponds to the input layer, with its pixels linked to

Table 4.1: Main characteristics of emerging NVMs (adapted from [169, 264, 271, 309–314]). **DRAM is volatile.*

Technology	DRAM*	NAND	ReRAM	PCRAM	FeCAP
Feature size (nm)		15–70	<5	20–45	130–180
Storage capacity		250 GB	0.064–1 GB	1–32 GB	1–128 MB
ON/OFF	102–103	105–107	104–106	102–104	102–103
Multilevel (bits)	1	1–5	2–8	2–4	1
Write (V)	<1	<10	<3	<3	<3
Read/write time	∼10 ns	10 μ/1 ms	<10/10 ns	10/50 ns	<10 ns
Write (J/bit)	5 fJ–1 pJ	50 fJ/1 nJ	0.1–1 pJ	2.5–10 pJ	50–100 fJ
Integration	High	High	High	High	Low (PZT)
Retention	∼64 ms	>10 years	>10 years	>10 years	10 years
Endurance (cycles)	$>10^{16}$	$>10^{4}$	10^{6}–10^{12}	106–109	10^{15}
DNN training	No	No	No	No	No
	[169]		Yes		
DNN inference	Yes	Yes	Moderate	Yes	No
	[169]	Yes	Yes	Yes	
SNN applications	No	Moderate	Yes	Yes	Yes
Cost ($/Gb)	0.6	0.03	High	0.16	10
	5.25	0.15		2.85	
		0.103 (QLC) 0.182 (TLC)			
Technology	**FeFET**	**FTJ**	**STT-MRAM**	**SOT-MRAM**	**ECRAM (Li)**
Feature size (nm)			28–130		
Storage capacity			10–100 MB		
ON/OFF	5–50	10^{2}	1.5–2	1.5–2	40–103
Multilevel (bits)	5	1–5	>1	1	10
Write (V)	<5	2–18	<1.5–2	<1.5	<1
Read/write access time	100 ns	600 ps–10 ns	10/5 ns	0.3–3 ns	1–10 μs
Write (J/bit)	<1 fJ	∼0.5 fJ	100 fJ–2 pJ	∼100 fJ	∼100 fJ
Integration	High	High	High	High	Low

Table 4.1 (continued)

Technology	FeFET	FTJ	STT-MRAM	SOT-MRAM	ECRAM (Li)
Retention	>10 years	>10 years	>10 years	Medium	Medium
Endurance (cycles)	$>10^5$	$>10^8$	10^{16}	$>10^{15}$	$>10^6$
DNN training	Moderate	–	No	No	Yes
		Yes			Yes
DNN inference	Yes	–	No	No	Yes
	Yes				
SNN applications	Yes	–	Moderate	Moderate	Moderate
Cost ($/Gb)			30–70		

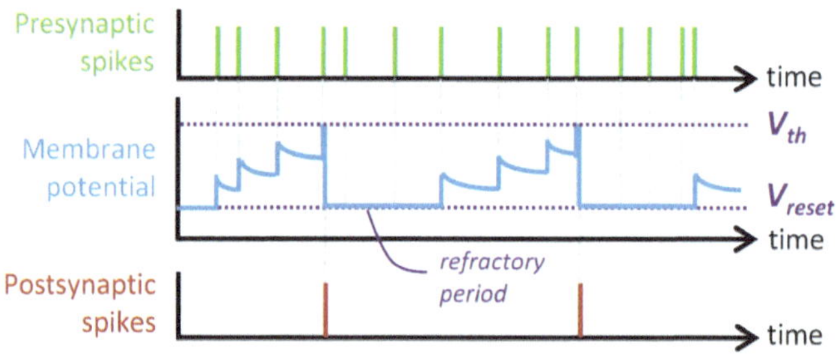

Figure 4.23: Illustration of the neuron dynamics of the LIF neuron model [68].

every excitatory neuron. Each excitatory neuron is tasked with identifying a specific class, and its output is linked to a designated inhibitory neuron. Each inhibitory neuron generates spikes that inhibit all excitatory neurons, except for the one that receives the input connection.

The neuron model illustrates the dynamics of a neuron. The leaky integrate-and-fire (LIF) model is preferred for its biologically realistic behavior while maintaining simplicity. The membrane potential (V_{mem}) increases in response to an input spike, while it decreases in the absence of one. A spike occurs when V_{mem} hits the threshold value (V_{th}), after which V_{mem} returns to the reset potential (V_{reset}), as demonstrated in Figure 4.23. For the synapse model, conductance increases with the corresponding weight (w) when an input spike is received, while w decreases otherwise [68].

The spike encoding converts the incoming data into sequences of spikes. Meanwhile, the learning principle determines the way a SNN acquires important characteristics from input data and adjusts the weights accordingly. Numerous biologically plausible learning principles have been created, yet spike-timing-dependent plasticity (STDP) remains the most commonly utilized learning principle [68].

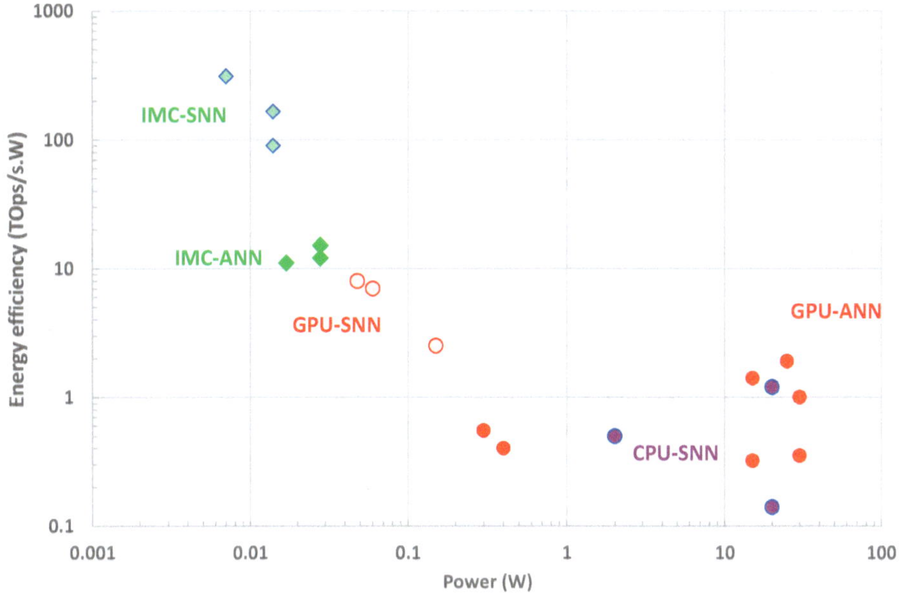

(a)

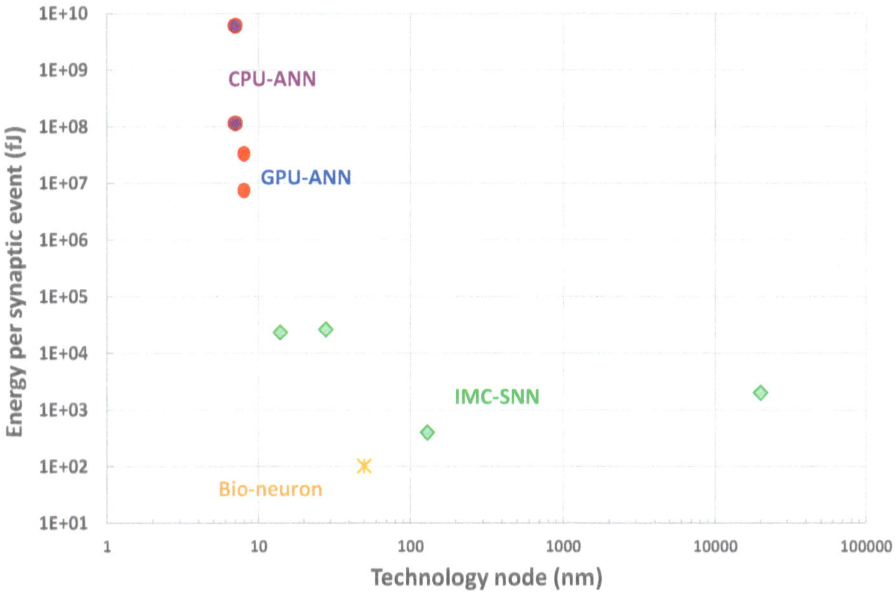

(b)

Figure 4.24: (a) Energy efficiency vs. power for different low-power edge AI accelerators – data from [314], (b) energy synaptic event vs. the technology node for digital ANNs and IMC SNNs, and comparison with human bio-neuron – data from [315].

As depicted in Figure 4.24, the use of SNNs instead of ANNs leads to a gain in energy efficiency of around 2 to 4 orders of magnitude for either training or inferencing such models. However, SNNs present many features unsuitable with a GPU execution. For example, spikes can be represented in the digital domain as single-bit events (high in the presence of a spike and low otherwise). In this case, the numerical representation used in the GPU is inefficient. Moreover, the computation is based on events: a neuron can react only in the presence of an active input spike, remaining in a quiescent state otherwise. Such an update policy would allow exploiting the sparsity of spikes typical of SNNs, with considerable savings in switching power.

In spite of significant advances in CMOS technology to emulate the logical processing and storage elements in SNNs, emulating complex neuronal and synaptic functionalities in CMOS technologies can lead to high Si area consumption. Moreover, there is a need for higher on-chip storage density. IMC using emerging NVM technologies allows component-wise direct emulation of neuronal/synaptic functionality at a

one-to-one level by the device characteristics. Such attractive characteristics, along with their high on-chip storage density and ability to perform massively parallel in-memory computations, make NVM technologies particularly suitable for neuromorphic computing systems, compared to digital accelerators (Table 4.1).

The AI computation schemes to define the synaptic weights in SNNs can be divided into two types: (i) supervised learning, where the inference process is based on training examples (like in ANNs and DNNs), and (ii) unsupervised learning, which does not use any feedback from an external teacher, but attempts to classify inputs based on the underlying statistics of the data.

Training SNNs is challenging due to the non-differentiability of the threshold function V_{th} of neurons. Classical back-propagation methods, used in ANNs, cannot be directly applied to SNNs, but other approaches have been developed, including supervised and unsupervised training.

In supervised training, the most common approach is to convert an ANN into an SNN, where the ANN's differentiable nonlinear function is trained using backpropagation, and the synaptic weights are used directly in the SNN. Unsupervised training uses a peculiar property of the brain: STDP. The assumption underlying STDP is that synapses tend to reinforce causal links. That means that when the presynaptic neuron spikes just before the postsynaptic neuron spikes, the synapse between the two becomes stronger. Therefore, if the presynaptic neuron spikes again, the synapse will allow the postsynaptic neuron to spike faster or with a higher occurrence probability.

SNNs have currently emerged as an energy-efficient approach to AI that leverages sparse and event-driven activations to reduce the computational overhead associated with model inference. While they have become competitive with ANNs for processing compact AI models, SNNs have proven to be more challenging to train, as aforementioned. As a result, their performance lags behind modern large ANNs, and until now, SNNs have yet to succeed at language generation on large-scale datasets.

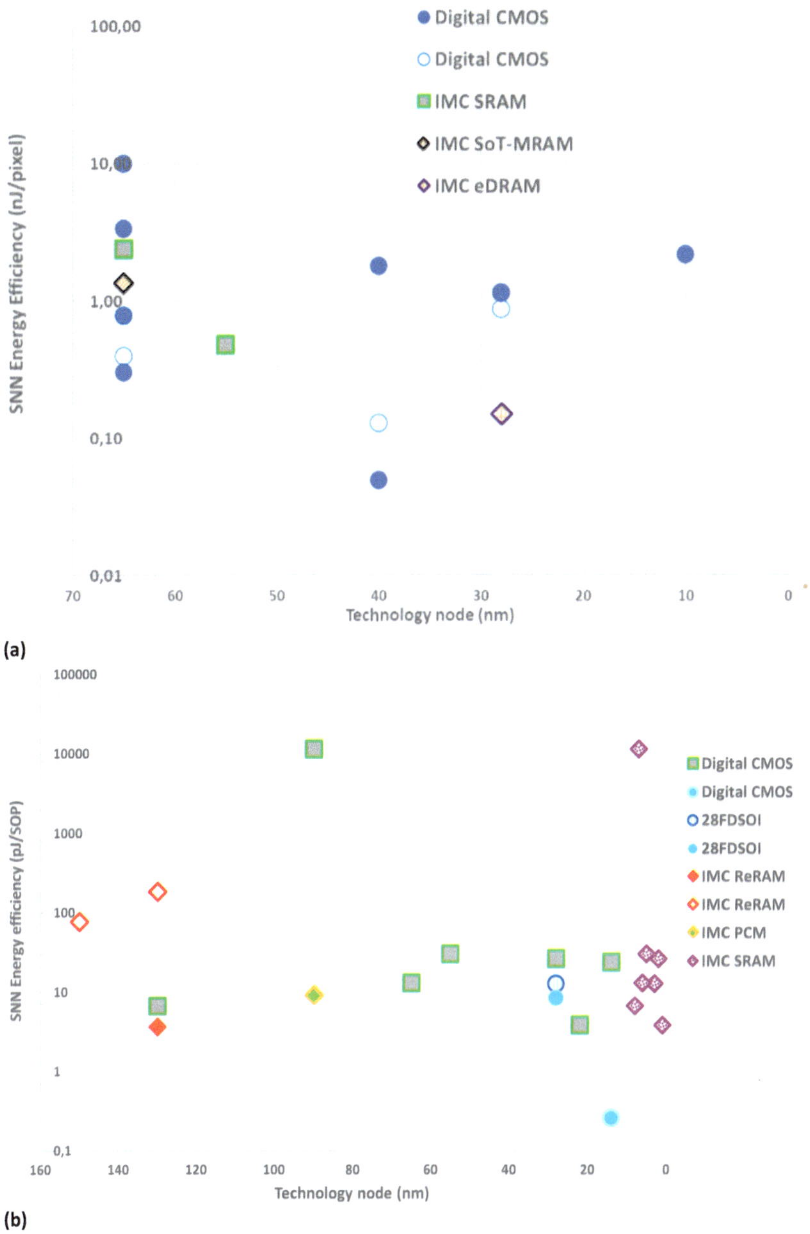

Figure 4.25: SNN energy efficiency expressed in (a) nJ/pixel and (b) pJ/SOP as a function of the technology node (filled symbols: training, empty symbols – data from [317–333].

In order to optimize performance and efficiency, one solution consists of using ANNs for training large AI models and determining the synaptic weights, then converting ANNs into SNNs. Using such an approach, generated SNNs are up to 38.7× and 12.2× more energy efficient, compared to ANNs with no significant drop in accuracy [316].

Since SNNs are more devoted to event detection, their energy efficiency is often expressed as energy per pixel for image recognition, and by energy per synaptic operation for other AI-related tasks. Figure 4.25 presents the SNN energy efficiency as a function of the technology node for digital CMOS chips (NPUs), and SRAM- and eNVM-based IMCs. The overall trend is a lower power consumption and thus a very slightly better SNN energy efficiency with a shrink in device dimensions. Furthermore, the energy efficiency of digital- and IMC-based SNN circuits is roughly similar.

The SNN power-efficiency (computing speed vs. power) of digital NPUs and IMC-based devices is shown in Figure 4.26 as a function of the technology node. This figure highlights that (i) the power efficiency of SNN devices is not well correlated to the technology nodes since 28 nm NPUs exhibit higher computing speed than the 7 and 14 nm NPUs, and (ii) the IMC-based computing devices globally present higher computing speeds than NPUs, and a better power efficiency.

In recent years, numerous emerging neuromorphic devices have been proposed to emulate the biological synaptic behavior for energy-efficient neuromorphic computing, including two-terminal memristors and three-terminal transistors. While two-terminal ReRAMs and FeCAPs possess a simple structure and advanta-

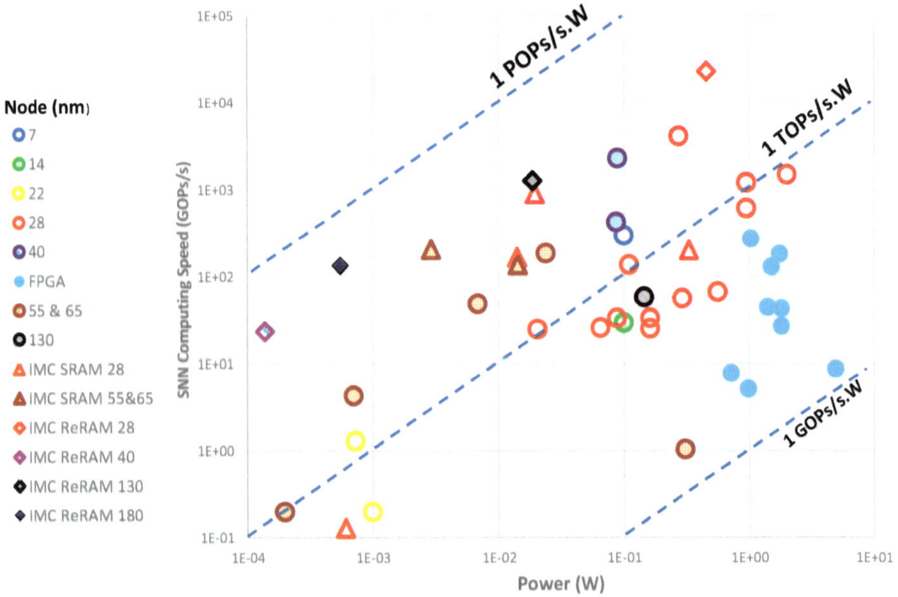

Figure 4.26: SNN computing speed as a function of power for digital AI chips and SRAM- and ReRAM-based IMCs as a function of the technology node – data from [149, 206, 317, 329, 324, 334–337].

geous NVM capabilities, they encounter challenges such as sneak current, variability, and randomness. Furthermore, the limitation of a single tunable conduction channel between the terminals hinders the simultaneous transmission of signals and learning processes. In contrast, three-terminal transistors can receive and process information simultaneously, replicating the information exchange and learning mechanisms of biological synapses. In these three-terminal devices, the source, drain, and gate electrodes can be selectively configured to act as either pre- or post-synaptic terminals (Figure 3.3, top), depending on the specific requirements of the neuromorphic system being implemented.

Additionally, three-terminal transistors demonstrate enhanced reliability, a superior signal-to-noise ratio, and improved signal modulation properties, contributing significantly to the substantial dynamic range and relatively high linearity essential for neuromorphic computing. Consequently, compared with two-terminal ReRAMs and FeCAPs, transistor-based devices exhibit significant potential for the development of neuromorphic platforms and intelligent systems [69].

To sum up the paragraphs on near-, in-memory, and neuromorphic computing, Figure 4.27 shows the current and future evolution of those concepts, to replace the energy-intensive von Neumann computing architecture.

Quantum computing could even provide a higher energy efficiency than the aforementioned computing architectures, for dedicated applications when it becomes effective.

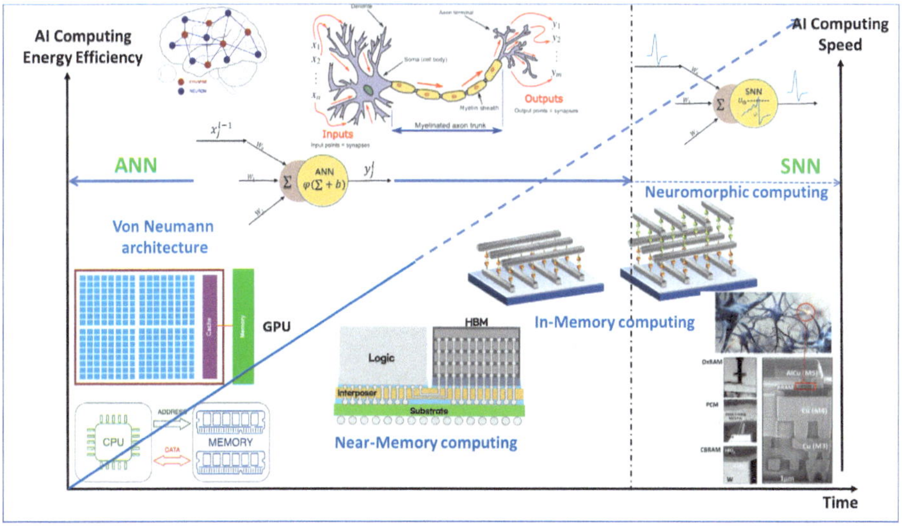

Figure 4.27: Evolution of the AI computing architecture and neural models.

4.5 Edge computing: edge AI

The integration of AI into edge devices offers significant benefits across various applicative sectors, contributing to a more efficient and resilient digital infrastructure. It also allows keeping privacy by processing sensible data locally. Computing now forms a continuum from edge to cloud, with edge computing processing data close to its source to improve performance, reduce data transmission, latency, and bandwidth, enhance safety and security, allowing real-time analytics and decision making, and decrease global power consumption (Figure 4.28).

As depicted in Figure 3.1, the global data volume is growing very fast. The gradual shift of cloud data and services to the edge is an integral part of the goal. Cloud computing needs secure and sustainable digital infrastructures, as well as safe data storage and transmission.

Whereas cloud computing happens mostly in large data centers, some years ago [338]. By 2025, this trend is reversed: 80% of all data is expected to be processed in smart devices closer to the user, known as edge computing or edge-AI. The availability of the combination of edge and cloud computing is vital in the spectrum of computing. to ensure that data is processed in the most efficient manner. Energy-efficient and trustworthy edge infrastructures and deep-edge (or end-point) devices (Figure 4.28), where data are processed close to their sources, will be fundamental for the sustainable use of computing and AI technologies.

At the core of this transition are intelligent sensors that embed novel AI compute directly within the sensing layer. Neuromorphic and IMC architectures make it possible to process data at the point of capture (deep-edge AI), dramatically reducing latency and energy use. By collapsing the traditional separation between sensing and computation, these systems enable always-on AI at the edge, making real-time responsiveness possible even in power-constrained environments.

Edge AI hardware was initially developed for perception tasks (vision, audio, signal processing) with high energy efficiency. But generative AI also emerges at the edge. First fueled by smartphones and computers with the need to be able to process most of the data locally (even if the network is not available or in a temporary outage) and adapting to the user's habits. It will progressively extend into other edge applications (robotics, faster patient diagnostics, high-level perception of the environment, etc.). Such an evolution drives new energy relative constraints not only for computing parts, but also to improve memory efficiency.

Running large AI models on the edge is a challenge, both from the compute intensity and memory bandwidth point of view. The solution most often adopted is to quantize the models to a lower precision. The quantization process always impairs accuracy. The key is to find the optimal precision that balances edge power and accuracy. Figure 4.29 shows the TOps/s requirements for the models of interest in the edge as a function of time [84]. The requested computing speed for edge-AI applications currently ranges from 1 to 100 TOps/s (Figure 4.29).

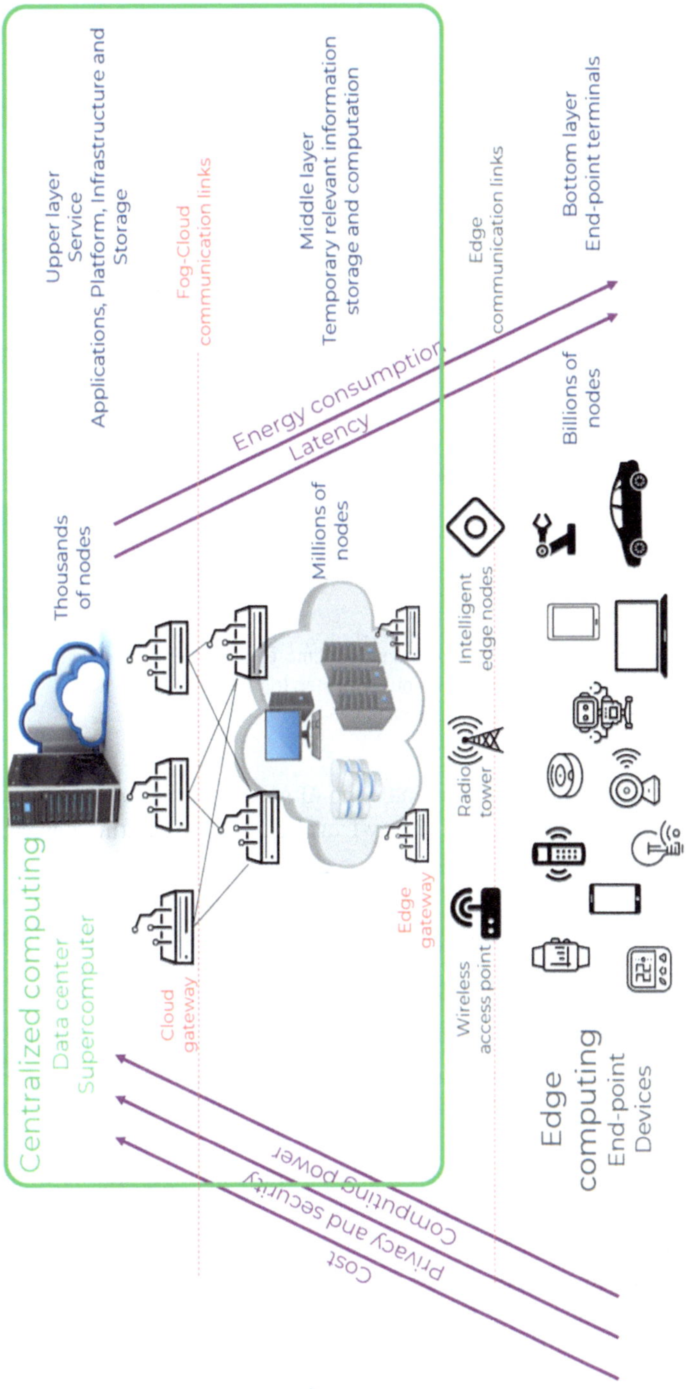

Figure 4.28: Cloud to edge computing continuum – courtesy of Yole Development – Generative AI 2025.

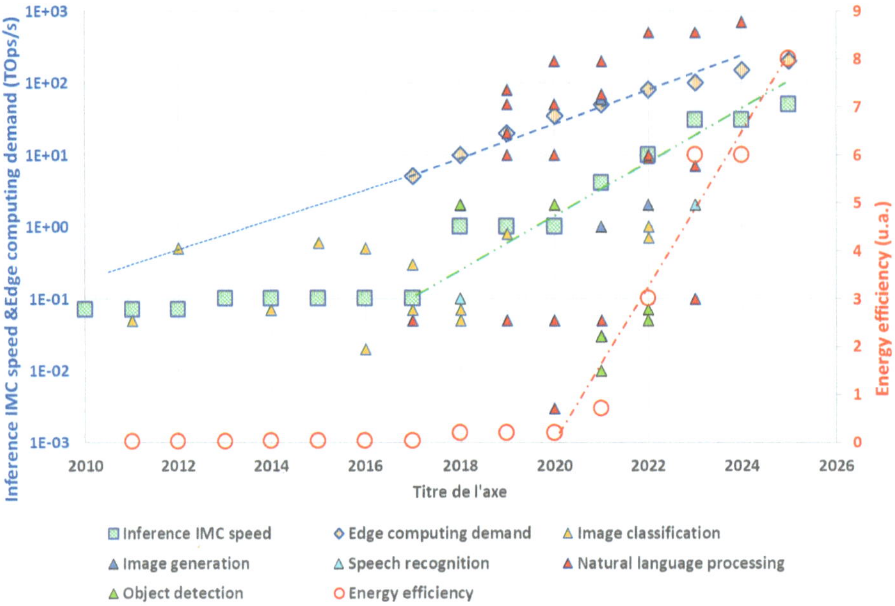

Figure 4.29: (left logarithm scale) AI edge computing demand over time, edge AI inference speed with time, and computing speeds required for inferencing various AI models (right linear scale). Power efficiency requirement for achieving a 10 W TDP – data from [84, 154].

However, since edge-AI applications will dominate the AI market in the coming years, the demand for more powerful and more energy-frugal edge-AI computing devices will increase in the upcoming years (Figure 4.29).

Edge devices in human proximity, such as earbuds, glasses, and even phones, will cause discomfort from temperatures over 40 °C. Therefore, effective heat extraction out of edge AI devices and limiting the heat generation are the biggest challenges facing the semiconductor industry. Moreover, technology scaling leads to higher junction temperatures both because of thermal conductivity degradation and higher self-heating (Figures 4.30 and 4.31). The bulk FinFET has a thermal conductivity that is approximately 75% less than that of a bulk planar device, and it diminishes further with miniaturization [339].

Edge AI devices face significant thermal management challenges due to their compact form factors and proximity to users.

Unlike server-grade hardware such as AI chips, edge devices cannot employ water cooling or advanced microfluidic systems because of constraints in size, cost, and user safety. As a result, air cooling remains the primary method for heat extraction, typically achieving rates of about 1.5 W per mm^2 (Figure 4.31). Therefore, edge devices do not need processors and accelerators produced using leading-edge technologies, which exhibit lower thermal conductivity (Figure 4.30). Edge AI devices cur-

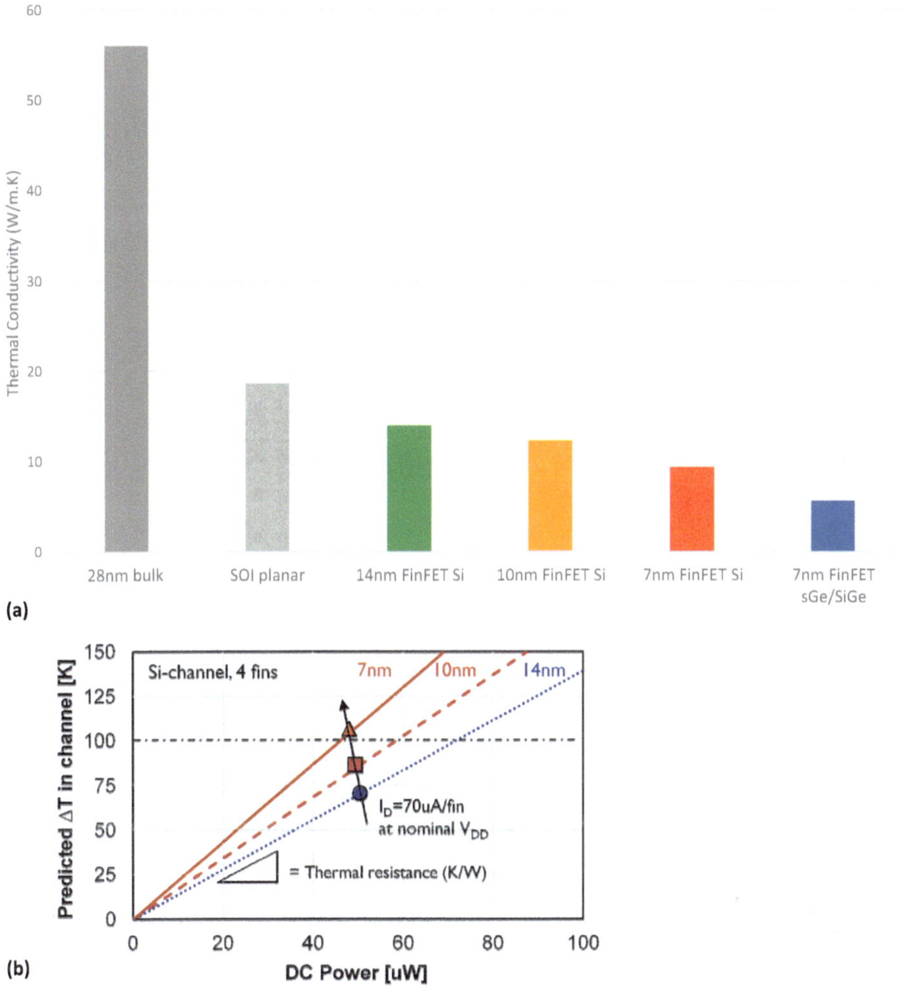

Figure 4.30: (a) Benchmark of thermal conductivity across different technologies – data from [339, 340]. (b) Predicted channel temperature rise of Si bulk FinFETs (4 fins) at given drive current [339].

rently typically require digital processors such as ASIC and FPGA, which are currently produced using relatively mature technology nodes like 28 or 22 nm, leading to computing speeds higher than 1 TOps/s (Figures 4.1 and 4.3). Moreover, eNVM-based IMC and neuromorphic devices can also lead to computing speeds higher than 1 TOps/s, with a better energy efficiency than digital devices (Figures 4.19 to 4.21), and so perfectly fit with edge-AI thermal constraints.

To maximize cooling efficiency, advanced packaging technologies must be engineered for optimal thermal conductivity, ensuring heat is effectively transferred from the device to the ambient environment [84].

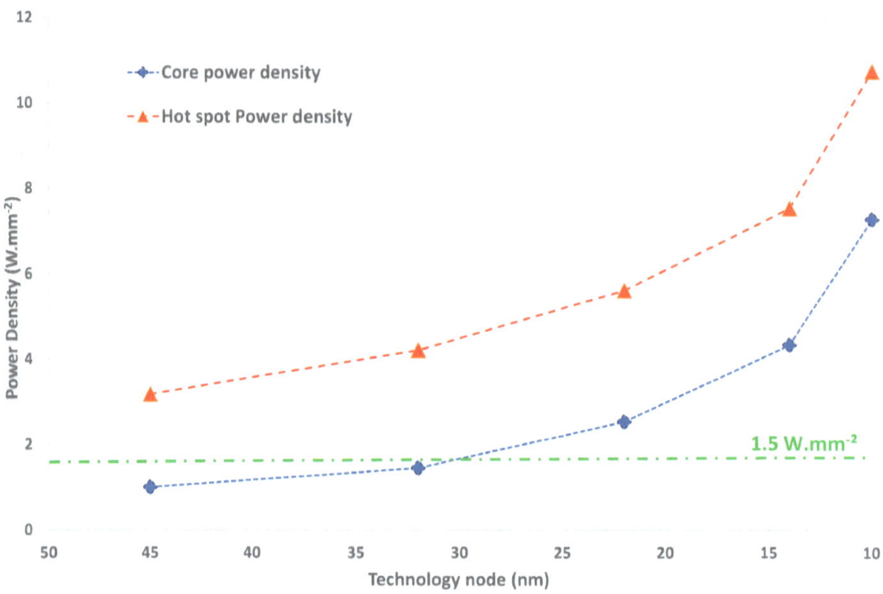

Figure 4.31: Evolution of the power density versus technology node – data from [84].

High energy efficiency of both processors and memories is compulsory for the so-called deep-edge computing and deep-edge AI, as they are included in systems, which are generally powered by batteries (vehicles, IoT, medical instruments, etc.). Zero standby energy and energy proportionality to load are essential for such edge devices. In order to reach a very high energy efficiency, the computing hardware is more specialized (ASIC, MCU, etc.), involving, for example, the use of NMC or IMC architectures, and models like convolutional neural networks [341, 342] or SNNs with far fewer neuron layers than DNNs. For very energy-efficient edge-AI devices, their hardware and the operating software must be codesigned and optimized for reducing their power consumption. Moreover, zero standby energy implies that NVMs must be embedded in such systems instead of SRAM and DRAM.

MCUs exhibit quite attractive properties for edge-AI applications. Microprocessor units like ASICs implement relatively high-performance CPUs and MAC units onto a single chip to improve processing performance. However, they typically require external system memory for work area of feature map, activations (intermediate information), and storage of weight data. In contrast, MCUs improve performance by integrating CPU, MAC unit, and memory onto a single chip. Even though the architecture is different, data transfer speed and memory usage efficiency are challenges for both embedded SRAM and eNVM (Figure 4.32a). In the case of microprocessor units, the data transfer speed within the chip exceeds 100 GB/s, but the transfer speed with external chips is about one-tenth of that. In contrast, MCUs have almost no external data transfer, but the data transfer capability of internal eNVM is significantly lower than that of system SRAM [343].

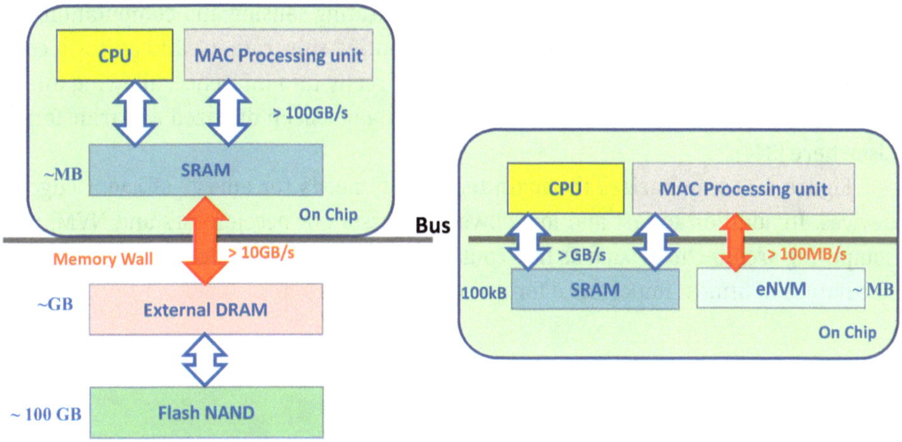

(a)

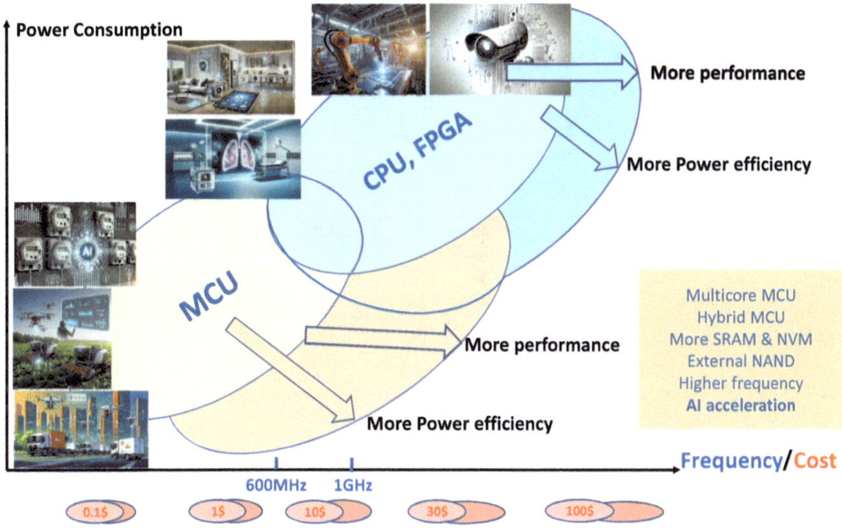

(b)

Figure 4.32: (a) Architecture differences between (left) MPUs or CPUs and (right) MCUs for edge AI applications, including approximate memory size and bandwidth for data transfer with each processing unit, and (b) relationship between performance and power consumption of MPUs and MCUs for edge AI applications. Adapted from [343] and from Yole Intelligence webminar on AI-enabled microcontrollers (November 2025).

However, due to their relatively poor computing performance, the edge-AI applications of MCUs are currently restricted to home appliances, medical devices, motor control, and smart monitoring (Figure 4.32b).

Furthermore, analog AI compute takes a fundamentally different approach from conventional digital processors. Rather than separating sensing and computation, analog AI uses the physics of the sensing and computing substrate itself to perform computation. In this model, energy is transformed directly into inference, allowing intelligence to begin at the sensor rather than after data has been digitized and transferred elsewhere [344].

Figure 4.33 summarizes the main technology needs for energy-efficient edge-AI devices. In addition to fast and low-power processors or accelerators and NVM, new computing architectures such as IMC could become mandatory. Moreover, secure connectivity is of utmost importance for preserving data privacy.

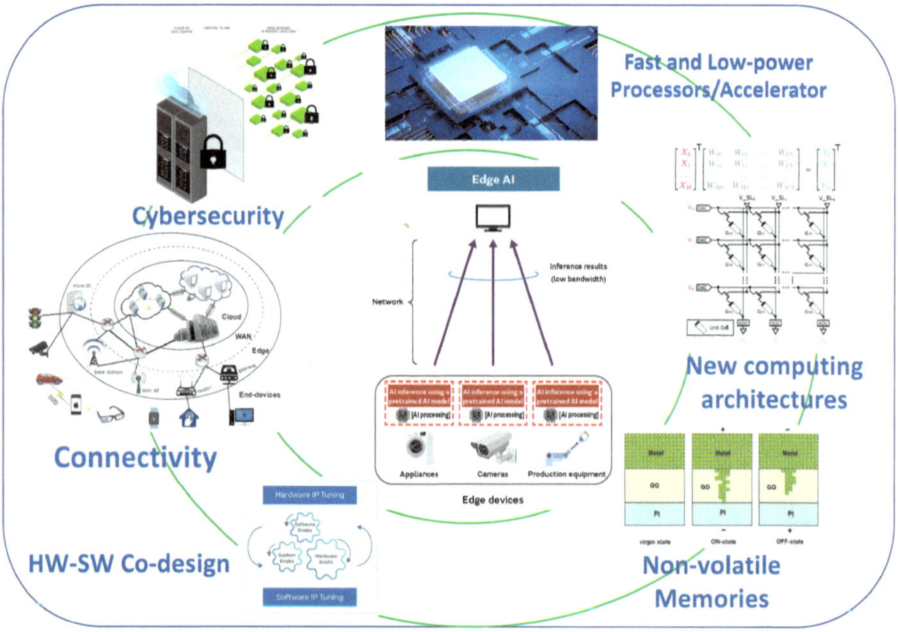

Figure 4.33: Basic technology needs for training and inference neuron models at the edge (Edge-AI).

5 Semiconductor technologies: future challenges

Following Moore's law and Dennard's rule is not only a matter of decreasing the transistor and memory dimensions. For achieving a significant improvement in their energy efficiency and performance (speed or storage density), the semiconductor industry has developed new architectures and introduced new materials. Such a trend will continue in the forthcoming years.

5.1 Logic devices

The integrated circuit semiconductor sector emerged with the self-aligned planar-gate silicon metal-oxide-semiconductor field-effect transistor (MOSFET) during the late 1960s. This design remained in use until the 28–22 nm technology stages. To manage the anticipated rise in performance while lowering power usage, new materials like copper and low-permittivity dielectrics (low-κ) were implemented at the 90 nm technology level to connect the transistors. Mechanical stressors were utilized to enhance carrier mobility in the transistor channels. High-permittivity dielectrics (high-κ) and metal gates were created to minimize gate leakage currents at the 45 nm node (Figure 5.1).

At the 28–22 nm technology nodes, for a better electrostatic control of the transistor, new fully-depleted transistor architectures have been introduced. They allow decreasing leakage currents and continuing transistor-switching speed improvements. Fully-depleted architectures include silicon-on-insulator (FD-SOI) for low-power devices, and nonplanar FinFET structures for high-performance devices (Figures 5.1 and 5.2).

The planar FD-SOI transistor is made up of a very thin silicon film (SOI) situated above a slender SiO_2 layer. The thickness of the SOI film is less than the device's depletion width. The narrow buried oxide layer restricts the flow of electrons from the source to the drain, which helps in minimizing leakage currents. In FinFET technology, the gate encircles the vertical thin channel or fin. The height of this fin is a key design element that enables enhanced current capability without affecting the device's active footprint. For fully depleted devices, precise control over the thickness of the ultrathin SOI film and the buried oxide is necessary for FD-SOI transistors, as well as managing the three dimensions of the fin in FinFET transistors, as illustrated in Figure 5.2. The FinFET architecture can be scaled down to the 5–3 nm technology nodes by keeping leakage currents under control. However, one of the FinFET shortcomings is that its gate surrounds the rectangular silicon fin on only three sides, leaving the bottom side connected to the silicon substrate. This pitfall allows some leakage currents to flow when the transistor is off. Lateral nanosheets (generally called gate-all-around (GAA) devices) have emerged as a new fully depleted transistor architecture offering improvement in both power consumption and performance over FinFET

 | https://doi.org/10.1515/9783112247952-005

at extremely scaled dimensions. GAA uses stacked horizontal nanosheets so that the gate surrounds the channel on all four sides. Those GAA transistors, which require dimensions accuracy around 1 nm (Figure 5.2e), will be introduced at the 3 or 2 nm technology node (Figures 5.1 and 5.2).

The GAA transistor critical dimension (gate and channel lengths) will continue to shrink in the forthcoming years to increase the device performance, the transistor density, and their energy efficiency (Figure 4.2). This shrink will be combined with innovations addressing advanced interconnects, middle of end interconnects (connecting transistors to the first metal layers), and backside power delivery networks [348] (Figure 5.3).

Continuous transistor scaling will require new EUV lithography tools with a higher spatial resolution and better overlay accuracy, which will be achieved through the increase in the numerical aperture (NA) of the optical pathway (Figures 5.3 and 5.4). Increasing NA allows finer designs to be patterned, allowing for increased density on a wafer. This increase in density during the manufacturing process requires fewer passes, thereby increasing production speeds and reducing the chances of defects occurring.

The 14 Å technology node will require an increasing need for double exposures with low-NA (0.33) EUV in a single processing step, raising costs. Multiple lithography exposures and the accompanying additional processing steps with other tools, such as etch, are also called multiple patterning, so there's double patterning, quadruple patterning, etc., depending on the number of repeated exposures the chipmaker has to do to complete the processing step. With high-NA EUV, on the other hand, double EUV exposures can be accomplished in a single step, which reduces the manufacturing costs and reduces the power consumption and GHG emission of the manufacturing line (see Sections 6.1 and 6.2). Consequently, high-NA EUV lithography tools are expected to be massively used at the 14 Å technology node (Figure 5.4).

The current low NA ASML's EUV lithography equipment requires over 100,000 parts from 5,000 suppliers. High-NA (0.55) and hyper-NA (0.75) EUV lithography tools will require even more parts, such as filters for managing the light polarization, for instance.

For extending future logic scaling, a new transistor architecture consisting of the complementary field-effect transistor (CFET) device is proposed. By stacking nFET and pFET vertically, CFET technology effectively doubles the transistor density, compared to GAA, without increasing the footprint. This leap in density will not only allow for enhanced functionality within the same area but will also bring improvements in power efficiency and device performance.

In addition to CFET, the continuous pursuit of improved performance and enhanced energy-efficient logic devices necessitates the exploration of channel materials that surpass silicon-based options. Low-dimensional substances like transition metal dichalcogenides (TMDs), exemplified by MoS_2 for nFET and WSe_2 for pFET, have at-

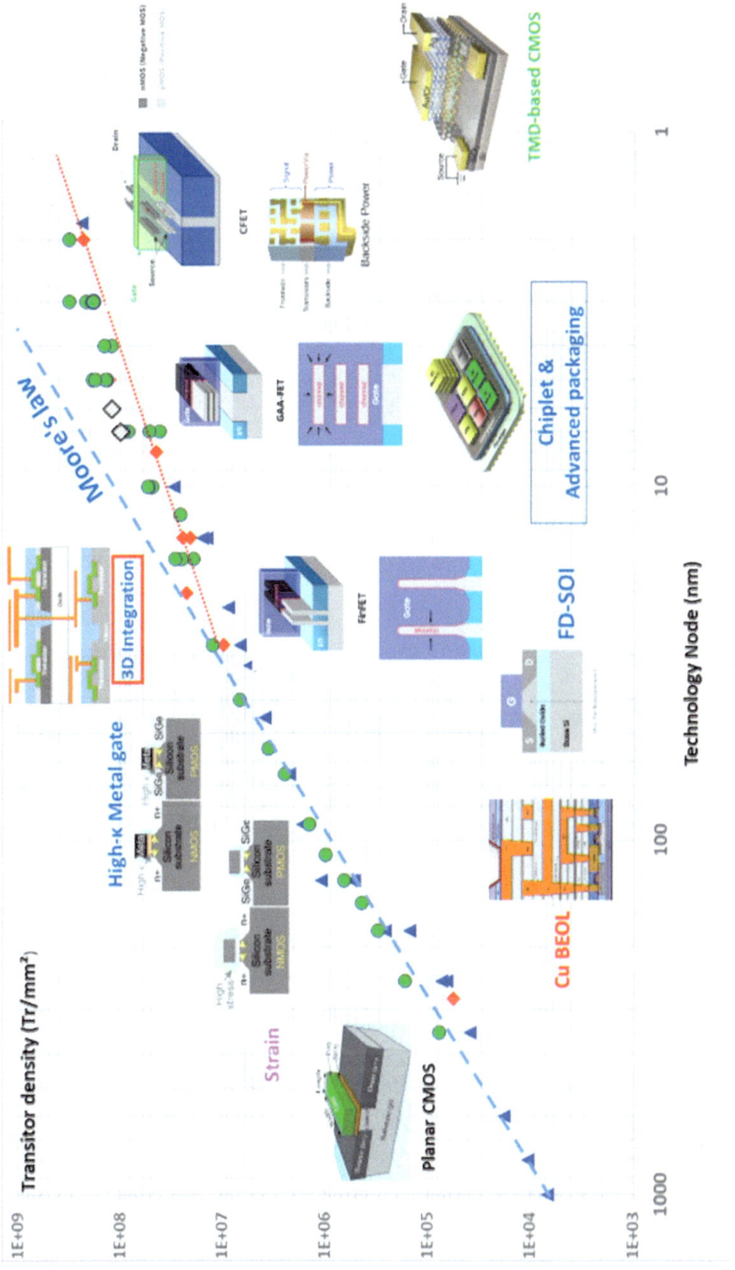

Figure 5.1: Evolution of the logic transistor density with the technology node updated from [1]. Sources: Intel, Samsung, TSMC, and SMIC.

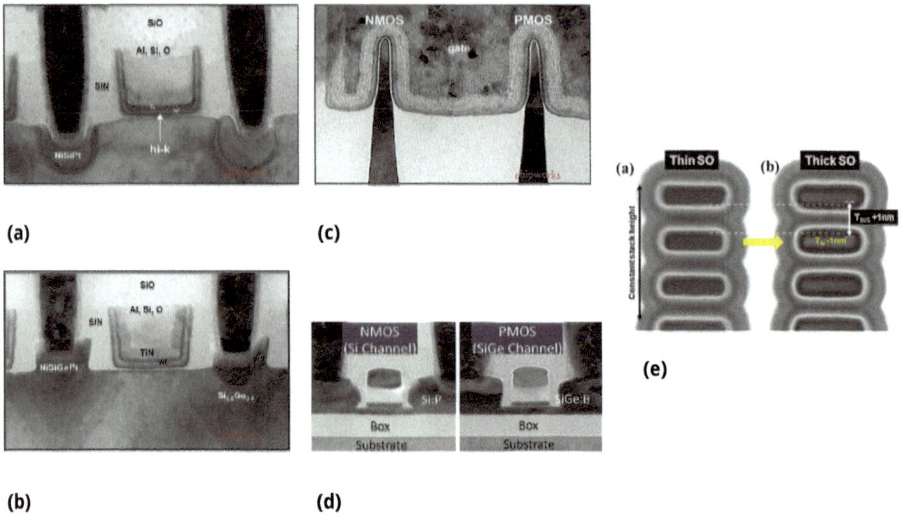

Figure 5.2: TEM cross sections of 20 nm (a) bulk NMOS and (b) PMOS [345], (c) 14 nm FinFET [346], (d) 22 nm FD-SOI [347], and (e) ~5 nm GAA [345] devices.

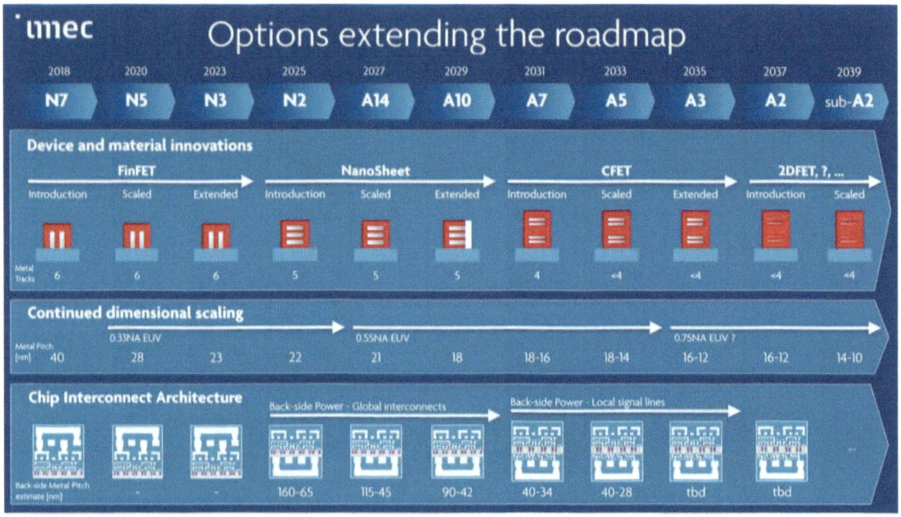

Figure 5.3: The imec logic technology roadmap – courtesy of imec [349].

tracted considerable attention owing to their unique physical and electronic characteristics.

Even in the CFET transistor era, the electrostatic control of the transistor channel will complicate further scaling. The continuous reduction of the transistor's gate and channel length requires ever-thinner semiconductor channels to restrict the pathway

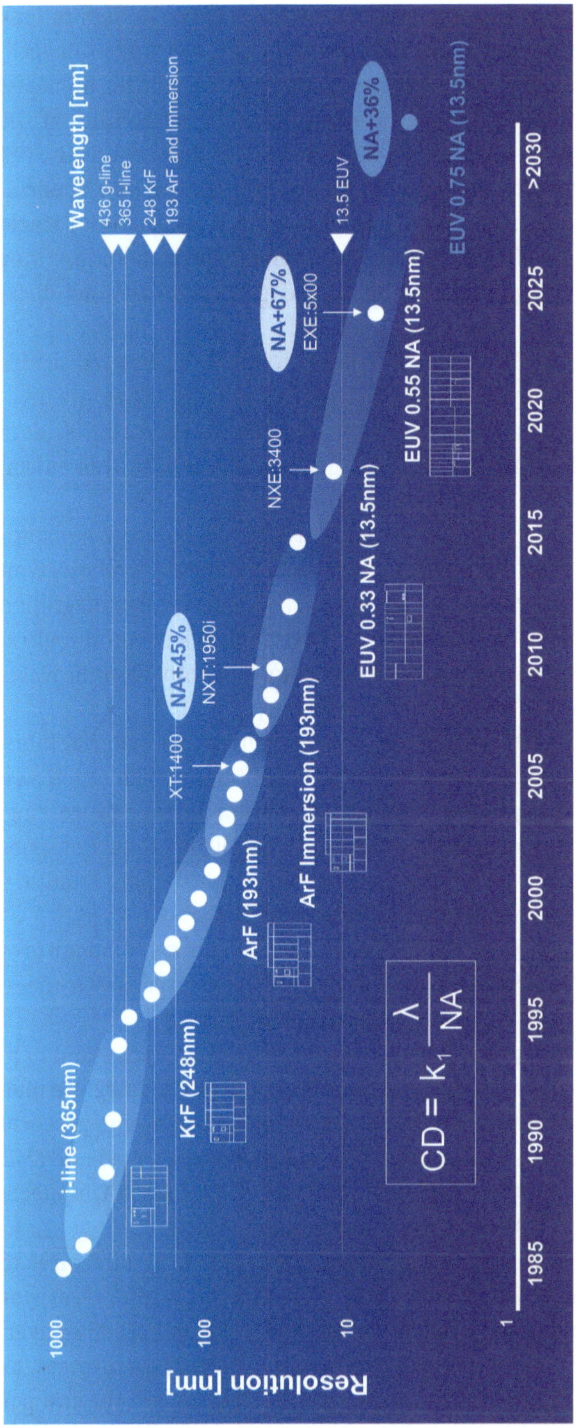

Figure 5.4: ASML technology roadmap – courtesy of ASML, November 14, 2024.

for the current to flow, hence limiting the opportunity for charge carriers to leak when the device is turned off. To move CFET transistors into the 2 Å transistor technology node with conduction channel lengths below 10 nm, the Si channel's thickness should also shrink below 10 nm. But in such thin Si channels, the mobility of the charge carriers and the transistor's on-state current begin to decline dramatically.

That's where 2D semiconductors, more particularly, TMDs, provide an opportunity. In these semiconductors, atoms are arranged in layered crystals with a single layer thickness of only ~0.7 nm, allowing for very thin channels. In addition, they promise to maintain a relatively high carrier mobility independent of channel thickness [349].

TMDs like MoS_2 and WSe_2 for nFET and pFET, respectively, have garnered significant interest due to both their physical and electronic properties. They are two-dimensional, meaning they have no out-of-plane dangling bonds and are therefore less prone to interface scattering that degrades carrier mobility in ultrathin silicon channels. This enables ultimate gate and channel length scaling (Figure 5.3).

However, several drastic challenges remain before using such innovative materials in logic devices for replacing Si, (i) achieving low-defect wafer-scale synthesis with a perfect control of their thickness (one or few atoms thick), (ii) gate-stack integration and dielectric deposition, (iii) developing low-resistivity contacts between the ultrathin TMD film and metals, (iv) doping TMDs, and (v) complying with 300 mm wafer integration in terms of reproducibility and economic viability.

One of the first and most urgent questions for TMD integration is material fabrication. Many studies of TMD-based transistors transfer these films, which are grown on dedicated substrates such as sapphire, to silicon wafers for further processing. Layer transfer techniques are very difficult to scale. Semiconductor production lines would prefer direct growth on the final substrate, avoiding the complex handling, contamination risks, and relatively high costs of transferred films.

Unfortunately, chemical vapor deposition of TMDs often requires temperatures exceeding 600 °C, which can degrade the underlying dielectric surface. TMDs themselves are also prone to plasma damage during subsequent processing.

TMD materials could lead to a post-silicon world. Theoretically, TMDs could enable ultrasmall logic transistors beyond 1 nm and have the capability to integrate photonics and electronics in a single chip. As of the time of writing, there are no practical and reliable manufacturable fabrication processes for semiconductors made of TMD material. But this may not be far away if recent investment and research from technology giants like Intel, TSMC, and others is any indication.

Despite the semiconductor industry mostly moving from planar architectures to nonplanar architectures at the 28–22 nm technology node, the increase in transistor density with the technology node slowed down at such nodes (Figure 5.1). Moreover, due to the complexity of the development and production of sub-22-nm devices, as well as the costs for building fabs and designing such devices (Figure 4.5), the cost per single transistor increases at the 28–22 nm technology node (Figure 5.5), which does not comply with Moore's law.

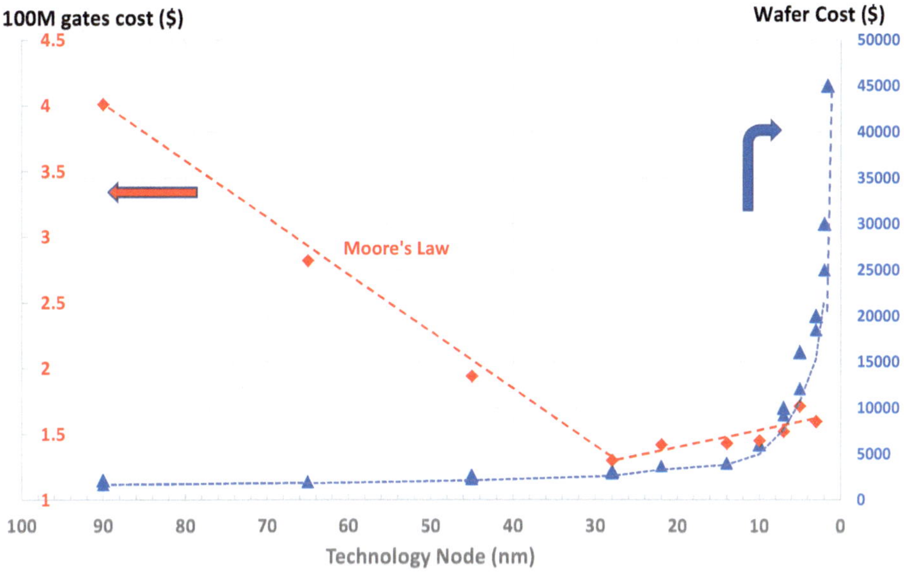

Figure 5.5: Evolution of the single transistor and 300 mm wafer costs with the technology node – data from [350–353].

5.1.1 Advanced packaging/3D integration

Power, performance, and area/cost are still the key metrics for chipmakers. The die cost does not scale linearly with the die area. This cost more than doubles with doubling the die area due to reduced production yields (number of good dies in a wafer). Therefore, it can be thriftier to disaggregate the IC or the SoC into smaller elements (cores for logic devices) and connect them inside the package, thanks to a Si interposer, for instance, to get higher core density at the package level. These packages with more than one logic die inside are sometimes called multi-chip modules. The dies inside the multi-chip modules are often referred to as chiplets.

This concept leads to the advanced packaging approach, which becomes central to performance, driven by AI and chiplet architectures.

Developed in the 1950s and still in use today, wire-bond technology is an interconnection technique that attaches the printed circuit board (PCB) to the die – the silicon square that contains the integrated circuit – using solder balls and thin metal wires.

The first major evolution in packaging technology came in the mid-1990s with flip chips, which use a face-down die, the entire surface area of which is used for interconnection through solder "bumps" that bond the PCB with the die. This results in a smaller form factor, or hardware size, and a higher signal-propagation rate. Flip-chip packaging is the most common and lowest-cost technology currently in use, mainly for central processing units, smartphones, and radio-frequency system-in-package solutions (Figure 5.13).

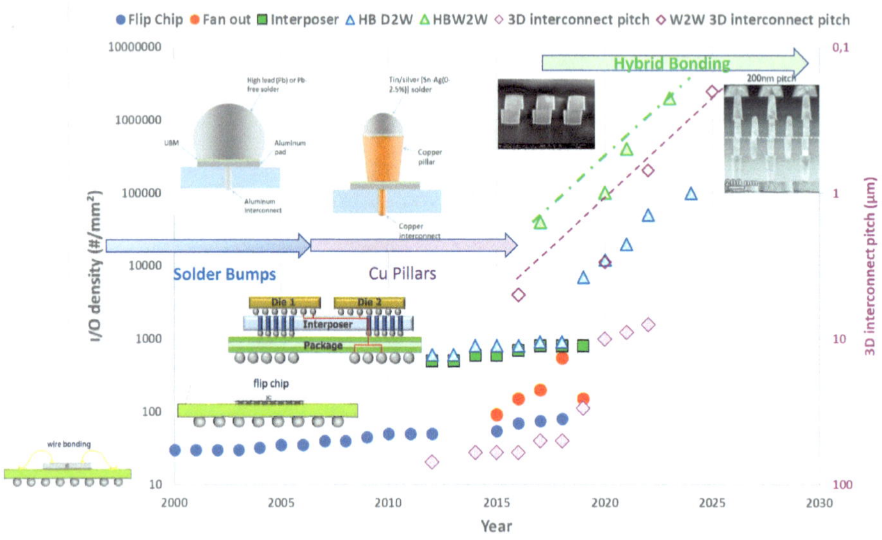

Figure 5.6: Advanced packaging technology roadmap – data from [354, 356].

While traditional packaging "dices" the silicon wafer into individual chips first and then attaches the chips to the PCB and builds the electrical connections, wafer-level packaging makes the electrical connections and molding at the wafer level, then dices the chips using a laser. The greatest difference between wafer-level chip-scale packaging (WLCSP) and flip chips in terms of chip configuration is that WLCSPs have no substrate between the die and the PCB. Instead, redistribution layers replace the substrate, leading to a smaller package and enhanced thermal conduction.

The past 15 years saw the development of stacked WLCSP, which allows for multiple integrated circuits in the same package and is used for both heterogeneous bonding, which integrates logic and memory chips, for instance, and memory-chip stacking. In 2.5D stacking, two or more chips are laid side by side with an interposer connecting one die to another. There are several categories of 2.5D stacking, based on the kind of interposer it uses.

Silicon interposers are the only type that requires TSV, or through-silicon via – a vertical electrical connection that passes through the silicon die or wafer. Silicon interposers use a stable technology that has been on the market for more than 10 years, but the cost of silicon is high and requires front-end technology and manufacturing capability [356].

The Si interposer, a thin silicon substrate with embedded metal interconnects, serves as the connection between the chiplets and/or modules and the wiring to the external pins. It allows increasing the energy efficiency compared to a monolithic IC or SoC. The use of an interposer leads to the so-called 2.5D packaging technology, which puts dies closer together and with small bumps, pillars, and hybrid bonding to connect to standard packaging technology, such as for NMC (Figure 4.27).

Silicon bridges are relatively new. Because they use smaller amounts of silicon than traditional silicon interposers, they are thinner, which reduces power consumption and increases design flexibility. Their advantage over traditional silicon interposers is that they can enable more advanced system-level integration, so they are used for high-performance computing (HPC) and AI [356].

Glass is also rising as a next-generation material for interposers. It offers low cost and low power loss in high-frequency bandwidths [356].

3D packaging takes the 2.5D packaging and goes further to stack the dies vertically for even shorter connections between dies. In 3D stacking, multiple chips are placed face down on top of one another, with or without an interposer. There are two main types of 3D stacking. The most common type is TSV with micro-bumps (μ-bumps) and/or Cu pillars (Figure 5.6).

The newer alternative, bumpless hybrid bonding, forms very short interconnections using a dielectric bond and embedded metal, ensuring a very high bandwidth between the logic device and the main memory. Moreover, hybrid bonding provides the highest interconnection density with the smallest 3D interconnect pitch (Figure 5.6). Therefore, this new technique is nowadays massively deployed in AI and HPC applications, as well as in CMOS image sensors.

Therefore, hybrid bonding often gets framed as the next logical step in density scaling, and in many ways it is. It offers the interconnect density and electrical performance needed for tighter die-to-die integration, especially as AI and HPC architectures continue to push for more bandwidth in less space. Hybrid bonding represents another approach to "shrink every dimension." It involves a perfectly flat surface where both the copper interconnects and the surrounding dielectric (usually SiO_2 or SiCN) are bonded together. The dielectric is bonded first, and a thermal cycle is used to expand the recessed copper to connect.

While hybrid bonding has the potential to offer the highest interconnect density, its requirements are extreme compared to conventional 2D packaging: leading-edge cleanroom environments, sub-nanometer surface roughness, and expensive chemical mechanical polishing steps. The nature and strength of the dielectric bond are just as important as the Cu–Cu bond. This puts restrictions on the nature of the dielectric, and processes need to be optimized for a specific dielectric. This makes the realization of heterogeneous hybrid bonding more difficult, especially if multiple foundries and material sets are involved. Hybrid bonding does have an advantage in alignment and overlay tolerance as the initial dielectric bond is performed at room temperature [357].

The hybrid bonding manufacturing challenges are changing as the pitch shrinks. At larger pitches, yield is still heavily influenced by defects and contamination. At smaller pitches, stress begins to dominate in ways that are less visible and often harder to control.

2.5D and 3D advanced packaging/integration technologies allow enhancing data transfer rates, reduce latency, optimize power consumption, and increase the overall performance of computing systems [358].

However, if moving from 2D to 3D packaging improves those metrics, it also increases the cost, as depicted in Figure 5.7. Therefore, 2.5D and 3D packaging/integration technologies are only implemented for applications that really deserve to get high memory bandwidth, like in HBM, to reach very high computing speeds and/or to reduce the overall device power consumption.

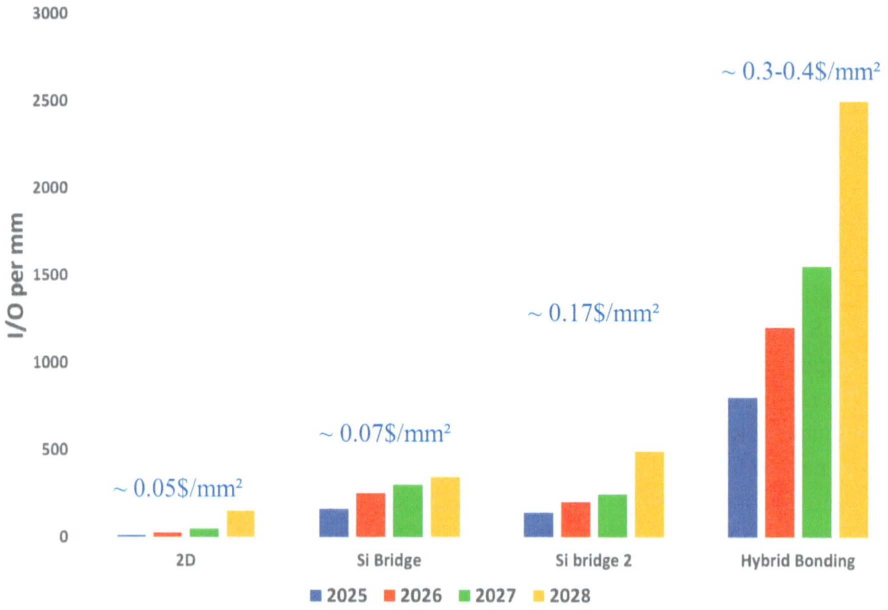

Figure 5.7: Cost of 2D, 2.5D (Si bridge), and 3D advanced packaging – data from [359].

All the constitutive elements of an IC and a SoC do not need the most advanced technology nodes. 2.5D and 3D integration offers the flexibility to best suit the needs of each partition. Disaggregation into multi-die assemblies enables different processors and functions to be prioritized, while simplifying contingency plans in case of less-critical component shortages. And rather than cramming every component onto a reticle-sized SoC at the most advanced node, different dies can be developed at whatever node makes sense.

Shrinking features are still important for some logic devices, but what gets scaled to the most advanced nodes is becoming a progressively smaller fraction of the overall design. The flip side is that more transistors in the form of chiplets can be added to boost performance, so long as data movement in and out of processors and memories is fast enough to handle the exploding volume of AI data.

This domain-specific technology optimization enables the creation of the most energy-efficient computing systems. Accordingly, if some elements can be manufactured with less expensive processes and lower mask costs, they should decrease the overall cost of the IC or SoC. As a result, some silicon development should be less expensive, although the additional effort necessary for an advanced-package design may consume some of those savings. 2.5D and 3D advanced packaging can also be designed to serve multiple products. They are now considered as the continuation of Moore's law.

However, if it is easier to design and manufacture a collection of chiplets than a full SoC, the integration of the various constitutive pieces is not easy.

Some 3D integration technologies can also be used for increasing the transistor density and so enhancing the computing performance. Utilizing the back side of the wafer for transistor powering and the front side for sending signals results in a smaller transistor footprint, and an amplification of the frequency signal is noted as well [348, 360]. In the coming years, a nano-electronic device would probably look like a skyscraper by stacking memories over the processor, and the development of 3D-NAND flash and 3D-DRAM.

If 3D integration is expected to become the bridge between the AI computing demand and the limitation of Si monolithic scaling, 3D and heterogeneous integration must be considered as a system and not only as a Si piece. Several optimized chiplets will need to be gathered using shrunk interconnects between those chiplets, starting from conventional solders to hybrid bonding.

5.1.2 Silicon photonics and co-packaged optics

The semiconductor industry now considers Si photonics as a substitute for Cu interconnects for transmitting signals (bits) from chips to chips, and even for intra-chip communications in the future (Figure 5.8). The main advantage of Si photonics, which deals with the manipulation of light in Si, lies in its ability to facilitate faster and more energy-efficient data processing, simultaneously mitigating latency issues. In contrast to the traditional copper-wired electronic systems, silicon photonics unleashes the potential of optical signals to propel data across the chip.

Introducing photonics inside the chip package is a very attractive trend for reducing the energy consumption of AI and HPC systems and chips. A data center includes switches, routers, storage solutions, etc. Their functions include the network of computing and storage resources that enable the delivery of shared applications and data. Today, more than 90% of transceivers are pluggable.

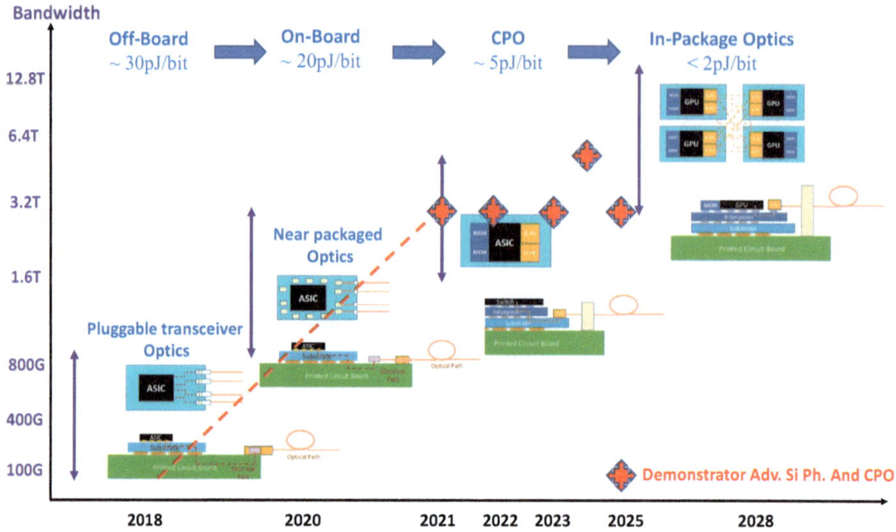

Figure 5.8: Evolution of silicon photonics from 2018 to 2028 – data from [361].

An optical transceiver defines the process of converting electric signaling towards the optical transmission with the help of the transmission optical subassembly or the Tx module, and performing the inverse action through the receiver optical subassembly or the Rx module. A Tx contains a semiconductor laser diode, laser driver, etc., while Rx contains a photodiode, an optical interface such as a lens, a transimpedance amplifier, a passive electrical interface, etc. The Tx module converts the electrical signal to the optical transmission light that lands on the optical fiber. The Rx is used to receive an optical signal from a fiber and convert it back into an electrical signal.

The pluggable transceivers are mounted on the edges of a PCB with the ASIC package attached to a package substrate. The distance from the photonic integrated circuit (PIC)/electronic integrated circuit (EIC) of the pluggable transceivers to the ASIC chip is the farthest, and therefore, the power consumption and electrical performance are the worst. For on-board optics, the PIC/EIC are mounted around the ASIC switch package, which is an improvement in power and electrical performance. However, the signals are still traveling on the PCB [362] (Figure 5.8).

In co-packaged optics, the PIC/EIC is placed side-by-side along the four edges of the ASIC chip on the same co-packaged substrate (Figure 5.8). In this case, the distance between the ASIC and the PIC/EIC is the shortest, and therefore, has the best electrical performance. CPO offers superior form factor, power consumption, and latency benefits compared to on-board optics [362].

The energy for transmitting one bit can be reduced from 30 pJ down to 5 pJ using CPO. However, since Si is an indirect bandgap semiconductor, it does not emit any photon with sufficient intensity, unless it is co-doped with Ge and Sn. Therefore, the integration of a III-V material such as AsGa and InP, or GaN, which are highly luminescent and the basic materials of lasers and optical amplifiers, as close as possible to the Si-based processor and accelerator is mandatory. Such an intimate 3D integration of the Si-based EIC with a III-V-based PIC and the laser diode is challenging, as it requires a perfect alignment with a sub-µm accuracy.

Micro-transfer printing (µTP) has emerged as a promising solution to these heterogeneous integration challenges, facilitating the precise placement of pre-fabricated, pre-tested III–V devices onto silicon substrates with high yield and minimal material waste [362]. µTP operates by first fabricating micro- or nanoscale photonic components, often including laser and amplifiers based on InP [363] or GaAs [364], or modulators based on $LiNbO_3$ [365], on a donor substrate. These components are then picked up using a soft elastomeric stamp and subsequently placed onto a receiving Si substrate. This decoupling of fabrication and integration allows significant flexibility in device design and process flow. One the most compelling advantages of µTP is its material agnosticism [366].

µTP offers a high-throughput potential, as the same donor wafer can yield thousands of microdevices, which are then selectively printed onto many receiving chips. Compared to wafer bonding, which generally integrates whole wafers, transfer printing allows selective and localized placement of components, with a placement accuracy better than 1µm [367]. This advantage is particularly useful for integrating only those active elements (e.g., lasers, amplifiers) that are not easily realized on silicon, while leaving the passive components to conventional silicon manufacturing processes. Hence, µTP allows coupling the emission, detection and optical modulation materials to the cost efficiency and manufacturability of Si devices [368].

CPO also introduces some severe issues in chip design for a successful coupling between the PIC and the EIC. Designers will need to use a chiplet-based architecture to create smaller, specialized dies for compute, memory, and photonics. Designers will need to create PICs that serve to integrate modulators, detectors, and waveguides onto silicon substrates using standard CMOS fabrication techniques. Photonic components are highly temperature-sensitive, so the designers must perform complex multiphysics simulations to manage heat flow from high-wattage compute EICs so they do not cause wavelength drift or alignment errors in the optics, etc.

CPO allows reducing the power consumption from 30 W down to 9 W compared to a pluggable transceiver solution because the optical conversion engine (OE in Figure 5.8) is embedded alongslide the switch ASIC. Hence, the signal (bits) is coupled to the optical fiber almost immediately instead of traveling over long electrical traces [369].

The near-future trend will consist of coupling the PIC near the processor accelerator or processor, and using the photons for transmitting the bits between the different processing cores constituting the computing chip (in-package optics in Figure 5.8). It would allow reducing the energy per bit down to 2 pJ and to strongly increase the bandwidth (Figures 5.8 and 5.9). Figure 5.9 shows an in-package CPO system for HPC driven by AI and low-power and high-bandwidth communication applications. It can be seen that the

GPU is surrounded by the HBM, and then surrounded by the Optical Elements (OE), such as the PIC and EIC. The ASIC switch is surrounded by the PIC and EIC. These two structures are connected through fibers and on a CPO substrate, which can be a TSV interposer, and then on a package substrate. Finally, both units are attached to a PCB [362].

While the CPO design achieves significant reductions in I/O power on the chip, transistor processing power remains unchanged. HBM has addressed memory power I/O concerns, and CPO now tackles the power and bandwidth challenges associated with I/O.

Hence, photonics has a great potential since lightwaves propagate and interfere without consuming energy, and thereby can be engineered to perform functions that scale without increasing the energy consumption. Silicon photonics has been extensively developed for the past two decades and is now fully capable of providing a nearly ideal platform to unleash such great potential. Indeed, silicon photonics can offer efficient high-density interconnects for high-bandwidth (Figure 5.8) and long-reach links, low-energy optical path switching regardless of signal bandwidths, and photonic neural networks capable of speed-of-light computations for the acceleration of AI computing in data centers [370].

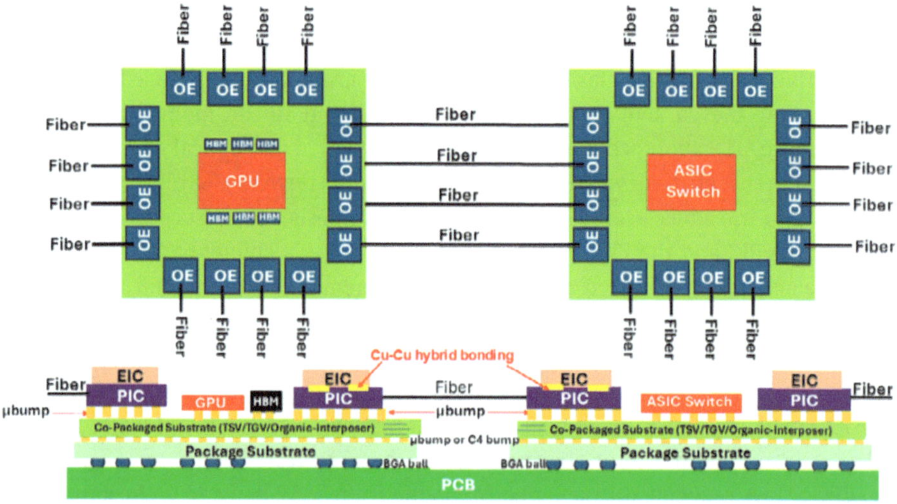

Figure 5.9: An HPC device driven by AI and a low-power, high-speed, and high-bandwidth communication system [362].

5.2 Memories

5.2.1 SRAM challenges and potential substitutes

SRAM shows high speed and enhanced endurance of read/write cycles with a low operation voltage and has been used as a crucial memory component in various hardware systems. However, SRAM technology faces drawbacks due to its volatile nature and the large cell size of a typical six-transistor (6T) design. The volatility leads to fixed power usage primarily influenced by the leakage current of transistors, which grows more prominent in advanced Si-CMOS technology nodes. Off-chip nonvolatile memory, e.g., embedded flash, is usually required to back up data before powering off SRAM for low-power artificial intelligence of things applications (e.g., mobile devices). Such a two-memory scheme increases the costs and causes extra delay and energy when powering ON/OFF.

While STT-MRAMs are not fast enough for performance-critical applications such as L1-L2 caches (Table 5.1 and Figure 5.10), they can still be a viable option for large-capacity last-level caches in SRAM-MRAM hybrid memory systems (NVSRAMs) or various emerging computing options such as neuromorphic or hyperdimensional computing systems [371, 382].

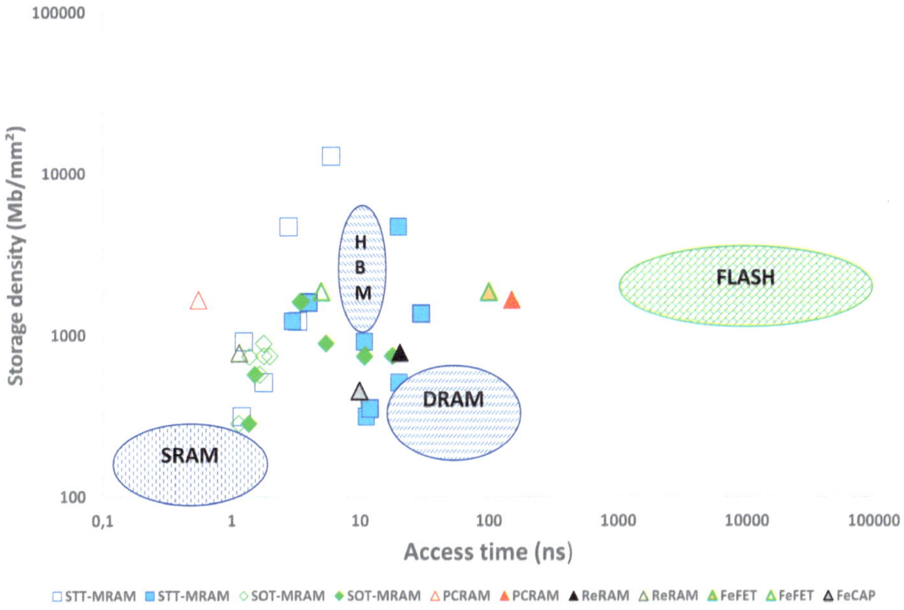

Figure 5.10: Storage density of stand-alone and embedded memory technologies with their access time. Open symbols: Read time; Filed symbols: Write time – data from [271–277].

SOT-MRAMs offer unmatched switching speed and endurance compared to STT-MRAM. In such devices, write (SOT) and read (TMR) paths are separated, and the three-terminal cell structure minimizes the risk of voltage breakdown, leading to reliable sub-ns and sub-volt operations. With these benefits, SOT-MRAM can potentially replace SRAMs at register, L1, and L2 levels (Figure 5.10) in CPUs, GPUs, and NPUs in the embedded domain (Table 5.1).

However, there are certain challenges facing SOT technology with regard to density (3T cell) and write current (material efficiency), as presently, the required SOT critical current for switching the SOT-MRAM cell is around one order larger than STT-MRAM [311].

By integrating nonvolatile modules onto the volatile SRAM cells, nonvolatile SRAM (NVSRAM) becomes one of the prevalent solutions for overcoming the above challenges. In this way, when powering ON/OFF is necessary, data can be transferred between nonvolatile modules and SRAM cells in a parallel bit-to-bit manner. Spintronic memories such as STT- and SOT-MRAM have been shown to be promising due to their long retention times, large endurance, fast read/write times (<10 ns), and compatibility with CMOS BEOL.

Various NVSRAM architectures have been proposed, including ReRAM-based 4T-two-resistor (4T2R) and 7T1R, magnetic tunnel junction-based 4T-two-magnetic tunnel junction (4T2M), ferroelectric field-effect transistor (FeFET)-based 8T, phase-change memory (PCRAM)-based 8T2R, charge trapping (CT)-based 8T, and ferroelectric capacitor (FeCAP)-based 6T-two-capacitor (6T2C) [374].

Although NVSRAM mitigates the volatility-related issue of SRAM, the challenge in density due to the large cell footprint remains, limiting capacity and further development. Since the SRAM bit density did not grow using advanced technology nodes (≤20 nm) (Figure 4.15), one alternative for increasing this density consists of stacking numerous SRAM and NVSRAM layers, thanks to monolithic 3D integration. Monolithic 3D integration that vertically stacks various functional layers onto the same chip can not only increase the bit density of SRAMs but can also shorten interconnect lengths with enhanced bandwidth and energy efficiency of SRAMs. For instance, monolithi-

Table 5.1: Comparison between the basic properties of SRAM, STT-, and SOT-MRAM as cache memories – data from [372, 373].

Technology	Level	Capacity	Area (mm^2)	Latency (ns)		Energy (nJ)		Energy/power
				Read	Write	Read	Write	Leakage
SRAM	L1	32 KB	0.0212	2.18	1.39	0.01	0.002	18.63 nJ
SOT-MRAM			0.0362	1.78	1.20	0.12	0.012	0.182 nJ
SRAM	L2	512 KB	0.3222	5.87	2.05	0.10	0.005	290.6 nJ
SOT-MRAM			0.3192	1.95	1.55	0.39	0.034	1.507 nJ
SRAM	L3	16 MB	11.448	6.589	3.274	0.191	0.182	462.9 mW
STT-MRAM			5.024	6.904	11.898	0.361	1.127	64.82 mW

cally 3D-integrated 4T2C NVSRAM through stacking of optimized IGZO thin film transistors and $Hf_{0.5}Zr_{0.5}O_2$-based FeCAPs with BEOL-compatible temperatures (≤400 °C) leads to a drastic increase in the bit density [374].

5.2.2 Flash NAND challenges and alternatives

Today, nonvolatile NAND flash memory is widely accessible to everyone, serving as the primary storage option found in smartphones and tablets, as well as in solid-state drives and USB drives. Additionally, they play a crucial role in cloud infrastructure.

The 2D-NAND flash dimension shrink should not continue in the forthcoming years. Accordingly, the NAND flash manufacturing would not require extensive use of EUV lithography (Figure 5.13).

Consequently, leading manufacturers of NAND chips are competing to boost the quantity of vertical 3D flash NAND memory layers. Currently, the layers range from 300 to 400 (Figure 3.25), but estimates suggest that this could rise to about 1,000 layers by 2030. Nevertheless, for 3D NAND structures exceeding 500 layers, technological and economic considerations for 3D integration solutions should also be taken into account. Future growth strategies entail logical scaling that seeks to increase the bits stored per cell, thereby improving data density, as well as physical scaling and the introduction of new cell designs. However, surpassing 1,000 layers and reaching 4 bits per cell (QLC), which is essential for sustaining bit areal density growth in 3D NAND (Figure 3.25), is proving to be increasingly difficult, with CT NAND devices encountering reliability issues even at their current scaled sizes [115, 375].

Moreover, there are significant hurdles to advancing 3D NAND technologies, stemming from CT flash cells and the severe issues linked to shrinking dimensions. Although CT NAND memory has its benefits, it encounters several restrictions that affect its scalability and operational efficiency. One major issue with CT NAND is the retention and durability of the memory cells. The charge-trapping layer, generally composed of silicon nitride, can experience charge loss over time, impacting the retention abilities of the memory cell. These reliability concerns are intensified by the strong electric fields necessary for programming and erasing cells, resulting in charge leakage in neighboring cells and a greater deterioration of the memory window even at the existing dimensions of 3D NAND. This scenario ultimately restricts logical scaling, as substantial write voltages are essential for TLC/QLC functionality.

Another key drawback of CT NAND concerning logical scaling is the diminishing number of electrons being trapped at each level. This, along with the uneven distribution of trapped charges, heightens the variability in threshold voltage (V_{th}). Such variability can lead to increased reading errors and diminished reliability of the memory array.

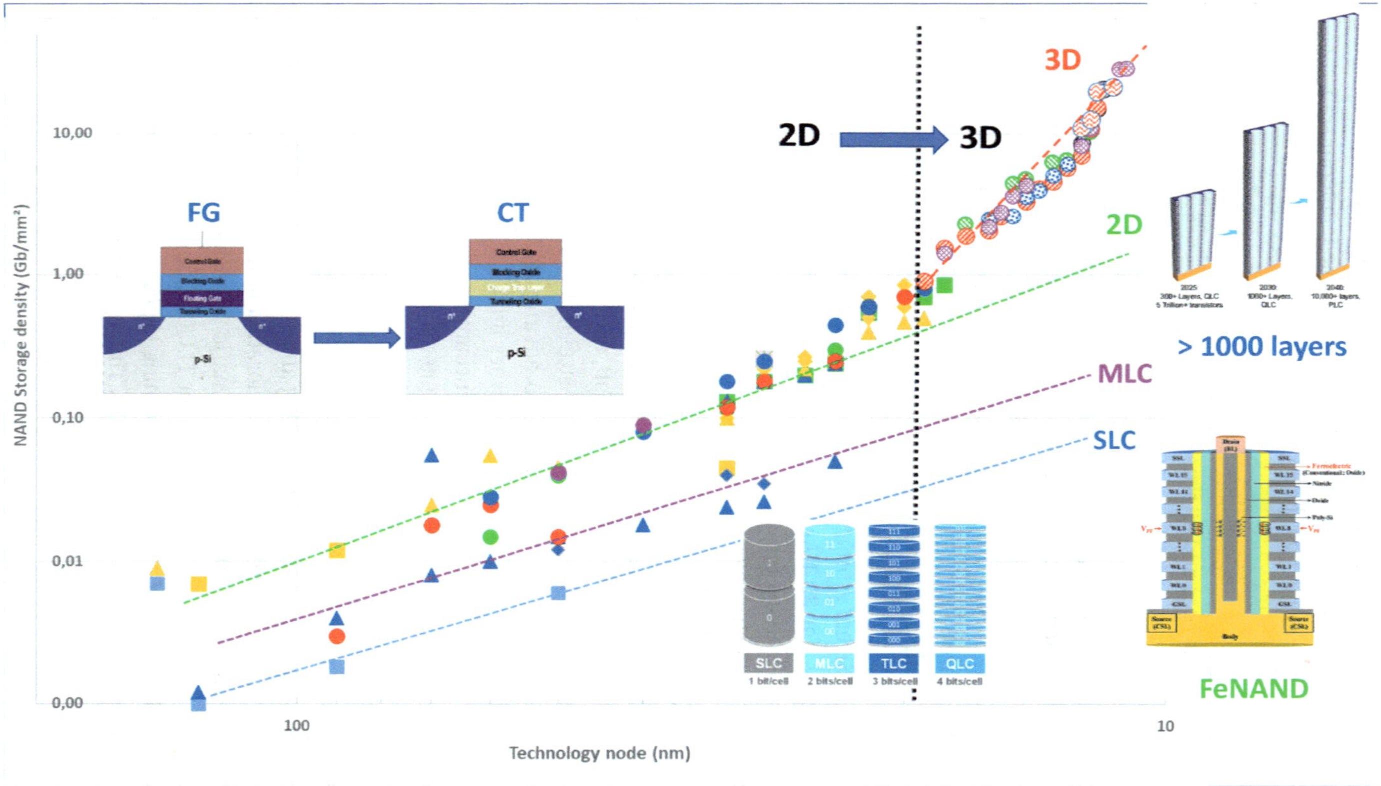

Figure 5.11: Evolution of the NAND flash storage density with the technology node updated from [1] – sources: Intel, Micron, Samsung, SK Hynix, Kioxia, Solidigm, and YMTC.

Additionally, the scaling down of CT NAND memory is hindered by the physical constraints of the charge-trapping layer. As memory cells become smaller, it becomes progressively more challenging to retain a high density of trapped charges without experiencing significant charge loss.

In recent years, hafnium oxide (HfO_2)-based FeFETs have surfaced as a potential alternative for charge trap layers in 3D NAND, facilitating anticipated data densities that exceed 100 Gb/mm^2 via z-scaling past 1,000 layers. FeFETs also function at lower voltages and nanosecond speeds, reducing endurance degradation at higher logic levels. Although the access time for 3D-FeFET stands at 100 ns, which is longer than that of planar FeFETs (10 ns), it remains significantly faster than NAND flash by a factor of 1,000 (Figure 5.17). The energy expenditure per write operation for 3D-FeFET, also called FE-NAND, is about 1 fJ, compared to 1 nJ for CT NAND flash, marking a six orders of magnitude improvement [376]. However, there are challenges faced by ferroelectric HfO_2-based FeFETs, as they tend to have restricted memory windows; the difference in V_{th} between the programming and erasing states [123].

Figure 5.11 presents the evolution of the NAND flash storage density as a function of the memory technology.

5.2.3 DRAM challenges and alternatives

DRAM works as the main memory in every modern computing system, from high-end server computers to simple hand-held devices. A DRAM cell has a simple structure comprising one capacitor (1C) connected by one transistor (1T) (1T1C) to the bit-line (the line through which information is written to or read from the memory cell) (Figure 5.12). This configuration allows DRAM to reach a very high storage density. But its pitfall is that since it stores data as charges, which leak, the charges need to be read and rewritten again every few milliseconds.

The DRAM readability is commonly gauged by the bitcell sensing margin (SM), i.e., the difference between the voltage development on the bitline (BL) of states "0" and "1." It is generally accepted that an SM of more than 0.1 V is the absolute minimum in DRAM in order to compensate for the cross-coupling and offset noises in array sensing. Meanwhile, the SM is governed by the capacitance ratio between BL parasitic capacitance (C_{BL}) and DRAM cell capacitance (C_S), where a rule of thumb of $C_{BL}/C_S < 4$ applies to ensure the desired >0.1 V SM under 1 V cell array operating voltage. This constraint implies that sufficient C_s needs to be preserved despite continued bitcell footprint reduction [284].

The DRAM cell capacitance C_s is the crucial parameter that determines the speed, data storage, data retention time, and power consumption. Shrinking the dimensions of planar DRAMs translates to a decrease in the capacitor area, and accordingly, in a decrease in the DRAM cell capacitances. In order to increase the capacitor area, the semiconductor industry moved to the trench and stacked capacitors concept (Figure 5.14).

The cell capacitor also evolved from planar to vertical structure and from cylinder to pillar shape to increase C_s on the same footprint (Figure 5.12). However, this solution also reaches its manufacturing limits, as the complexity of the manufacturing process is largely increased, such as a high aspect ratio.

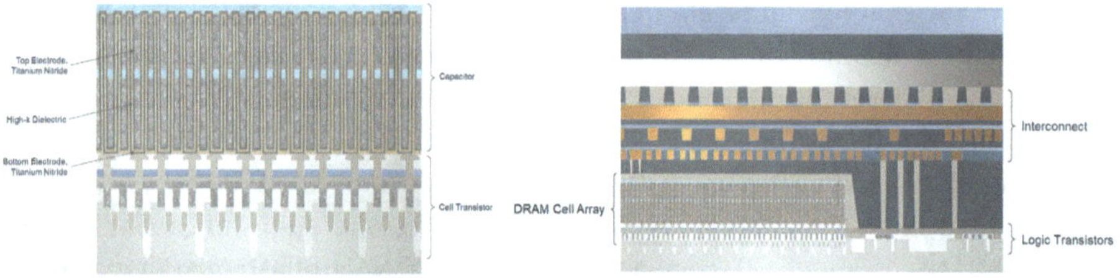

(a) **(b)**

Figure 5.12: Schematic of (a) DRAM showing an array of cell transistors (bottom) and tall capacitors (top), (b) DRAM cell array, peripheral logic transistors, and interconnect wiring [377].

The first DRAM generations used a silicon/insulator/silicon (SIS) film architecture, in which doped poly-Si layers act as electrodes and a deposited $SiO_2/Si_3N_4/SiO_2$ (ONO) film stack as the dielectric. In order to increase the capacitor area, the poly-Si deposition process was finely tuned in order to form hemispherical-poly silicon grains (HSG) in the trench and stacked DRAM capacitors. Since SiO_2 and Si_3N_4 films present relatively low dielectric permittivity, which is not sufficient enough for reducing the leaking current and for increasing the storage capacitance, the DRAM industry introduced high-κ materials, including Al_2O_3, Ta_2O_5, HfO_2, ZrO_2, and TiO_2 films for replacing the ONO dielectric. Since these high-κ materials are generally hypo-stoichiometric and tend to create a low-κ interfacial layer with Si (mostly SiO_2), DRAM manufacturers shifted the bottom electrode material from heavily doped poly-Si to metal electrodes such as TiN, WN, TaN, etc. This shift to a metal-insulator-metal capacitor structure also enhances the DRAM capacitance (Figure 5.14).

Through extensive work, it has been proven that none of the emerging NVM studied technologies can directly replace DRAM from a product standpoint due to the strict latency, power, and bandwidth requirements of DRAM. STT- and SOT-MRAM and ferroelectricity (FE)-based memories can achieve near-DRAM access times (Figure 5.17). But the scalability and storage capacity of STT-MRAM are severely limited due to the high current requirements for cell operation and the low sense margin of scaled STT cell architectures. The SOT-MRAM is under development and currently not mature enough as a DRAM substitute.

FE-based nonvolatile DRAM (NVDRAM) technology is the only technology that provides the combined capacity, scalability, performance, power, reliability, and cost to suit the requirements of high-capacity AI/ML applications requiring near-DRAM performance. Additionally, the nonvolatile nature of NVDRAM allows for reduced refresh overhead commands relative to conventional DRAM, leading to power reduction and reduced memory access collisions [286].

Recently, CMOS-compatible, HfO_2-based ferroelectric capacitors (FeCAPs) have emerged as a potential solution to the scaling limitations of DRAM. Due to their high spontaneous polarizations, FeRAMs promise to provide the necessary SM with smaller Aspect Ratio (AR) below 50 nm technology nodes, thus retaining the desired cost, area, and layer thickness scaling [284]. The so-called nonvolatile DRAM (NVDRAM), combining DRAM and FeRAM, recently achieved a bit density of 0.45 Gb/mm^2, similar to the industry-leading 1β planar DRAM technology (Figure 3.19) [286].

Scaling the 2D-DRAM memory cell and structural changes would continue in the forthcoming years, thanks to the use of EUV lithography for patterning the DRAM cells. The EUV lithography request is expected to rise more than 20% node on node (Figure 5.13). However, as DRAM memory cell dimensions decrease beyond the 10 nm process node, the leakage current of the access transistor becomes the foremost challenge.

In terms of cell geometry scaling, efforts are underway to introduce a vertically stacked DRAM, commonly referred to as 3D-DRAM, to achieve a higher integration density compared to the current cell architecture, like its flash memory counterpart.

1T1C 3D-DRAM cell array is currently investigated in two configurations: the vertical and the horizontal bit-line. In the vertical bit line scheme, bit-lines run vertically through stacked tiers along the z-axis, while word-lines extend laterally within each layer. Conversely, in the horizontal bit-line configuration, the array layout is rotated so that bit-lines run horizontally and word-lines vertically. In both cases, the plate line will be vertical. Comparative analyses have reported that the vertical bit-line architecture can achieve more uniform and roughly twice the SM of horizontal bit-line designs at comparable stack heights, indicating that the vertical bit-line is a more promising foundation for future 3D-DRAM scaling, despite its increased technological complexity [168].

One of the main technology issues for monolithically stacking several DRAM layers is the isotropic etching of the full stack for embedding the access transistor, and the integration of the capacitor as close as possible to the access transistor. DRAM manufacturers used the selective etching of Si versus SiGe epilayers, which is used to manufacture the GAA transistors (Figure 5.2e), or the selective etching of amorphous SiO_2 versus Si_3N_4 layers in an oxide/nitride stack to create holes in the vertical stack for embedding capacitors. The GAA nanosheet can be favorably used in the development of DRAM for upcoming technological nodes. The GAA capacity to vertically arrange several nanosheets enables the potential for developing a DRAM structure that is both easily expandable and economically efficient.

Using a single-crystal Si channel for the access transistors (3 nm FinFET) of 3D-DRAM, simulation results show that a stack of 64 DRAM layers leads to a bit density of 0.43 Gb/mm^2, which is similar to that of 2D DRAM (Figure 3.19). A 128 DRAM layer stack increases the bit density up to 0.66 Gb/mm^2. Replacing the Si channel with an

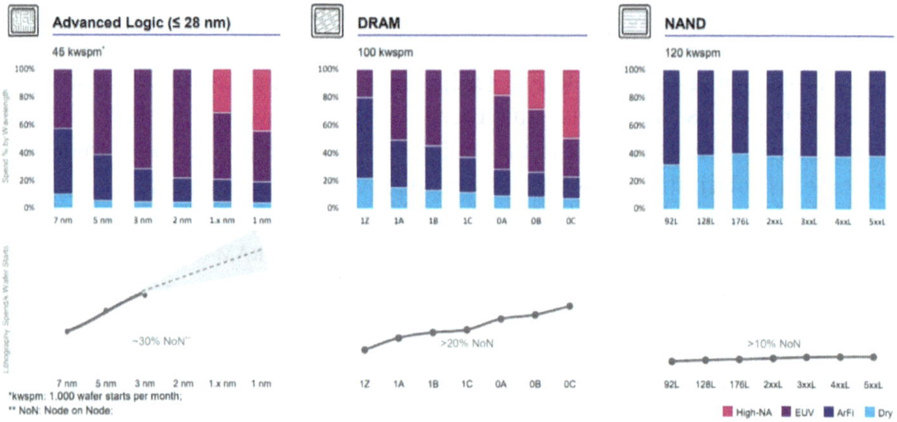

Figure 5.13: Lithography needs for manufacturing advanced logic devices, DRAM, and NAND memories – courtesy of ASML (November 2022).

amorphous W-doped In_2O_3 semiconductor channel allows enhancing the bit storage density up to 0.74 and 1.08 Gb/mm^2 with 64- and 128-DRAM layers, respectively [378].

Amorphous semiconductor oxides such as W-doped In_2O_3 and IGZO are very attractive for 3D-DRAM since this can reduce refresh power consumption, which is still mandatory for 3D-DRAMs.

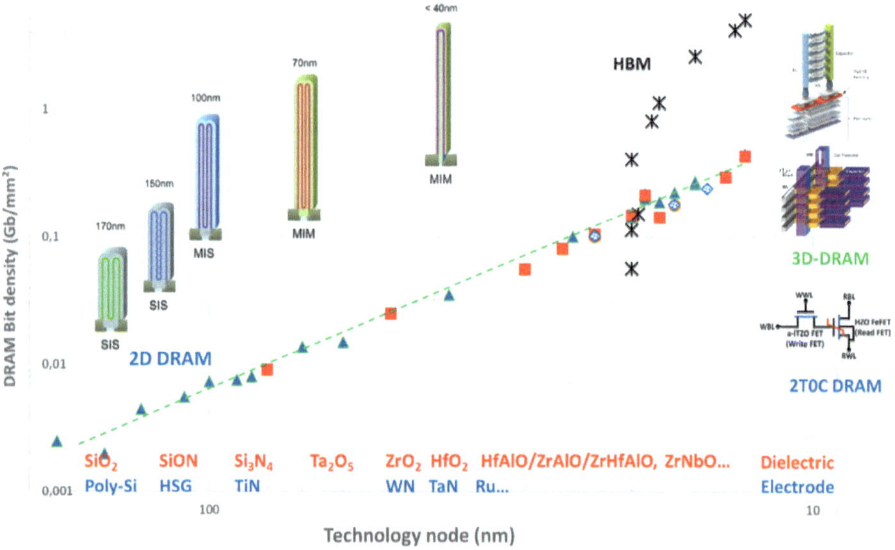

Figure 5.14: Evolution of the DRAM storage density with the technology node updated from [1] – sources: Micron, Samsung, SK Hynix, and CXMT.

3D-DRAM can also be produced by bonding the 3D-DRAM cells to the access transistors, thanks to hybrid bonding [379].

HBM is already the dominant solution for ultrahigh-density and ultrahigh bandwidth memory for AI acceleration and HPC. Today's HBM (i.e., HBM3E – Figure 3.19) stacks multiple DRAM dies – typically eight 24-Gb dies – on a base logic die using TSVs. While this provides higher than 1 TB/s bandwidth per package, it may face fundamental limits in stack height, die yield, and base-die scaling. Instead of eight separate base dies, a single 3D-DRAM die (~74 stack layers × amorphous oxide semiconductor channel) provides equivalent capacity (24 GB per HBM cube). In a 3D-DRAM configuration, signals traverse only ~7 μm vertically (versus >360 μm through eight stacked dies for HBM3E), drastically reducing interconnect delay, TSV resistance, and energy per bit transferred. The simplified bonding interface – one base die bonded to buffer die instead of eight – also improves yield, reduces assembly cost, and enhances thermal management [378].

The configuration known as two-transistor-zero-capacitor (2T0C) DRAM has been examined as a potential option for IMC applications due to its alignment with logic operations and its ability to perform read actions without causing damage. A 2T0C DRAM cell consists of two transistors, namely a write transistor (WTr) and a read transistor (RTr), in which the source of the WTr is connected to the gate of the RTr, thereby forming a storage node at the RTr side. This capacitor-less structure simplifies the fabrication process while enabling more aggressive scaling. In addition, the independent operation of the WTr and RTr allows nondestructive read operation. This type of DRAM also brings benefits like rapid operation speed, reduced energy use, and long-lasting performance. Traditional 1T1C DRAM struggles with parallel processing and computation, which are essential for IMC applications, as the data stored cannot be retained during read processes. In contrast, the 2T0C DRAM unit cell features distinct pathways for writing and reading, with the drain current from the reading transistor acting as the output current. This property of 2T0C DRAM allows for nondestructive reading, facilitating efficient parallel processing and computation.

However, the volatile characteristic and capacitorless structure of 2T0C DRAM cause challenges, including a larger unit cell area and shorter retention time. Consequently, 2T0C DRAM has a low memory density currently and necessitates frequent refresh operations, potentially leading to considerable energy consumption [380].

As an example, an amorphous oxide semiconductor FET based on Zn-doped $(In_{1-x}Sn_x)_2O_3$ for writing data with a FeFET based on $Hf_{0.5}Zr_{0.5}O_2$ for reading data leads to an excellent multibit retention time of more than 2,000 s and low power consumption of a 2T0C DRAM. Since they can be embedded in the CMOS BEOL, such devices could also be used for 3D-DRAM [381].

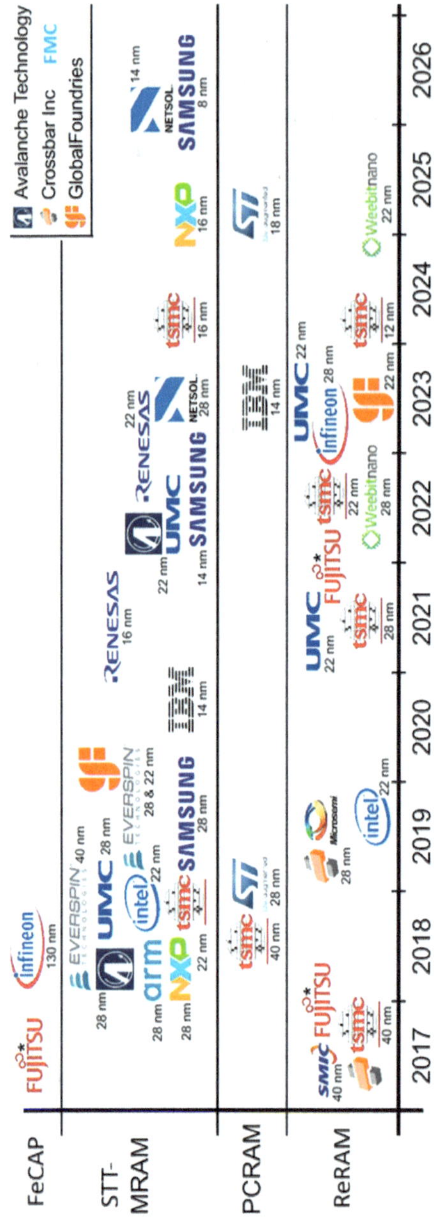

Figure 5.15: Timeline of recent technology demonstrations of embedded NVM with respective CMOS nodes, adapted from [384].

5.2.4 Embedded memory challenges

Emerging NVM already has a strong foothold in the semiconductor industry, with the main target of replacing embedded flash memory, and soon possibly DRAM and even SRAM [383]. Various emerging NVM are already available as consumer products at a large scale and at very advanced CMOS nodes down to 22 nm, with even smaller nodes at qualification and development stages (Figure 5.15).

Longer latencies of resistive RAM, phase change memory, and 3D XPoint mainly limit their use to storage-class memory.

STT-MRAMs, FeRAMs (FeCAPs and FeFETs), and ReRAMs are the current frontrunners among emerging NVM for embedded flash and DRAM replacement [383]. Indeed, their storage capacity has considerably increased during the last two decades, and they are close to the storage density of NAND flash and DRAM memories. Moreover, their storage capacity tends to increase using more advanced technology nodes (Figure 5.16).

The access times of some emerging NVMs like STT-MRAM and spin-orbit torque MRAM (SOT-MRAM) range between those of SRAM and DRAM (Figure 5.10). Other emerging NVMs like PCRAM, FeFET [384], and ReRAM exhibit access times ranging from those of DRAM and NAND Flash memories (Table 4.1 and Figure 5.10). Furthermore, their read/write energy per bit ranges from 5×10^{-3} pJ (FeFET) to 50 pJ (ReRAM), which is quite lower than the typical energy per bit of flash memories (10^4 pJ) [384].

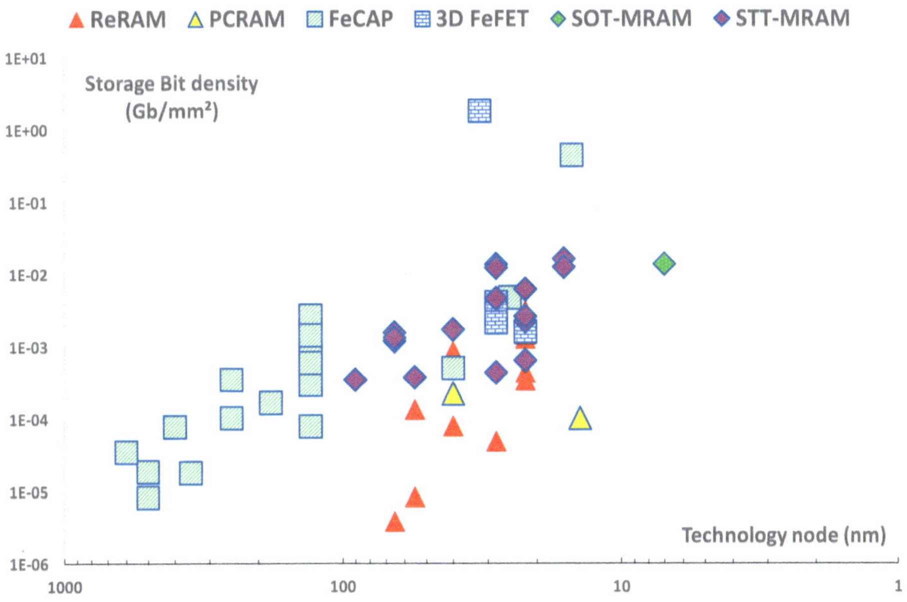

Figure 5.16: Scaling trend of memory capacity for the main emerging NVM technologies [189, 192, 248, 271–285, 287–297].

Therefore, emerging NVMs present electrical properties that could also compete with SRAM (see Section 5.2.1), DRAM, and NAND flash as stand-alone memories (Figure 5.17).

Furthermore, the radiation hardness of emerging nonvolatile memories, such as STT-MRAM, ReRAM, and PCRAM, is much higher than that of flash memories. These memories are relatively insensitive to radiation, which is of utmost importance for embedded memories that are used in applications in which the devices are exposed to cosmic rays, such as automotive, plane, etc. [385]

However, their production costs are still too high compared to the manufacturing costs of DRAM and NAND flash, which use mature production processes (Figure 5.18). Therefore, emerging nonvolatile memories are currently only dedicated to niche markets as embedded memories, and are mainly used in MCUs. However, they represent a very attractive substitute for GPUs as AI accelerators in terms of energy efficiency. Such an important application could probably lead to high-volume manufacturing of the most interesting memories, such as FeFETs, FeRAMs, and ReRAMs in the forthcoming years (Figure 4.19).

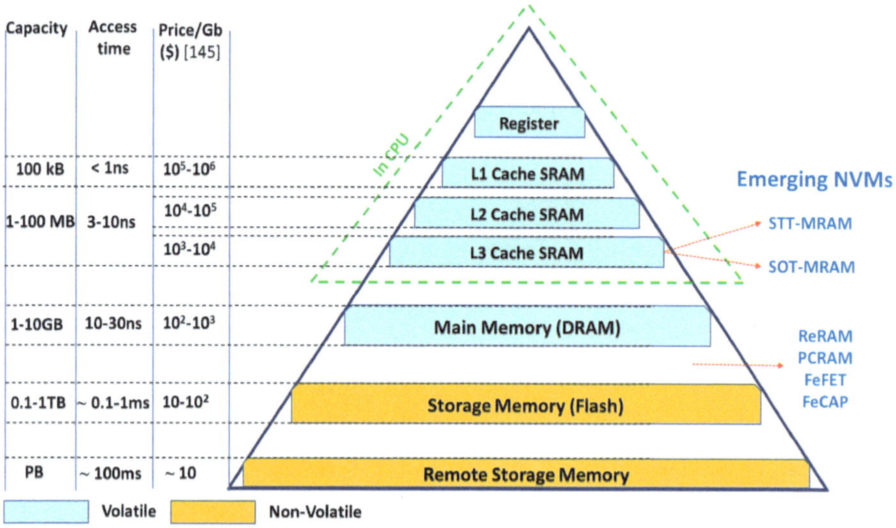

Figure 5.17: Memory hierarchy for computing.

5.3 Power and RF devices

Table 5.2 presents the main physical and electrical properties of different materials for power electronics. UWBG materials, such as β-Ga_2O_3, AlN, and diamond, present a higher bandgap E_g, and, above all, much higher electrical breakdown field and figure of merit [FOM = $\varepsilon\mu E_g^2$ (2)] than Si, in which ε is the dielectric permittivity and μ the carrier mobility. They could exhibit higher breakdown voltages (BVs) and lower elec-

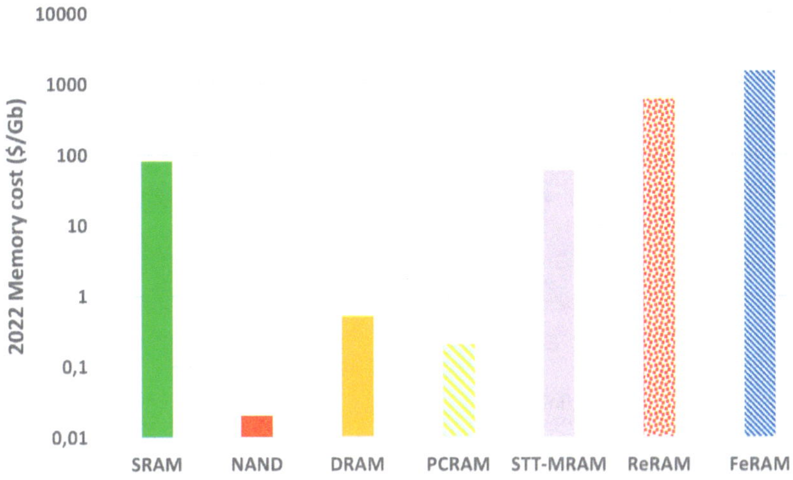

Figure 5.18: 2022 Memory cost expressed in $/Gb – courtesy of Yole Development Emerging Nonvolatile Memory 2023.

trical losses as power devices than their Si counterparts (Figure 5.19). Therefore, they could be used as very high-power devices at voltages around 10 kV for railway applications, for instance. However, obtaining large crystals and doping of those materials remains a major challenge [386]. The use of UWBG materials could be required to develop improved power devices in the forthcoming years.

Table 5.2: Main properties of WBG and UWBG materials.

Properties	Si	4H-SiC	GaN	β-Ga_2O_3	Diamond	AlN
Bandgap E_g (eV)	1.12	3.2	3.4	4.9	5.5	6
Electron mobility μ (cm^2/Vs)	1,240	930	1,000	180	2,000	426
Relative permittivity ε	11.7	9.7	8.9	10	5.7	9.76
Thermal conduct. (W/mK)	145	350	140	27	3,450	319
Breakdown field (MV/cm)	0.3	3.1	4.9	10.3	15.4	17.5
FOM (2) (V^2/Wcm^2)	8.8	6,270	27,900	36,300	554,000	336,000
Melting point (°C)	1,414	2,730	2,500	1,700	3,500	2,830

Due to their better electronic properties than Si for power devices, and as 200 mm in diameter SiC single crystals are currently available, while GaN epilayers can be grown on SiC wafers and on 200 and 300 mm Si wafers [387], SiC and GaN will partially replace Si power devices for numerous applications (Figure 2.3). The main applicative domains for the deployment of WBG devices are the automotive sector (Figure 5.20), energy grids and data centers for reducing their energy consumption. The trend for larger SiC single crystals will continue up to 300 mm in diameter, owing to the huge demand for SiC-based devices (Figure 5.20).

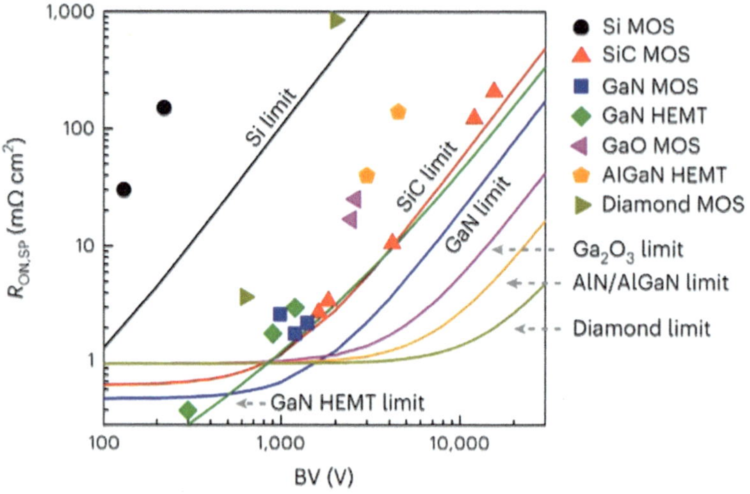

Figure 5.19: Electrical losses versus breakdown voltage (BV) for WBG and UWBG devices [**49**].

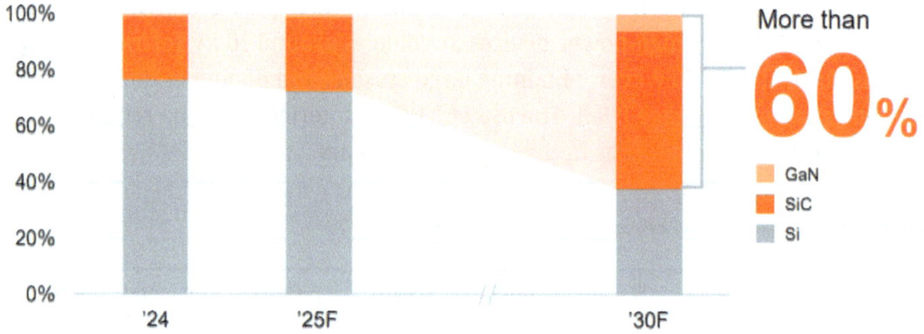

Figure 5.20: Share of SiC and GaN power semiconductor in automotive [388].

The mismatch between the crystal lattices of GaN and Si (or SiC) degrades the electrical properties of GaN power devices and affects their reliability. However, when GaN transistors are grown on GaN substrates or on substrates that are perfectly matched to GaN crystal lattice parameters and thermal expansion coefficients (CTE), GaN layers with a thickness higher than 1 µm can be epitaxially grown on this type of substrate. The several micrometers-thick GaN layers represent typical drain (n^+-GaN) and drift (n-GaN) regions in vertical transistors [389, 390]. Since the thicker the GaN epilayer, the higher the BV, producing GaN devices on substrates that exhibit identical crystal lattice parameters and CTE, allows increasing the BV of GaN devices, and widens their applications (Figure 5.21). Even, if new GaN transistor architectures on Si (red diamonds in Figure 5.21) begin to present similar BV than GaN on GaN devices.

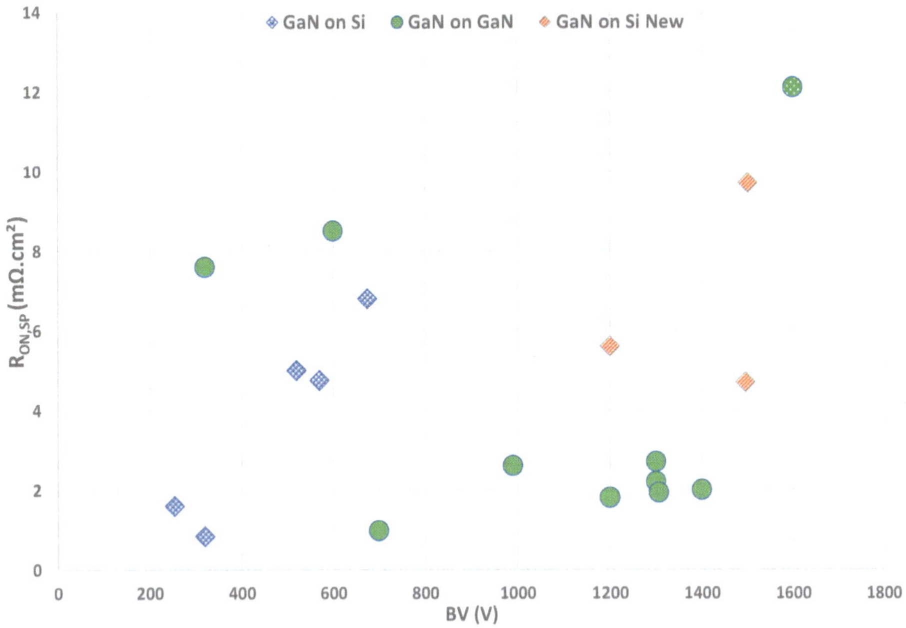

Figure 5.21: Electrical losses versus breakdown voltage (BV) for GaN epilayers grown on Si and GaN substrates – data from [390, 391, 405].

Thick and high-quality GaN epilayers enable the manufacturing of components capable not only of sustaining very high voltages but also of leading to vertical transistors. Vertical GaN technology really unleashes the full potential of GaN's superior material properties, such as higher BV by increasing the drift layer thickness and low electrical resistance by increasing the device surface area. Therefore, vertical GaN devices are very promising for numerous power device applications [393–404].

Power device improvement is envisioned to no longer solely rely on material innovation, but also to multidimensional architectures such as FinFET, multichannels, and stacked channels, like the advancements that have been achieved for improving the performance of logic devices [49].

Technological challenges associated with data- and tele-communications include:

- The expansion of wireless and wireline technologies (amplifiers, filters, switches, etc.) from 5G to 6G and the THz band.
- New approaches to network densification that take into account the growing importance of security.
- New device architectures for millimeter waves.
- Device technologies to maintain high bandwidth: heterogeneous integration of III-V materials (GaN, GaAs, InP) on SOI wafers, chiplet design, Si photonics, etc.
- A greater shift toward edge computing and local data storage.

6G will use the spectrum above 100 GHz and will eventually extend to the THz band. The advantages of such a high frequency are obvious: a wide bandwidth can be used, enabling peak data rates in the order of Tbps with latency in the order of microseconds. However, using such a high spectrum presents several limitations.

Selecting the most suitable semiconductors to increase the communications range is the most critical factor in the development of 6G. SOI-based CMOS can cover devices operating below 150 GHz, particularly for short-range communications. For longer ranges, the use of other semiconductors such as SiGe or III-V materials for power amplifiers may be necessary. The state-of-the-art frequency performance of InP, GaAs, GaN, SOI, and SiGe transistors in terms of communications shows that InP transistors are by far the fastest and the only two transistor types with maximum oscillation frequencies above 1 THz. They are followed by SiGe transistors, GaN transistors, and then by SOI MOSFETs (Figure 5.22).

Figure 5.23 shows a compilation of the output power of integrated power amplifiers realized in various semiconductor technologies as a function of frequency. It clearly highlights the degradation of the output power at increasing frequencies. At 200 GHz, a frequency within the spectrum for 6G, InP heterojunction bipolar transistor (HBT) power amplifiers, and most likely soon, power amplifiers based on GaN HEMTs and InP HEMTs deliver the required output power. Power amplifiers using SiGe HBTs, GaAs HEMTs, and Si MOSFETs with an output power of 200 mW, which corresponds to the power required for power amplifiers, at 200 GHz could also be realized in the future [41].

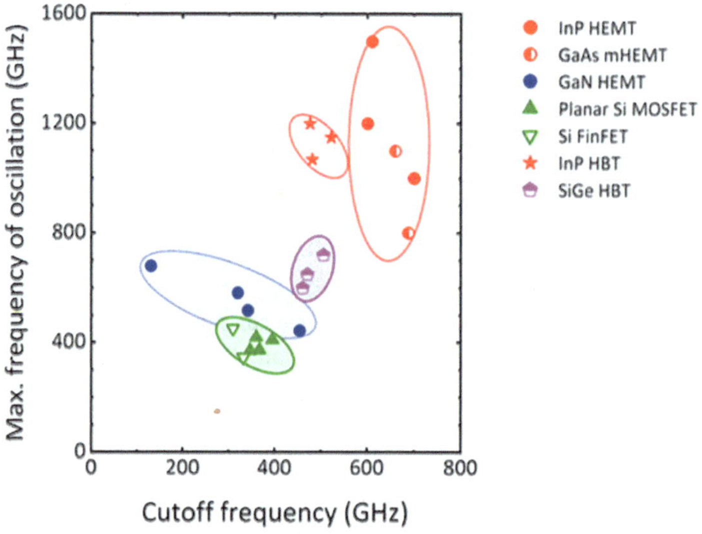

Figure 5.22: Best reported f_{max}-f_T data of GHz-THz InP HEMTs, GaAs HEMTs, GaN HEMTs, Si MOSFETs, InP HBTs, and SiGe HBTs [41 and references therein].

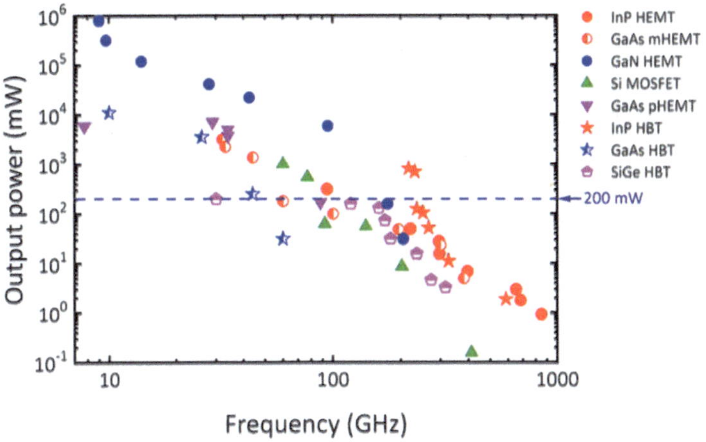

Figure 5.23: Output power of integrated power amplifiers based on different technologies versus frequency [41 and references therein].

As compute solutions scale to higher power for graphics and server platforms, and data rates continue to increase through emerging 5G/6G communications, semiconductor technologies such as gallium nitride (GaN) and advanced 3D packaging are gaining an increasingly important role in offering better performance, efficiency, integration, and density beyond current silicon- and III-V-based technologies. The 300 mm GaN-on-silicon technology is emerging as an attractive technology for high-density, high-performance power, and high-speed/RF electronics due to its ability to integrate low-voltage to 48 V GaN HEMT with silicon CMOS [406].

5.4 Sensors

A first future trend overall of sensor technology will be the increase in their spatial resolution, sensitivity, and integrability, and the reduction of their power consumption. Sensors will be integrated close to AI devices (logic devices and memories), thanks to 2.5 and 3D integration for achieving edge-AI, thus allowing real-time decision-making and convenient privacy.

The improvement of sensor performance would also allow extending the applications of various sensors. As an example, the pixel size of CMOS image sensors (CIS) is forecasted to be reduced down to 0.4 μm in the forthcoming years versus the current 0.6–0.8 μm pixel size [407] for enhancing their spatial resolution, while keeping their light sensitivity (Figure 5.24). The pixel size reduction results in a smaller CIS footprint. Therefore, it allows their use in healthcare devices, especially in smaller flexible video endoscopes. The next frontier for compact CIS is cardiovascular monitoring, where CIS technology is poised to enable the delivery of direct and real-time color images.

Extending the “spectral” range of the sensors is a second major trend of sensor technology. Hyperspectral imaging is an example of such a trend. Hyperspectral imaging integrates imaging and spectroscopy, leading to a fast, nondestructive, and environmentally friendly technique. Hyperspectral imaging involves the acquisition of an image of a material system at different spectral bands, simultaneously providing spatial and spectral information about the material. Its applications range from remote sensing, extending into fields like airborne sensing, water quality, waste management, pharmaceutical and medical fields, agriculture, and the food industry.

THz imaging provides a detailed resolution at the sub-millimeter level that is on par with human vision, making it possible to capture images through nontransparent substances, and responds to the moisture levels and conditions in systems that contain water, such as biological tissues.

Quantum sensors are currently being developed as they offer better sensitivity or reduced volume, weight, energy consumption, and cost, or even longer stability compared to conventional sensors. Quantum sensors are based on qubits similar to those used in quantum computers. They essentially include superconducting circuits, photonic systems, cold atoms, and nitrogen vacancies in diamonds. But while the extreme sensitivity of qubit quantum states to their environment creates challenges for quantum computing, the interaction of quantum states with properties such as motion, magnetic fields, and gravity facilitates quantum sensing.

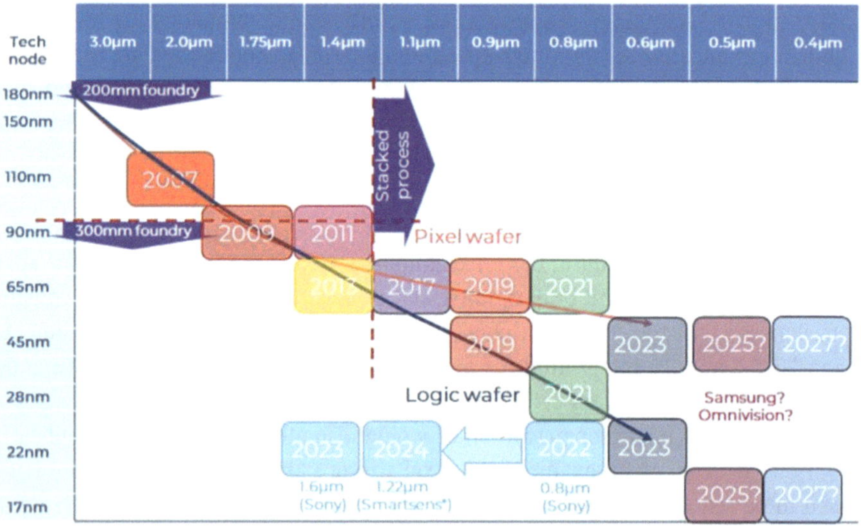

Figure 5.24: CMOS Imagers roadmap – courtesy of Yole Intelligence – status of the CMOS Image Sensor Industry report 2024.

Additionally, integrating multiple sensor types and data fusion techniques can improve overall data accuracy and enhance the reliability of sensing systems in agriculture, automotive, and medical applications. One of the most promising advancements is the development of AI-enhanced multi-sensor fusion systems. These systems can combine optical, acoustic, electromagnetic, and chemical sensors into integrated monitoring platforms capable of providing highly detailed, real-time insights into various applications. These applications include, but are not only soil and crop health, plant physiology, medical diagnosis, ADAS, and defense applications. Instead of relying on individual sensors with isolated functionalities, next-generation multimodal IoT networks will use AI-driven data fusion to extract deeper insights and optimize decision-making.

In order to fully benefit from AI potentials, which requires relatively advanced technology nodes, and to ease the 3D integration of sensors with AI chips, the manufacturing of sensors will progressively move from 200 to 300 mm Si wafers. Such a transition occurred for the manufacturing of CIS around 2010 for stacking logic AI devices, memories, and photodetectors [408], and will continue in the near future.

6 Semiconductor manufacturing sustainability challenges

For many years, the semiconductor sector has dedicated significant resources to improving the energy efficiency of electronic devices used in computing and sensing (Figure 4.2). Key factors driving these advancements include substantial enhancements in circuit architecture, integration of components, software development, power management strategies, and the incorporation of specialized accelerators.

Paradoxically, the components enabling the transition to a green economy – semiconductors – have historically been among the most resource-intensive to produce. The manufacturing process for these ubiquitous chips consumes huge amounts of electricity, often from fossil fuels, and billions of liters of water annually, while also generating hazardous waste and greenhouse gases (GHGs).

As outlined in the GHG protocol [409], there are three categories of GHG emissions, which are depicted in Figure 6.1 for the semiconductor sector:

- Scope 1: Emissions that occur directly from company activities and production processes
- Scope 2: Emissions that are indirect, encompassing those arising from power generation to fulfill the energy needs of the business
- Scope 3: Includes all other indirect emissions within a firm's value chain; upstream emissions originate from suppliers or their goods, whereas downstream emissions pertain to the use of semiconductor-inclusive products throughout their operational lifespan. It is important to note that scope 3 downstream emissions are largely unaffected by the phase of semiconductor production.

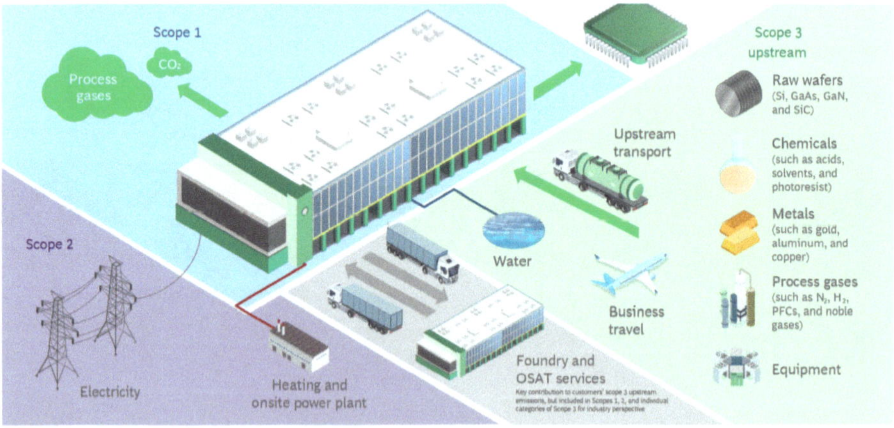

Figure 6.1: GHG emission from the semiconductor industry [410].

 | https://doi.org/10.1515/9783112247952-006

GHG emission of large semiconductor foundries has almost doubled during the last decade (Figure 6.2). Due to the growth of demand for AI chips such as logic devices and memories, chip makers' GHG footprint is expected to increase even more significantly in the next few years as manufacturing capacity expands. At the same time, production complexity is increasing as innovative advanced nodes, particularly in the logic and memory segments, decrease in size – an improvement that requires more electricity and uses more process gases and other raw materials that further fuel global warming [410].

Chip customers, such as electronic, computing device makers and automotive suppliers, are setting ambitious emissions-reduction goals (often spurred by their own customers who favor green products) and are asking their suppliers to help in this effort. Therefore, the semiconductor ecosystem must reduce its GHG emission in the forthcoming years for not increasing the semiconductor-related climate change trend.

In a standard semiconductor manufacturing facility, 35% of GHG emissions are classified as scope 1, while scope 2 accounts for 45% and scope 3 upstream represents 20% (Figure 6.2) [411].

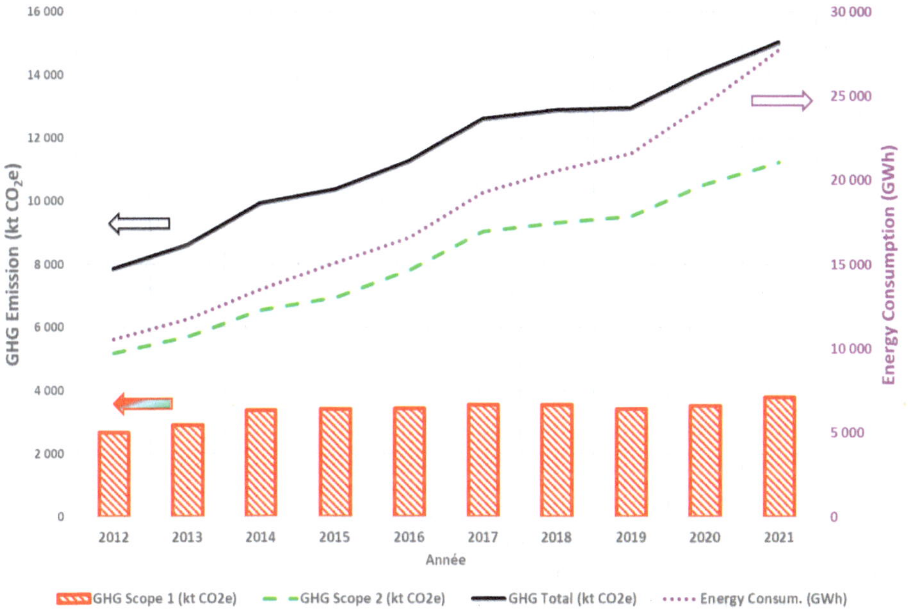

Figure 6.2: Evolution of the annual GHG emissions and energy consumption of large semiconductor foundries [411].

6.1 Scope 1: GHG emission and PFAS

In contrast to the overall GHG emissions of the semiconductor industry, which drastically increased during the last decade, scope 1 GHG emissions did not increase significantly during this period (Figure 6.2). A sharp decline from 2021 to 2023 was observed and was attributed to increased efficiency in gas abatement systems. However, scope 1 direct GHG emissions from chip production have been rising again since 2023. This phenomenon is explained by the surging demand for high bandwidth memory; vertically stacking more memory layers requires more frequent etching and cleaning, which results in higher fluorinated gas use [412]. This trend will need to be strongly reversed in the near future.

Between 85% and 90% of scope 1 emissions stem from the utilization of fluorinated process gases during wafer etching, cleaning, and chamber cleaning processes. In contrast, 10–15% result from the application of fluorinated heat transfer liquids. These gases comprise compounds like perfluorocarbons (PFCs), hydrofluorocarbons (HFCs), nitrogen trifluoride (NF_3), sulfur hexafluoride (SF_6), and nitrous oxide (N_2O), all of which possess high global warming potential (GWP) and endure in the atmosphere for extended durations. Furthermore, their emissions increase as technology node sizes decrease. The sixth assessment report from the intergovernmental panel on climate change lists the GWP100 values for CF_4, C_4F_8, NF_3, and SF_6 as 6,630, 10,200, 16,100, and 23,500, respectively [413, 414].

To lower PFC emissions effectively, various approaches must be rapidly developed, including alternatives, optimizations in processes, recycling, recovery, and emission reductions. However, finding suitable substitute materials for PFCs will necessitate considerable research and development. Enhancing the efficiency of processes that consume PFCs by modifying parameters such as temperature and chamber pressure can lead to a decrease in both PFC consumption and emissions. Additionally, the recycling, recovery, and treatment of PFCs from exhaust emissions can mitigate their release.

PFAS, which stands for per- and polyfluoroalkyl substances, are referred to as "forever chemicals" due to their prolonged existence, spanning thousands of years. They are synthetic chemicals characterized by multiple fluorine atoms bonded to an alkyl chain. The carbon-fluorine chemistry inherent in PFAS materials provides a unique combination of surface tension, stability, and chemical compatibility essential for many semiconductor applications. These applications primarily involve: (i) lithography that utilizes photoacid generators in chemically amplified resists, bottom antireflective coatings, top antireflective coatings, surfactants, barrier layers, and photoimageable polybenzoxazoles; (ii) wet processes employed in semiconductor manufacturing for tasks such as cleaning, stripping, wet etching, chemical mechanical planarization, metal plating, and aiding other processes. Specific wet chemistry applications depending on PFAS materials for certain performance criteria may encompass actions like post-plasma photoresist stripping, mitigation of high aspect ratio collapse, selective film inhibition, wetting low surface energy substrates, and cleaning particular components; (iii) dry etching and deposition processes that involve PFCs and HFCs classified under the PFAS category, including compounds like CF_4, C_2F_6, C_4F_8, and C_3F_8.

Considering the extensive variety of individual PFAS and their specific roles within the semiconductor manufacturing process, the timeline for identifying and implementing alternatives can be quite extended, even under the most favorable circumstances. Currently, there are no substitutes available for PFAS in photoresists, developers, antireflective coatings, rinsing agents, or wet and dry etching. Consequently, the immediate strategy to diminish or possibly eradicate PFAS emissions from the semiconductor sector involves eliminating materials containing PFAS from fabrication water, solvent waste, gases, and chemical vapors [415].

For instance, reactive ion etching (RIE) of Si is performed using SF_6 gas, which is not a PFAS. However, because the majority of photoresists used to pattern wafers are organic compounds, which means that when fluoride reactive ions bombard the surface of a patterned wafer, they not only react with the silicon in order to etch it, they also react with the photoresist and etch it as well. The reactive fluorine ions form PFAS gases or polymers with the carbon atoms that are present in the photoresist. Ion beam etching (IBE) is a similar process to RIE, but it is a PFAS-free process. In a typical IBE process, a plasma is created from an inert gas such as argon by electron ionization, and the positively charged ions are accelerated to form a concentrated ion beam. The ion beam is then directed at a silicon wafer, which can be moved or rotated according to the necessary etch pattern or angle of incidence of the beam. Both RIE and IBE succeed in etching large patterns (200 nm) on Si. IBE consumed more energy than RIE for the same process [416].

Figure 6.3 presents the annual scope 1 emission of some of the most important chip manufacturers and designers in 2020 and 2021. As expected, the five listed fabless companies disclose much less GHG emissions in their scope 1 than the IDMs. Memory chip manufacturers, such as SK Hynix, Micron Technology, and Kioxia, tend to have larger scope 1 carbon intensity than manufacturers concentrating on other activities [413].

6.2 Scope 2: power and water consumption

Scope 2 emissions, constituting the largest share of GHGs from semiconductor manufacturers (Figure 6.2), are associated with the power needed to operate their vast production plants. The origins of these emissions comprise:

- Numerous manufacturing devices, including lithography machines, ion implantation systems, and high-temperature kilns. The energy usage from these tools accounts for a significant portion (39%) of a typical semiconductor fabrication facility.

- Spacious cleanroom environments that necessitate control of temperature and humidity, along with over-pressurization and filtration of particles.
- Comprehensive sub-fab setups designed for gas removal, exhaust systems, water cooling units, and water treatment.

As the size of chip nodes progressively reduced, energy demands at production sites soared markedly over the past ten years, thus greatly elevating the scope 2 GHG emissions (Figure 6.4).

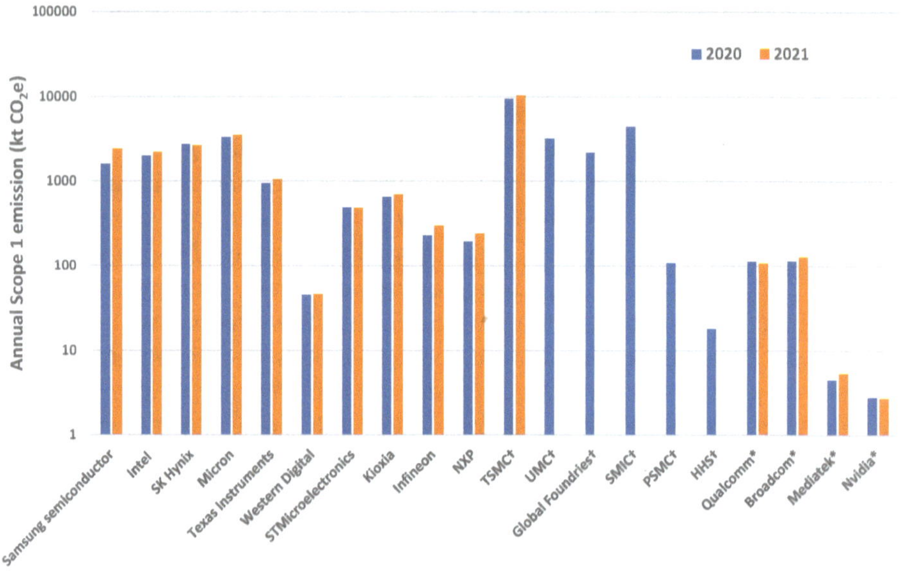

Figure 6.3: GHG scope 1 emissions of the 15 main semiconductor market entities in 2020 and 2021. Companies marked by † and * are foundries and fabless, respectively – data from [411, 413].

The energy consumption for manufacturing chips is closely correlated with the number of manufacturing process steps. As the number of steps increases with the advanced technology nodes, the energy consumption for producing logic devices and memories strongly increases (Figure 6.4). Consequently, as the node size of chips continues to shrink, energy requirements at production facilities are expected to increase significantly. The energy consumption for manufacturing a 300 mm wafer has been modeled at 400 kWh for a 28 nm node and 1,400 kWh for a 3 nm node (Figure 6.4).

Figure 6.5 shows the evolution of cumulative energy demand (CED) as a function of the dynamic random-access memory (DRAM) technology node. CED considers the whole energy demand during the manufacturing process at the level of wafer and packaging, including raw materials. CED is directly proportional to the volume of materials and energy consumed; these, in turn, are directly proportional to the number of photomasks and are therefore inversely proportional to the technology node size. Within each node, the utility (electricity, natural gas) had the largest CED share of nearly 75%, and the remaining share was attributed to materials (nitrogen, natural gas, Si wafer, etc.) [418]. The CED associated with the Si wafer is nearly constant with the technology node, while the CED for nitrogen and natural gas continuously increases.

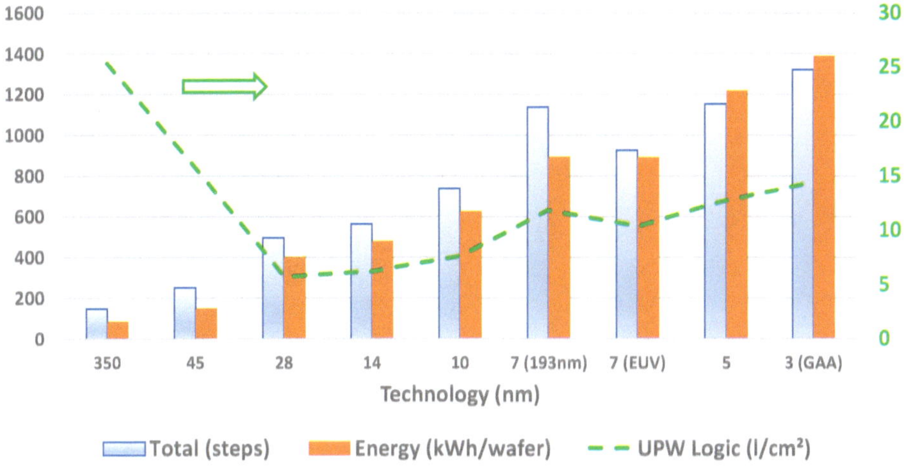

Figure 6.4: Correlation between energy and ultrapure water (UPW) consumption and the number of steps for manufacturing logic components. Data at 350 and 45 nm are experimental [417] and simulated from 28 to 3 nm [155].

CED for manufacturing DRAM and logic devices is equivalent and significantly higher than that for producing of 2D and 3D-NAND flash memories. It is also much higher than the CED required for packaging those chips (Figure 6.6).

Wafer-level chip-scale package (WLCSP) exhibits 30-fold lower CED than advanced flip-chip ball grid array (FCBGA) chip packaging, because the FCBGA possess 40 times more connections than WLCSP. While the older conventional packaging (quad flat package (QFP) and thin shrink small outline package (TSSOP)) has less electrical connections, which translates into a lower CED (Figure 6.6) [418].

The use-phase CED of DRAM was computed to be 19 MJ/cm^2, assuming the device is being used for 8 h/day in a 70% active state for 5 days/week and 50 weeks/year with 70% power supply efficiency. The share of the use-phase CED footprint in the total life cycle was nearly 28%, whereas the manufacturing and assembly shares were estimated to be 66% and 5%, respectively, for a representative 2 GB DRAM. The IC use-phase energy share is obviously very sensitive to the number of daily operating hours of the device [418].

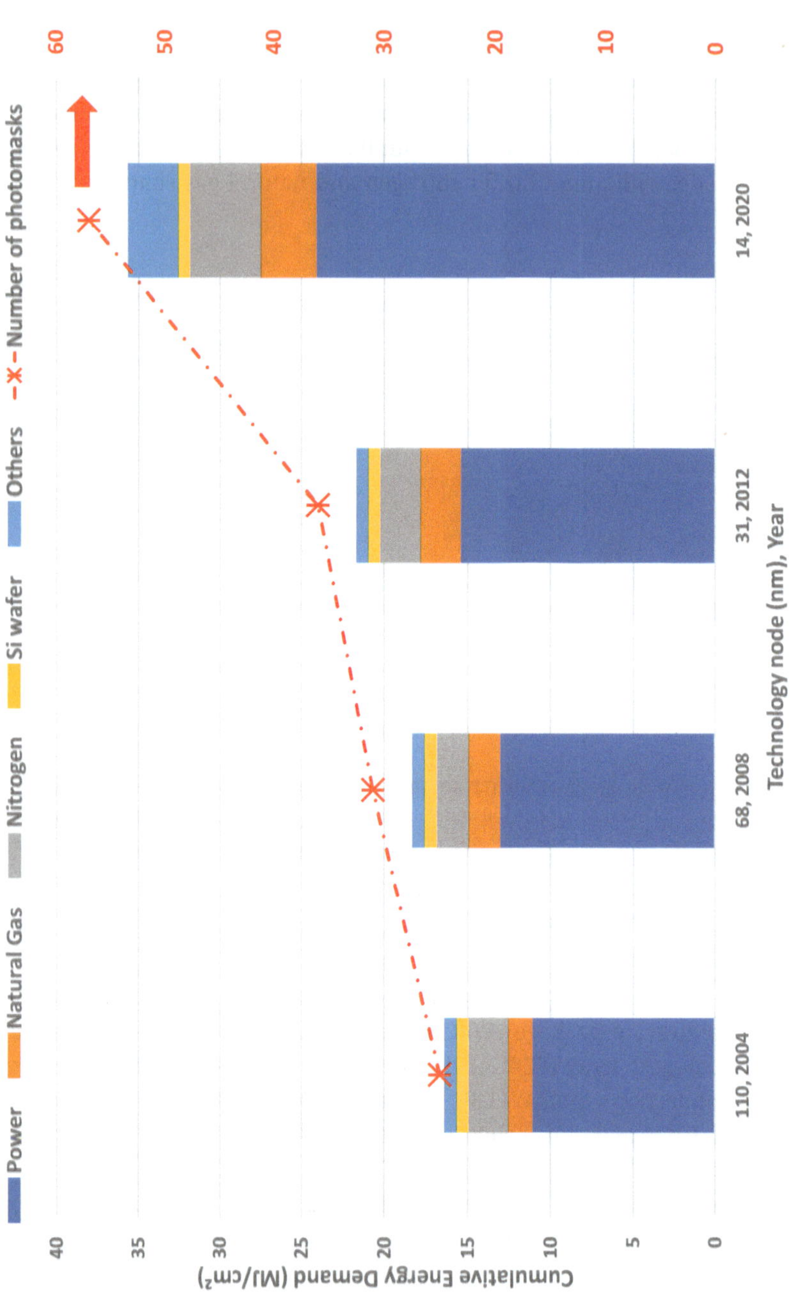

Figure 6.5: Variation of cumulative energy demand – CED (MJ/cm^2) with respect to DRAM technology node and manufacturing year – data from [418].

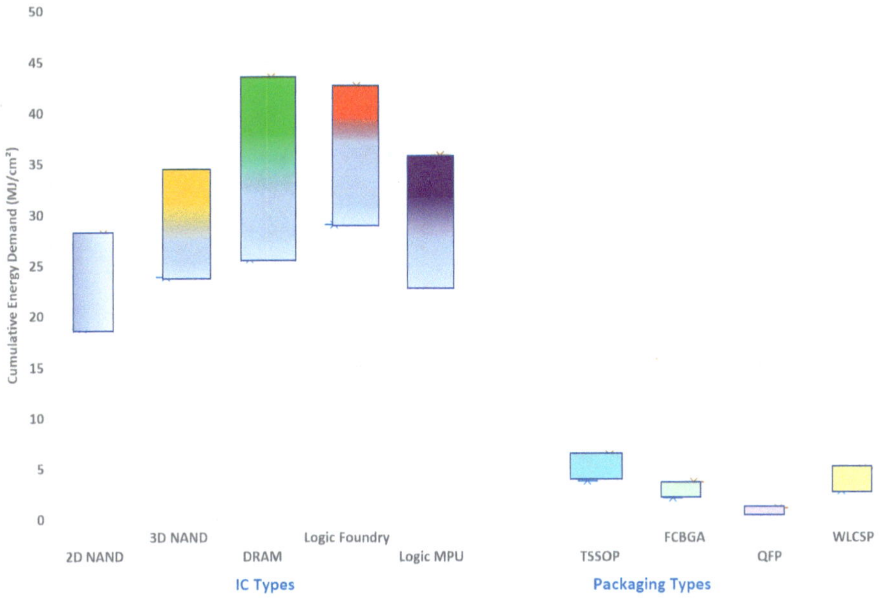

Figure 6.6: Variation of cumulative energy demand – CED (MJ/cm^2) of various chips at the wafer and packaging levels – data from [418]. *TSSOP: thin shrink small outline package, FCBGA: flip-chip ball grid array, QFP: quad flat package, WLCSP: wafer-level chip-scale package.*

Figure 6.7 shows the annual scope 2 GHG emission of the same chip manufacturers and designers, as shown in Figure 6.3. Scope 1 and 2 emissions are strongly related to the annual semiconductor manufacturing and the associated energy consumption [413]. Scope 2 emissions reflect the impact of the energy consumed by the production facility, and the investment made by the manufacturing company in renewable energies. As an example, a Chinese company SMIC did produce far less chips in 2020 and 2021 than TSMC. However, their scope 2 emissions are much larger than those of TSMC. This results from the Higher use of fossil fuels in China compared to Taiwan (Figure 6.8). The energy efficiency for producing chips of TSMC looks better than the one of SMIC. On the contrary, the European manufacturers (STMicroelectronics, NXP, and Infineon), primarily utilize renewable and nuclear energies (Figure 6.8), which leads to lower annual scope 2 emissions (Figure 6.7).

Using nuclear and renewable energy sources as the main energy providers for manufacturing chips is currently the best way for reducing the scope 2 GHG emission.

Reducing scope 2 GHG emissions presents a significant challenge for the semiconductor sector, largely because of the energy required for chip production and the specific needs of the manufacturing methods. This difficulty is further intensified as the semiconductor industry strives to grow to satisfy the tremendous demand for chips, particularly those used in AI applications.

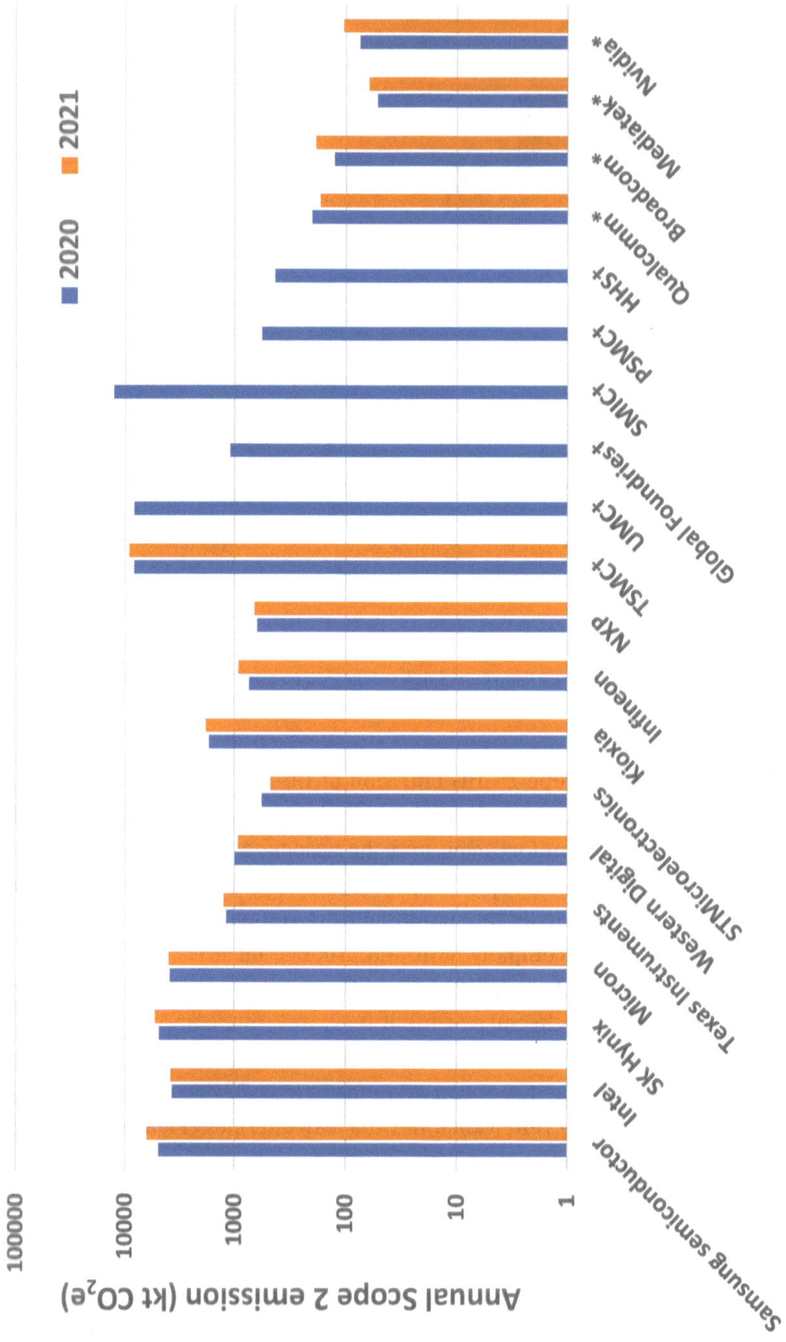

Figure 6.7: GHG scope 2 emissions of the 15 main semiconductor market entities in 2020 and 2021. Companies marked by † and * are foundries and fabless, respectively – data from [411, 413].

The production of semiconductors is also highly reliant on water and necessitates substantial amounts of ultra-pure water (UPW) to prevent the contamination of electronic components. Typically, a large semiconductor facility that utilizes 300 mm silicon wafers uses between 38 to 40 million liters of municipal water daily, which is comparable to the daily water usage of 300,000 households in the West or 500 pools. To produce 3,800 L of UPW needed for chip production, approximately 6,000 L of city water are required. Like the energy consumption, the water consumption of 300 mm manufacturing plant is strongly correlated with the number of process steps (Figure 6.4). The water consumption per Si area was higher for 200 mm wafers than 300 mm wafers; this transition occurred at 90/130 nm technology node. At those nodes, the front-end of line (FEOL) lithographic resist removal steps gradually moved from stripping using a solution containing per-oxidized sulfuric acid to a dry plasma process, thus gradually reducing the number of "wet" steps, in which the wafer is immersed in an UPW-based solution. This change explains the reduction in water consumption between the 350 and 45/32 nm technology nodes (Figure 6.4).

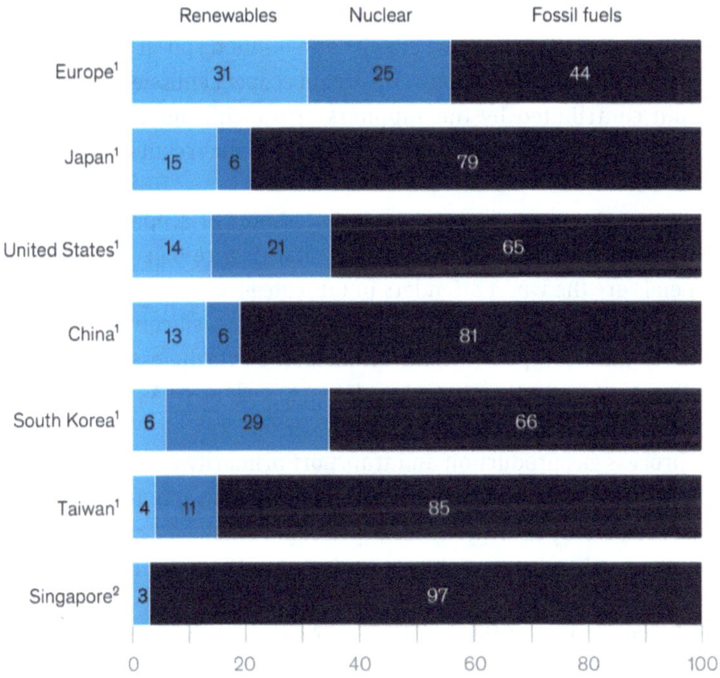

Figure 6.8: Share of electricity generation type, by region, % – [1]2020 data, [2]2019 data from [414].

In light of prolonged dry spells and a tremendous rising need for chips, the semiconductor industry is concentrating on innovative methods to reclaim, decrease, and repurpose the water utilized in their manufacturing processes. Recent progress in water purification technologies has been developed to enable semiconductor producers to treat and reuse wastewater while eliminating specific pollutants. This progress facilitates achieving a recycling water percentage, which is calculated as the ratio of recycled water to the total water consumption, comprising both the recycled water usage and the water intake, ranging between 12% and 70% [419, 420]. Nevertheless, given the anticipated surge in semiconductor demand in the upcoming years, the semiconductor industry needs to enhance this recycling water percentage, particularly in regions facing water scarcity.

6.3 Scope 3: GHG emissions

Scope 3 GHG emissions from the semiconductor emission are split in two categories: downstream and upstream. Downstream scope 3 emissions, covering the use of the final electronic device over its lifetime as well as its end-of-life disposal or recycling, will be partially described in the next chapter. Upstream scope 3 emissions primarily encompass CO_2 output contributed by the suppliers of the chip manufacturer; in other words, raw materials, manufacturing, logistics, and the like related to supplier production.

Purchased goods and services are the largest contributor of scope 3 upstream emissions, accounting for approximately 60% of the total (Figure 6.9). Raw wafers, process gases, and metals are the worst offenders in this category.

Raw wafer production-related emissions are driven mostly by silicon because it is used in higher volumes than compound semiconductors. Nevertheless, during the production of one kilogram of compound semiconductors, such as GaAs or SiC, more GHG are typically emitted than for the same amount of silicon.

Emissions from process gas production and transport primarily come from nitrogen and argon but also from PFCs. Other chemicals such as acids, photoresists, and solvents were found to have only limited scope 3 upstream emissions associated with them, because they are used in lower volumes and more efficiently than process gases.

Metal-related CO_2 emissions are the result of emissions-heavy production processes (including mining and multiple remelting steps to reach the required high purity) and the transport of these high-density materials.

In the category of capital goods, front-end and back-end manufacturing tools and equipment are estimated to contribute another 20–30% to scope 3 upstream emissions, according to reports from equipment manufacturers. Notably, emissions are driven almost entirely by the equipment manufacturers' own supply chains.

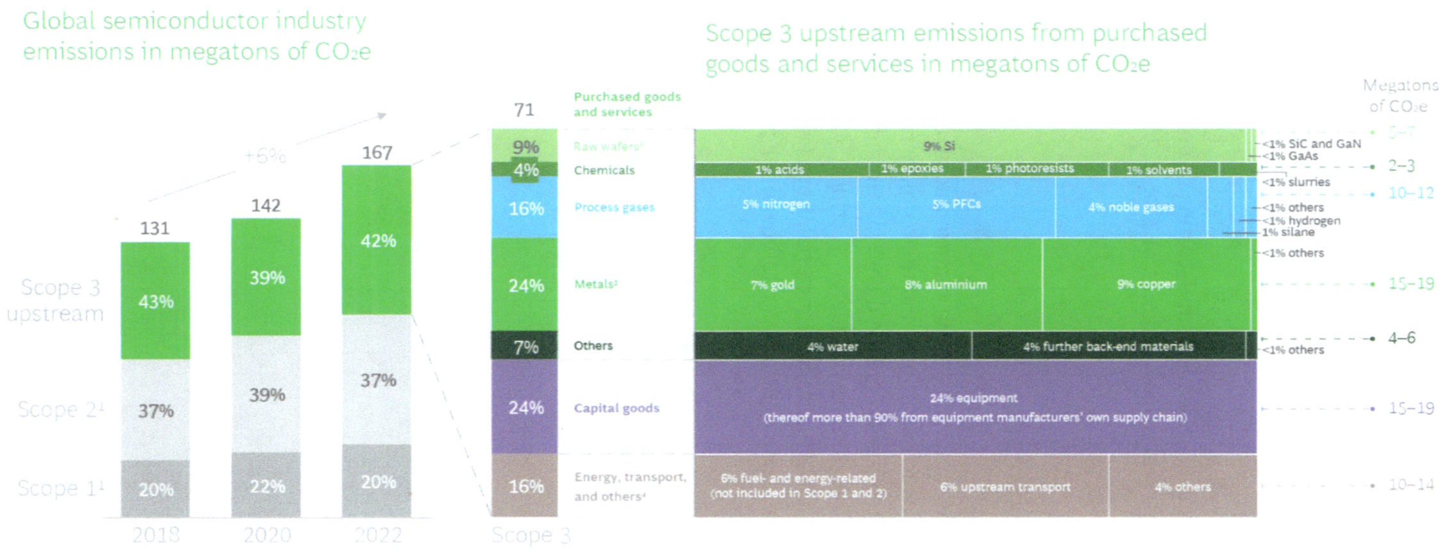

Figure 6.9: Evolution of the GHG emissions by the semiconductor industry, and distribution of the upstream scope 3 emissions [410].

Upstream scope 3 GHG emissions also depend on the number and complexity of process steps for producing the chips. Each additional process step involves more photolithography, process gases, and energy, thereby increasing GHG emissions across scopes 1, 2, and also 3 upstream. Further, advanced node manufacturing requires a greater number of state-of-the-art wafer fab equipment, expanding the environmental footprint even more.

Therefore, the production of leading-edge logic devices leads to an upstream scope 3 GHG emission around 32 Mtons CO_2e, while manufacturing of memories and discrete components results in an upstream scope 3 emission of 22 and 17 Mtons CO_2e, respectively [410].

6.4 Critical raw material consumption

Figure 6.10 gathers the numerous raw materials used in semiconductor manufacturing classified according to their economic importance and their supply risk for their extraction and processing stages. Raw materials in blue diamonds are not considered as critical, while the raw materials in red dots are considered as critical [421]

High-purity Si metal is considered as a critical raw material. However, its supply risk essentially results from the huge demand of polycrystalline Si for photovoltaic

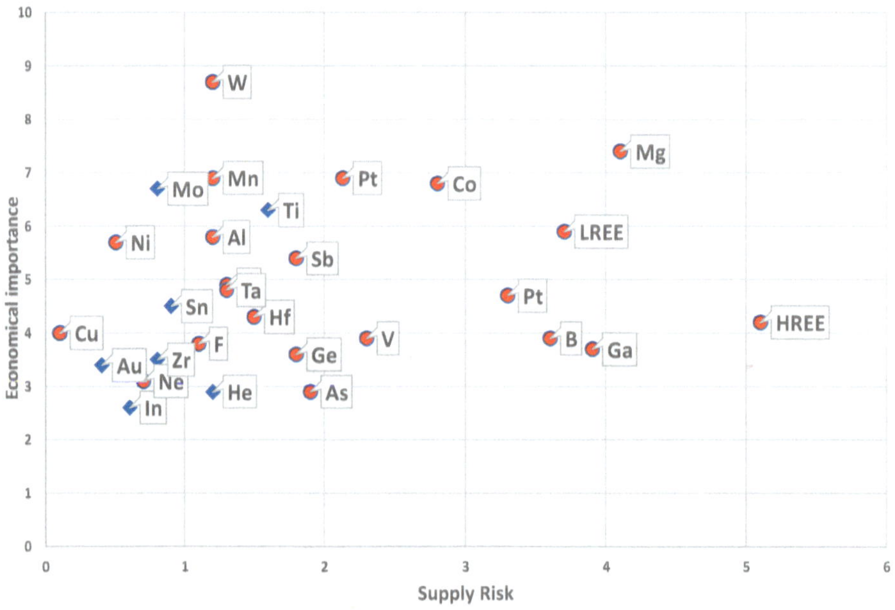

Figure 6.10: Assessment of the criticality of raw materials in EU – data from [421]. *Heavy rare earth metals (HREE: Dy, Er, Eu, Gd, Ho, Lu, Tb, Yb, Y), light rare earth metals (LREE: Ce, La, Nd, Pr, Sm).*

cells and batteries, while the semiconductor industry only consumes a very small part of the Si metal consumption.

Figure 6.11 gathers the main producer countries of selected critical raw materials. They are divided in two categories: (i) the mining countries (Figure 6.11a) and (ii) the processing or refining countries (Figure 6.11b).

The main country for extracting critical raw materials involved in the semiconductor from the ores is China, with a monopoly for Ga, and a large dominance for rare-earth elements, W, Si, P, As, and F (Figure 6.11a). Other main countries are South Africa (Pt-group metals), Congo (Co) – even if the Congolese mines mainly belong to Chinese companies – Australia (Al and Hf) and Brazil (Nb). The contribution of industrialized countries (the USA, Europe, Japan, etc.) in extracting critical raw materials for nanoelectronics is rather weak.

The dominance of China is even stronger for refining the raw materials. It has a quasi-monopoly for Ga, Ge, rare earth metals, and Si refining, and dominates the refining market for Al, Co, Cu, In, Ni, Sb, and W metals. The industrialized countries are more active in the field of refining metals that they are in the mining domain but at a much lower level than China (Figure 6.11b).

Due to its high dominance in the extraction and, above all, refining critical raw materials, in December 2024, China announced a complete ban on exports of Ga, Ge, and Sb to the USA. China extended its export controls in February 2025 to a range of materials, including W, Te, Bi, In, and Mo, and then in April 2025, its export controls concerned seven medium and heavy rare earth elements.

Mineral mining is a water- and energy-intensive industry with a complicated and negative reputation due to frequent incidents of environmental degradation, social unrest, and workplace accidents. Therefore, it also contributes to the upstream scope 3 GHG emission of the semiconductor supply chain.

One possible way to reduce the mining-related GHG emissions and the dependence of Western and Eastern countries to China in this sector is recycling metals and/or other elements from electronic waste (e-waste), electric vehicle batteries, and other scraps from the overall industry.

E-waste, technically known as waste electrical and electronic equipment (WEEE), includes nonfunctioning and discarded electrical or electronic equipment and its discarded components, ranging from smartphones and laptops to circuit boards and televisions. The economic value of metals contained in e-waste produced worldwide in 2022 was estimated 62 million tons (7.8 kg per capita) corresponding to a potential value of $91 billion after recycling. It is predicted that this amount will exceed 82 million tons by 2030, with an estimated annual increase of 2–2.5 million tons (Figure 6.12).

About 62 million tons of WEEE were produced in 2022, of which 13.6 million tons were formerly documented as formally collected, for a global recycling rate of 22.3% [425].

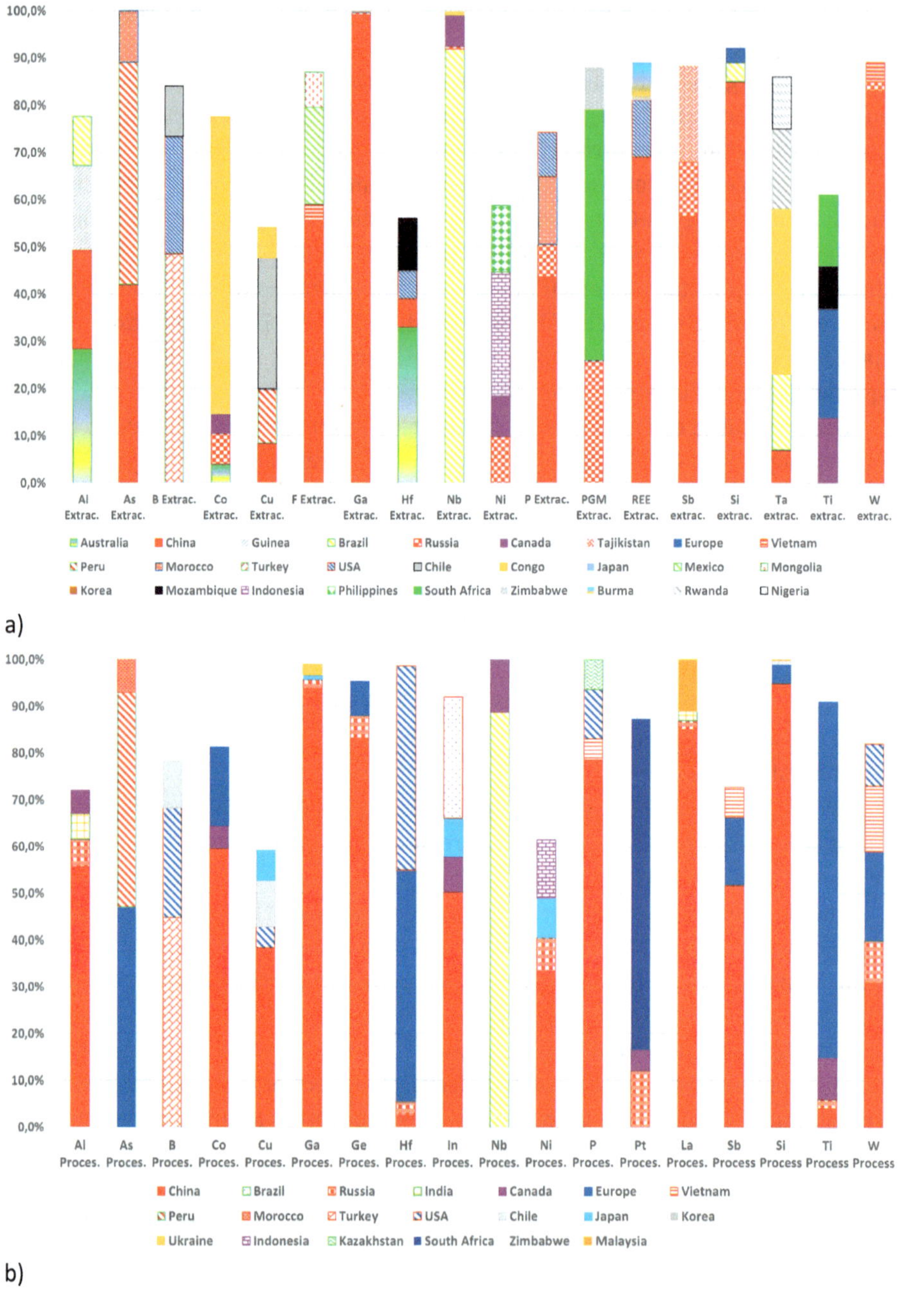

Figure 6.11: Share of producing countries in (a) mining and (b) refining of selected minerals – data from [414, 422–424]. *PGM: platinum-group metals, REE: rare earth elements.*

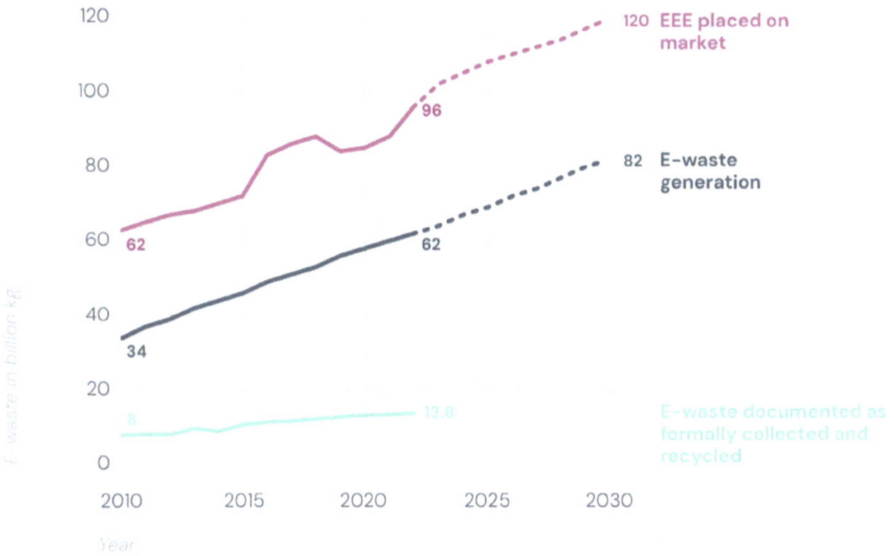

Figure 6.12: Evolution of e-waste generation [425].

The most valuable recovered raw materials are copper ($19 billion), gold ($15 billion), and iron ($16 billion), as these metals have high recycling rates using current technologies (Figure 6.13). Improving collection rates could therefore significantly increase recovery rates. Currently, e-waste management generates $28 billion in recovered raw materials out of a maximum of $91 billion, due to the poor recycling efficiency and very low economic viability of numerous elements. The main costs for this recovery were $78 billion, including $10 billion for the treatment and $78 billion as external costs for the population and the environment [425].

The end-of-life recycling input rate (EoL-RIR) reflects the total material input into the production system that comes from recycling of post-consumer scrap. Eol-RIR can be expressed using the following expression:

$$\mathrm{EoL\_RIR} = (B\_1 + B\_2)/(C\_1 + C\_2 + C\_3 + G\_1 + G\_2) \quad (6.1)$$

where B_1 and B_2 are the production of primary material as main product in a country sent to processing in the same country and the production of primary material as by-product in country sent to processing in the same country, respectively. C_1, C_2, and C_3 are the imports to country of primary material, secondary material, and processed material, respectively. G_1 and G_2 correspond to the production of secondary material from post-consumer functional recycling in a country sent to refining and processing in the same country, and sent to be directly used for manufacturing in the same country, respectively [426].

Figure 6.13 gathers the EoL-RIR of several elements. The highest EoL-RIRs are observed for metals, which present a high economic value (noble metals, especially)

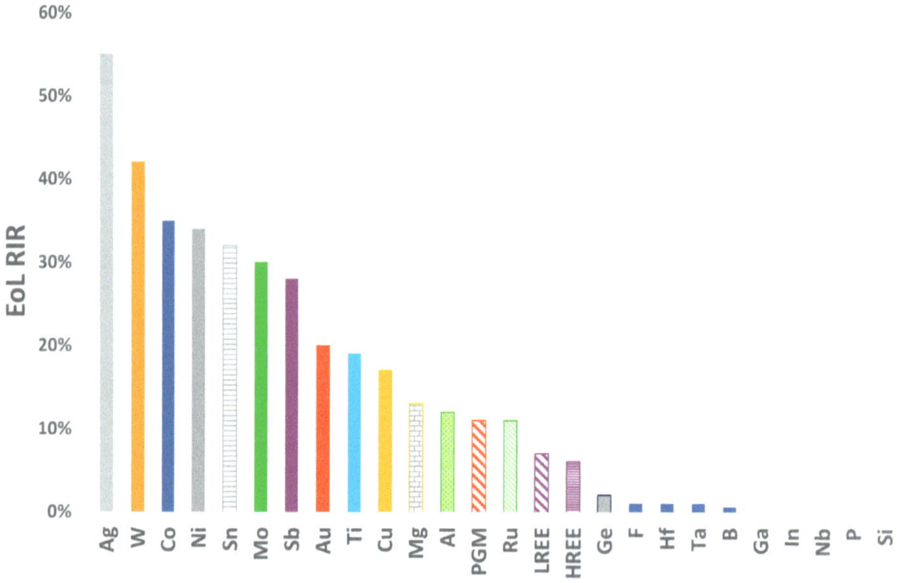

Figure 6.13: EOL-RIR for Europe of different elements that are used by the semiconductor industry – data from [426].

and/or which are not too difficult to recycle such as Cu, W, and Co. The Ga and Ge EoL-RIRs are rather poor and would not currently overcome the export controls setup by China.

Recycling platinoid metals shows the highest economic value in consolidated applications such as automotive, catalysts, and jewelry (Figure 6.14). The recycling rates of lithium from battery metals and of rare earth metals are markedly low, but still slightly higher than that of semiconductors such as Ge and Ga [427] (Figure 6.14). For recycling rates to increase, greater waste availability, improved end-of-life recycling, and profitable business models for recyclers must emerge.

The energy and water consumption for producing metals from WEEE are quite lower than those used for extracting and refining the same metals from ores (Figure 6.15) [426, 427]. This low energy and water consumption allows decreasing the recovery costs between \$10,000 and \$20,000/kg of Au using advanced WEEE recycling technologies, which is significantly lower than the \$30,000–\$50,000/kg in primary mining [429].

Recovering critical metals from electronics offers multiple advantages. It allows reducing wastes by preventing Pb, Hg, and Cd from contaminating soil and water. It also allows saving natural resources; it takes 1 ton of ore to yield the same amount of gold as one ton of recycled circuit boards. It lowers the mining impact, since ore mining consumes vast amounts of water and energy (Figure 6.15). Recycling is also

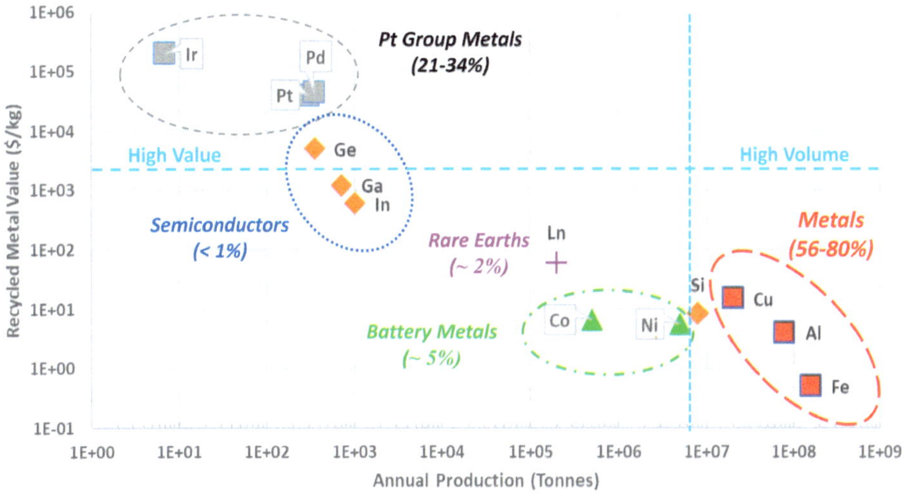

Figure 6.14: Recycled metal value versus the worldwide annual production of various metals (% represents the global recycling rate) – data from [428].

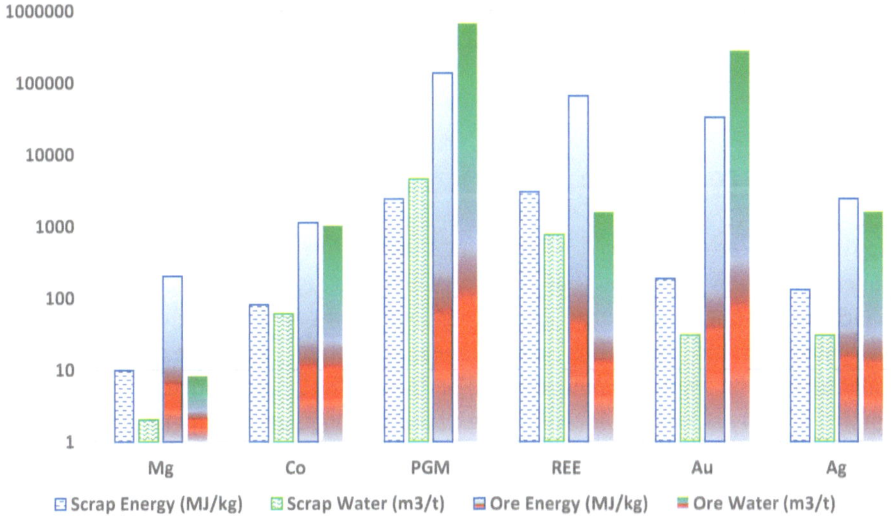

Figure 6.15: Energy and water consumption in production of metals from scraps and ores – data from [426, 427].

a billion-dollar industry providing skilled employment in sorting, refining, and technology development [430].

The semiconductor industry usually needs numerous metals (Figure 6.10), and with a very high and reliable purity level. Such a very high-purity level still remains an important technical challenge for using recycled metals, for instance. Therefore, eco-friendly innovation in recycling critical raw materials from WEEE looks paramount for reducing the dependency of industrialized countries to Chinese exports.

To reduce the downstream scope 3 GHG of the semiconductor ecosystem, moving from ore mining to WEEE recycling could be also an important step in terms of sustainability of the whole semiconductor supply chain. Current industrial WEEE recycling practices primarily employ pyrometallurgical and hydrometallurgical techniques or a combination of both. Pyrometallurgical recovery involves energy-intensive thermal treatment through incineration, smelting, or roasting. Hydrometallurgical recovery is a less energy-intensive alternative to the pyrometallurgical processes, involving a leaching process with various chemical agents (Figure 6.16) [431].

Contemporary metallurgical practices and studies that are aimed at reducing energy and chemical consumption should be explored, thereby enhancing overall efficiency, effectiveness and practical applicability of WEEE recycling [432]. Emerging bioleaching and phytomining technologies as alternative selective recovery methods that adhere to sustainability criteria without excessive capital investment, are attractive processes to comply with forthcoming environment and cost requests [433].

In parallel, the dismantling of printed circuit boards (PCBs) (Figure 6.16) could be performed automatically in order to adjust the recycling process as a function of the electronic components present on the PCBs. Such an adjustment strongly limits energy and chemicals usage [434].

6.5 Workforce shortage

The worldwide economy faces a significant shortage of skilled workers, which poses challenges to their economic growth and their technology leadership. This shortage impacts all technology industries, but strongly affects the semiconductor industry and its supply chain due to the huge current demand for chips, which is expected to drastically increase in the forthcoming years. Moreover, the semiconductor supply chain requires a highly skilled workforce. Researchers and engineers in the R&D and design phases require highly advanced degrees (top talents) and workers in fabrication facilities operate complex manufacturing equipment and require significant training.

This shortage is mainly induced to some complex societal factors such as (i) an insufficient number of students pursuing science, technology, engineering, and mathematics (STEM) degrees, (ii) among students who pursue STEM fields, too few pursue advanced degrees in these fields, and many students with STEM degrees find employment in non-STEM occupations (e.g., finance and business), and (iii) the semiconductor industry has developed a notorious reputation for tough working conditions full of long hours and intense pressure. Therefore, students are no longer willing to pursue masters and doctorates in semiconductor-specialized fields when there are more appealing opportunities in other job markets.

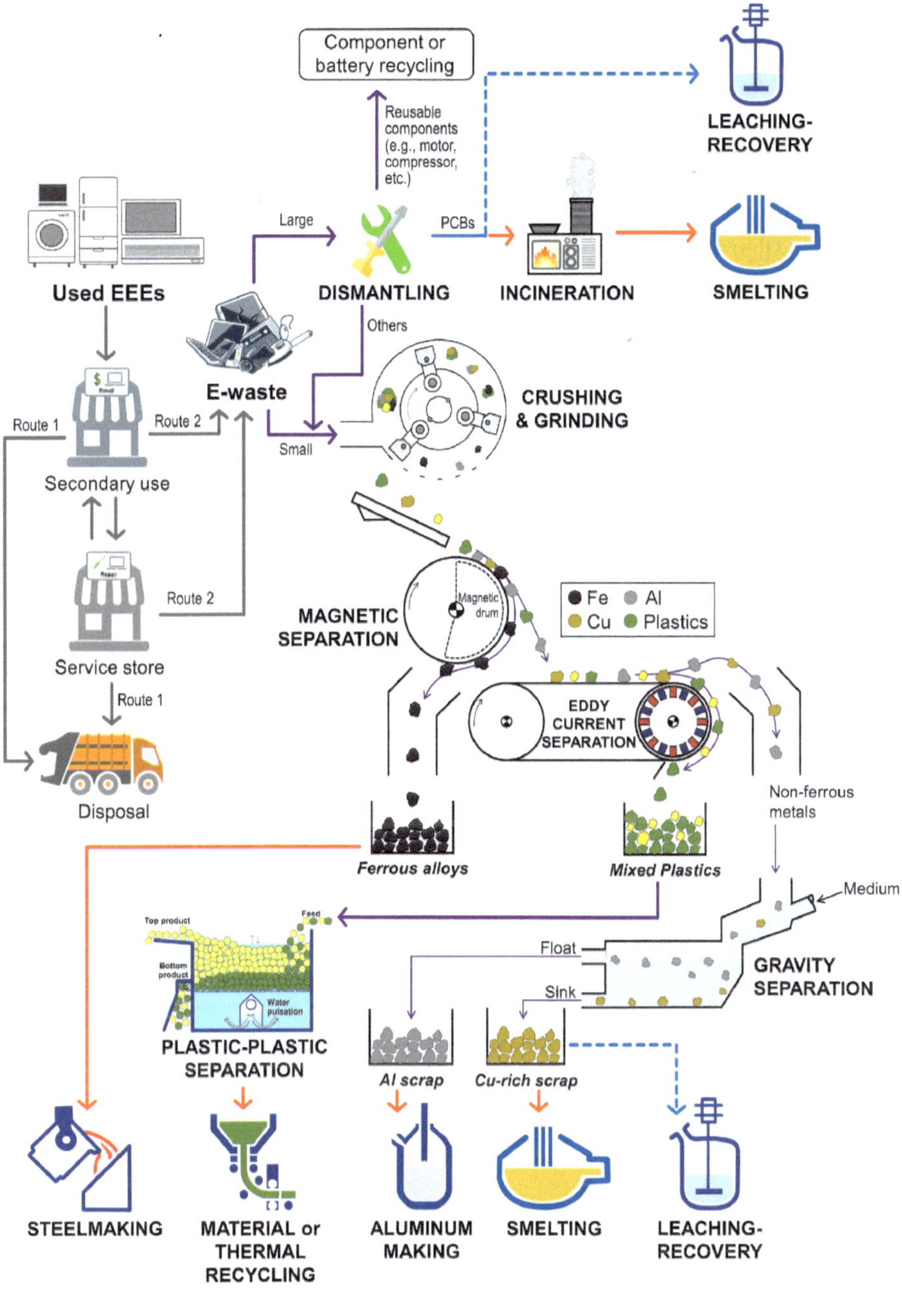

Figure 6.16: A generalized schematic flowchart of WEEE during recycling. *Note: purple arrows = crushing, sorting and preconcentration; blue arrows = hydrometallurgy; red arrows = pyrometallurgy [431].

The semiconductor sector is currently experiencing a notable shortfall in labor, caused by a rise in job vacancies that exceeds the rate of new graduates entering the field. From 2017 to 2023, job vacancies in the semiconductor sector grew at an compound annual growth rate of 11%, fueled by robust job growth, while the number of graduates in related disciplines remained fairly constant.

This labor shortage will drastically increase in the forthcoming years. In the USA, the workforce gap in the semiconductor industry between 2024 and 2029 is estimated between 59,000 and 146,000 engineers and technicians [435]. A recent report mentioned that 67,000, i.e., 58% of new jobs in both manufacturing and design will go unfilled by 2030 in the US semiconductor industry [436].

In Europe, the gap ranges from 76,000 to 350,000 employees between 2024 and 2030 [437, 438], while in Taiwan, this gap ranges between 16,000 and 34,000 skilled workers per year [439, 440].

7 The semiconductor supply chain and its vulnerabilities

The semiconductor supply chain, schematically illustrated in Figure 7.1, is highly fragmented and globalized. It starts with the raw materials, which are extracted from ores, then refined to deliver the convenient materials, gases, and molecules. A significant and quite important part of this supply chain is constituted by the gas, wafer, lithography, and equipment suppliers. Thanks to the beneficial help of electronics and design automation (EDA) tools and software, semiconductor chip designers use intellectual property (IP) licenses and design verification to provide designs to wafer manufacturers. Semiconductor suppliers use single crystal silicon or other substrates (SiC, GaAs, InP, etc.), lithography masks, and equipment to produce chips. Integrated device manufacturers (IDMs) and outsourced semiconductor assembly and test (OSAT), package, and assemble those chips with PCB substrates for delivery to end customers.

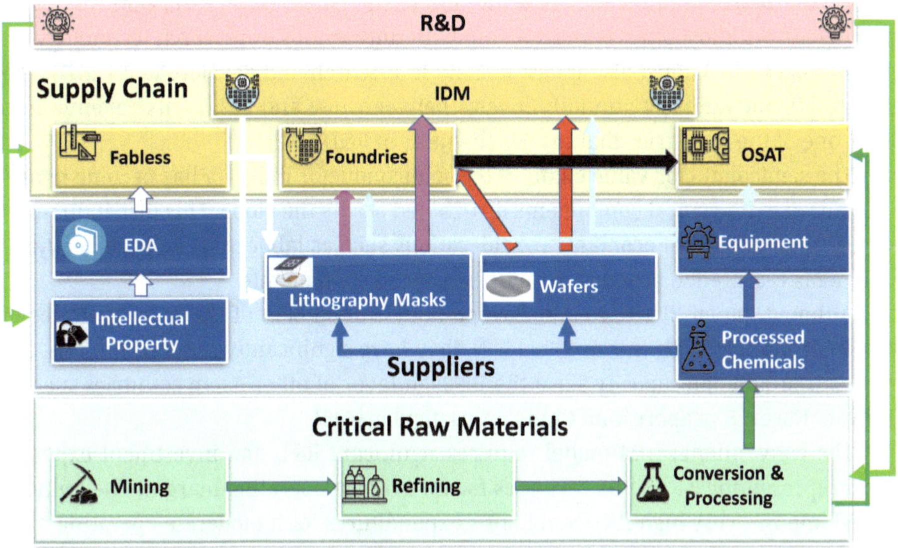

Figure 7.1: Schematics of the global semiconductor supply chain.

Varying levels of specialization and functional division within the semiconductor value chain have led to two primary operating models: Integrated Device Manufacturers (IDMs), which both design and manufacture chips, and the fabless-foundry model.

In the IDM model, a manufacturer performs all stages of production, including design, manufacturing, assembly, electrical testing, and packaging. Memory manufac-

 | https://doi.org/10.1515/9783112247952-007

turers such as SK Hynix and Micron, alongside European manufacturers ST, NXP, Bosch, and Infineon, operate as IDMs. Intel and Samsung were vertically integrated IDMs. Intel could probably disaggegate in the forthcoming years, with the production side expected to become a foundry on one side and the design side (fabless) on the other. Samsung is simultaneously a memory manufacturer, a foundry, and an IDM.

In the fabless-foundry model, design and manufacturing are not handled by the same manufacturers. Design firms/houses such as Qualcomm, Broadcom, Nvidia, AMD, Huawei/Hisilicon, Apple, Meta, Google, Tesla, and Microsoft focus on research, design, and outsource manufacturing, and do not have a production line. Foundries (TSMC, GlobalFoundries, UMC, SMIC, Intel Foundry Services, etc.) focus on developing new manufacturing processes and producing chips (Figure 7.2).

A third group of companies, although not strictly part of the fabless-foundry model, performs assembly, testing, and packaging (ATP). This third group is generally known as OSAT companies (Figure 7.2). ATP can also be defined as the back-end-of line (BEOL) process steps of the semiconductor manufacturing process, while front-end-of line (FEOL) refers to the process steps that are performed on the semiconductor wafers.

Furthermore, in order to stay at the top of the global chips race, semiconductor manufacturers, most often in partnership with universities or research technology organizations (RTOs), must also invest heavily in research and development (R&D). The semiconductor industry annually invests between 15% and 20% of its revenue into R&D, one of the industries that invests the most in R&D [442].

The semiconductor value chain in the semiconductor industry has become hyperspecialized, with critical components distributed across the globe. This distribution resulted from significant economic considerations such as labor costs and the heavy investments required to produce the most advanced components.

Although Japan, the USA, and Europe hold around 68% of the market shares of the semiconductor industry in 1990 [442], they have significantly reduced their investments in production capacity since the 1990s in favor of other Asian countries such as Taiwan, Korea, Singapore, and China, in particular [443].

The conventional IDM model requires significant R&D and investment expenditures representing 34% of its revenues for an average operating margin. The foundry model imposes very high R&D and CapEx expenditures for a moderate operating margin. In contrast, the fabless model allows for reduced R&D and CAPEX expenditures and leads to a higher operating margin. The OSAT model is inexpensive in terms of R&D and CAPEX, but labor-intensive, and leads to the lowest operating margin.

Given their low operating margin and to increase their profitability, ATP activities have been largely outsourced to Asian OSATs by Japanese, American, and European chip manufacturers. The current development of advanced packaging, which partially constitutes the future of the semiconductor industry, is primarily developed and marketed by foundries, reducing OSATs' revenue. Furthermore, given the massive investments required to manufacture integrated circuits and memories using the most advanced technology nodes, numerous IDMs have given up the race for advanced ICs

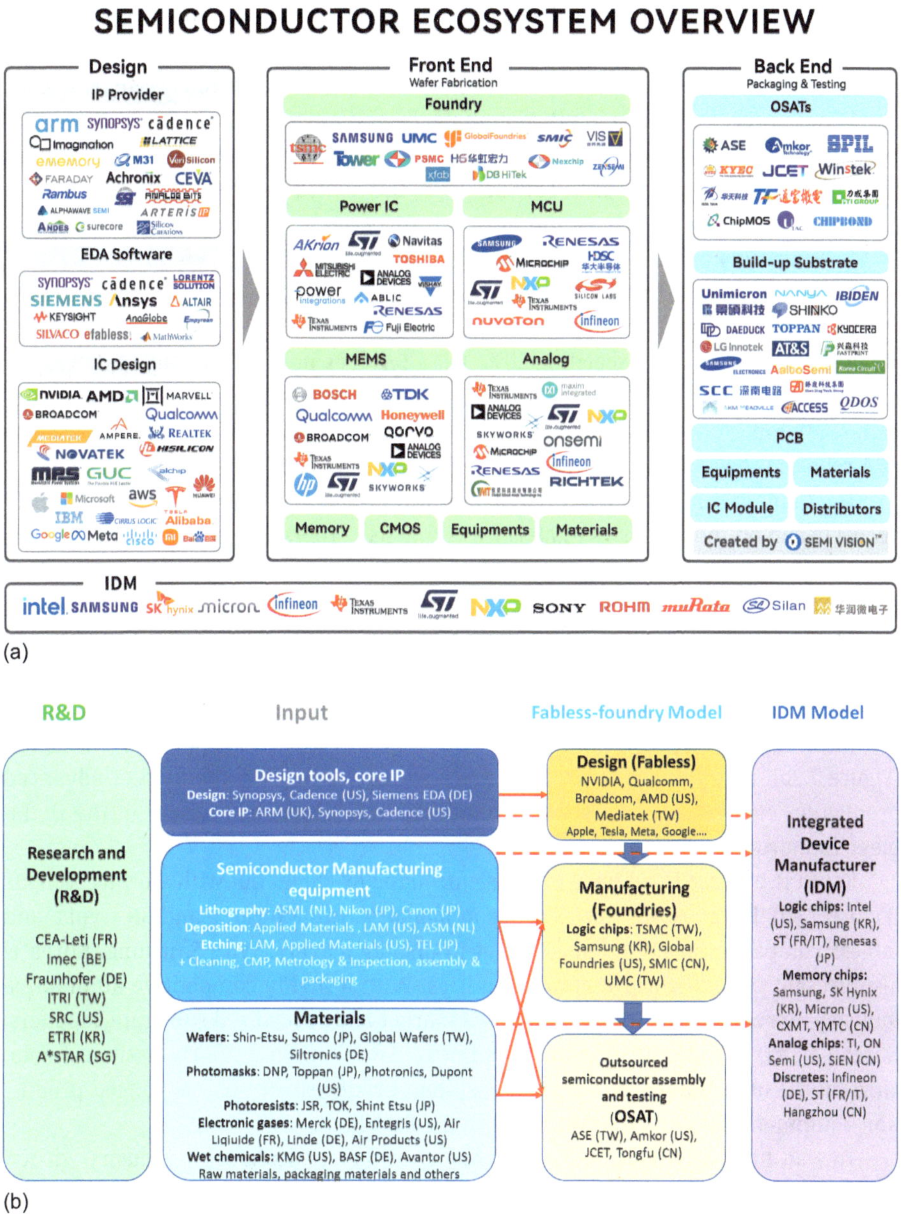

Figure 7.2: Main players in the semiconductor ecosystem (b) [441] – (a) courtesy of SemiVision.

and memories for the benefit of a handful of technology behemoths that can only afford the financial and technical demands of leading-edge semiconductor research.

The semiconductor supply chain is thus organized according to geographic specialization. Different regions have strengths in different areas of the semiconductor industry. For example, US-based companies are leaders in design, core IP, and EDA. The USA, Europe, and Japan jointly dominate the equipment market. Companies headquartered in China, Japan, Taiwan, and South Korea lead in materials. Companies headquartered in Korea and Taiwan lead the world in electronic component manufacturing, particularly in the most advanced technology nodes (≤10 nm), and OSATs are mainly concentrated in China and Taiwan (Figure 7.3).

In deeper details, the USA dominates (Figure 7.3a): (i) the EDA, core IP, and fabless sectors, with a 70% global market share, (ii) the advanced production equipment market, with market shares exceeding 50%. The USA does not have a single lithography equipment manufacturer, and (iii) the IDM sector, with the presence of Texas Instruments, Micron, and Analog Devices, to which should be added Intel. In contrast, the USA relies on the global ecosystem for silicon wafers, masks, foundries, and OSATs.

The Chinese semiconductor ecosystem relies primarily on its expertise in the field of OSATs and on the foundries established within its borders. China has a growing number of IDMs present in the More Than Moore market (power, discrete, and optical devices). It is also home to a large number of fabless companies. Finally, China is present in the memory segment (Figure 7.3). Its main current weakness lies in the lack of advanced production equipment, for which it relies heavily on Japan, the USA, and Europe.

The strength of the Japanese semiconductor industry lies in the production of silicon wafers and photomasks, for which it holds nearly 50% of the global market share (Figure 7.3b). Japanese manufacturers are also very present in the field of advanced production equipment. On the other hand, Japan is practically absent in the IP, fabless, foundries, and OSAT sectors.

Korea is present in all links of the semiconductor chain, but with varying weights (Figure 7.3b). Korea is the world leader in memory with Samsung and SK Hynix, and fabless companies are often specialized in display drivers. With Samsung, Korea is also present on the world map of foundries. With a highly competitive automotive industry, Korea also has IDMs serving this market and specialized substrate suppliers. Korean OSATs are relatively numerous but remain focused on fairly basic technologies. On the other hand, the Korean ecosystem produces very few advanced production equipment and remains very weak in the EDA field.

Taiwan has based its political autonomy on the semiconductor industry (silicon shield). The bulk of its ecosystem (300 companies) is based not only on its foundry activities, the production of silicon substrates, and OSATs, but also on the vitality of its fabless companies (Figure 7.3b). In contrast, the Taiwanese ecosystem does not produce any advanced production equipment and remains absent in the fields of EDA and core IP.

European competitiveness in the semiconductor sector is primarily based on its equipment suppliers and its sub-systems equipment producers, its materials manufacturers (Si substrates, gases, and chemicals), and its R&D ecosystem (Figure 7.3a). However, the European semiconductor ecosystem has numerous weaknesses in lithography masks, EDA tools, and the small number and competitiveness of its fabless companies, foundries, and OSATs. It is also worth noting that Europe is currently absent from the memory market, and that the European automotive brands currently support the demand for integrated circuits. However, Europe has committed to deeply

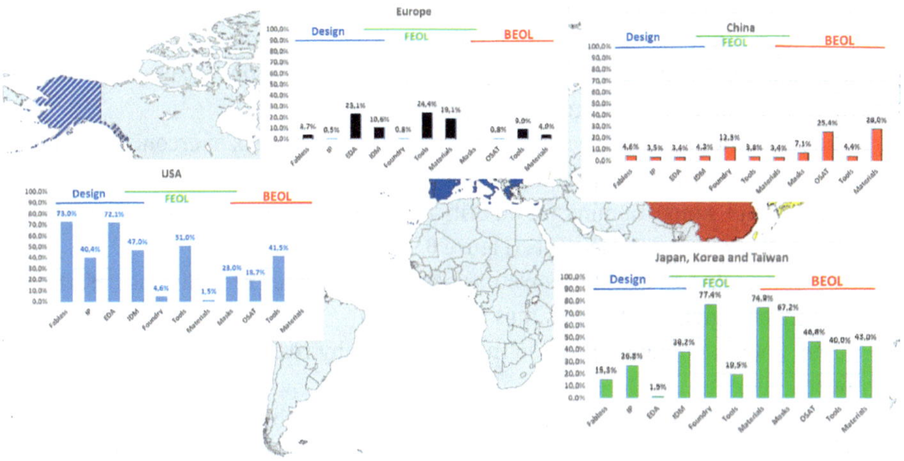

(a)

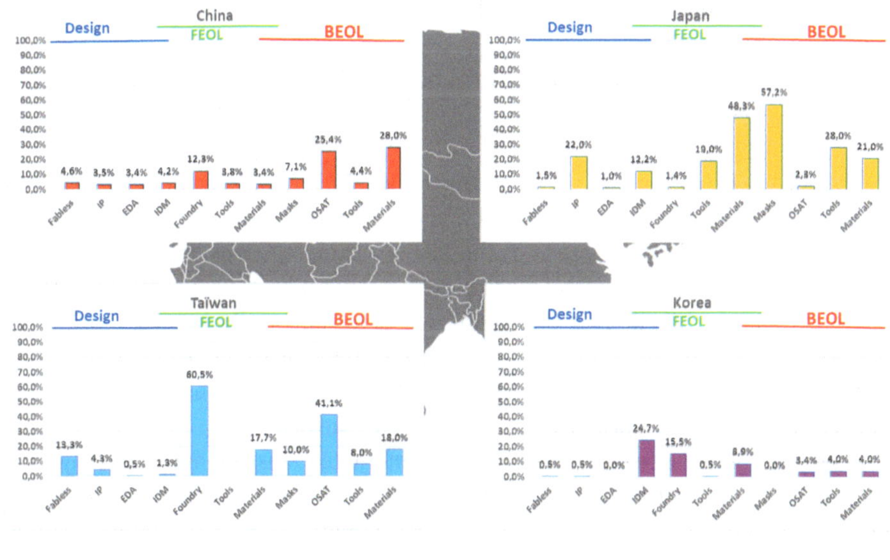

(b)

Figure 7.3: (a) A global overview of the semiconductor supply chain in 2023 as a function of the location of the company headquarters, with (b) a focus on major Asian countries updated from [1] – data from [20] – *The distribution of BEOL activities between Japan, Korea, and Taiwan is approximate.*

invest in the semiconductor supply chain for enhancing its competitiveness (EU Chips Act).

Any disruption of the global semiconductor supply chain, i.e., any arrow in Figure 7.1 that could be broken, induces chip shortages, as the one that occurred during the COVID-19 pandemic, highlighting the vulnerabilities of this value chain. The COVID-19 pandemic delivered a definitive operational lesson on the dangers of overconcentration. China's stringent zero-COVID policy led to sudden lockdowns, factory closures, and logistical chaos, exposing the acute vulnerabilities in the supply chains heavily reliant on a single geographic region.

The semiconductor supply chain is currently exposed to six overconcentration regions, which can strongly impact its resilience:

- China with its huge dominance on critical raw materials, and especially on Si, Ga, Ge, As, and rare earth elements (Figure 6.9)
- The EDA tools and core IP, as well as design of GPUs for AI, which are essentially or only provided by US companies (Figure 7.3a)
- Japan holds a dominant position in the lithography resist processing market, with over a 90% share
- The significance of Taiwan in strategic terms arises from its possession of 92% of the globe's leading semiconductor production capabilities, along with the availability of the most cutting-edge parts (≤10 nm) (Figure 7.3b)
- Europe, which welcomes the sole extreme UV (EUV) lithography equipment supplier that enables the patterning of nanostructures for the most advanced technology nodes (≤5 nm)
- Taiwan and China that concentrate nearly 70% of the OSAT market shares (Figure 7.3b)

In addition to the risks associated with this regional overconcentration, the semiconductor industry faces other serious risks that could significantly disrupt its activities.

Events from recent years such as the Red Sea shipping crisis, the Russia-Ukraine war impact the production of neon gas used in lithography tools, for instance. Ongoing conflicts in the Middle East reduce the access to some critical species such as He and bromine, and considerably disturb the semiconductor supply chain. Geopolitical tensions between China and Taiwan could lead to a conflict or a blockade of the Taiwan strait disrupting the production of leading-edge devices.

The semiconductor supply chain is also vulnerable to disruptions caused by natural disasters, such as earthquakes, freezes, heat waves, and flooding, which will be enhanced by the climate change. Since major suppliers of the semiconductor supply chain are concentrated in geographical areas prone to extreme weather events and natural disasters (Figure 7.3), such risks cannot be overestimated.

Furthermore, the semiconductor industry has become a critical battleground in the ongoing economic and technological rivalry between the USA and China. As the world's largest consumer of semiconductors, China has made it a strategic priority to reduce its dependence on foreign semiconductor technology and build up its domestic industry. Its initial goal was to become to become self-sufficient in all aspects of semiconductor manufacturing, from R&D and design to manufacturing and packaging around 2025–2030. Although this objective was not be fully realized by 2025, the aspiration has faced opposition from the United States, which considers China's progress in technology as a possible danger to its economic and national security concerns.

Therefore, the USA strategically limited the Chinese AI, supercomputing and quantum technology sectors by restricting their access to high-end chips, to strategic equipment set (in collaboration with Japan and Europe) for producing advanced ICs (≤5 nm), and to advanced EDA tools and software. This reflects the belief held by top leaders in both nations that dominance in AI technology is crucial for future global military and economic power. In reaction to the export control rules set-up by the USA, China has aggressively invested in building domestic chip capabilities and intents to invest more heavily in its domestic semiconductor industry within its new strategic plan to be announced in 2026. China also deployed export restriction on critical raw materials such as Ga, Ge, and Sb in December 2024; W, In, and Mo in February 2025; and rare earth elements in April 2025.

Shocks like the COVID-19 pandemic, the Russia's invasion of Ukraine followed by the escalating trade and technology rivalry between Washington and Beijing have unveiled the fragility of the semiconductor supply chain and refocused worldwide attention on the critical materials, equipment and processes of manufacturing semiconductors. Previous semiconductor shortages, access to the most advanced chips, associated with the ongoing huge demand for AI chips have become a critical technological vulnerability and a potential national security threat for major economies including the USA, Japan, China, and Europe. Furthermore, many industries only rely on Taiwan for leading-edge semiconductor devices, and on China for critical raw materials.

Semiconductor and AI technological leadership is now defined along the overall supply chain from critical raw materials to chip packaging. Those who control key nodes shape both the pace and the direction of innovation.

8 Semiconductor-related government incentives

In response to the semiconductor supply chain vulnerabilities, multiple regions around the world are currently seeking to build resilient, reliable, and largely self-sufficient sources of supply for semiconductors. Although the complexity of the semiconductor supply chain ensures that they will remain only part of a globally integrated industry, governments around the world have competed fiercely to secure advanced semiconductor R&D, domestic manufacturing for their national security and/or sovereignty.

Therefore, large-scale government intervention has become the new normal. In a context of heightened power competition, both the USA and China treat control over the semiconductor supply chain as a matter of national security. Consequently, both countries have crafted powerful industrial strategies that are to some extent comparable in their stated goal to secure and expand their own semiconductor industrial base [444].

Before the COVID-19 pandemic, the geographic distribution of roles between IDM, fabless, foundries, and OSAT offered enormous value for the semiconductor industry and consumers who benefit for reasonable chip costs. Fully localizing the end-to-end control of a supply chain for advanced semiconductors in a single country or a geographic region, even in the USA or China, remains economically unfeasible without significant cost increases and production delays. A hypothetical alternative with entirely "self-sufficient" parallels, i.e., local supply chains in each region, and, given current and future levels of semiconductor demand, would have required at least $1 trillion in initial investment, ultimately resulting in a 35–65% increase in semiconductor prices.

As previously mentioned, since 2014, China was the first country to massively invest around $150 billion for its self-sufficiency in the whole semiconductor supply chain (Figure 8.1) by launching the "Guidelines to Promote [the] National Integrated Circuit Industry initiative." China's self-sufficiency rate in semiconductors has increased from 15% in 2019 to about 25% in 2025, which is still below its initial 70% target by 2025. For instance, Chinese subsidies to Semiconductor Manufacturing International Corporation allow this company to report in 2024 operating profit of $473 million, of which $411 million came from government subsidies [445].

Moreover, China very recently pledged to use "extraordinary measures" to support the country's bid to become a global leader in artificial intelligence (AI), quantum technology, and other cutting-edge technological fields, according to its 15th 5-year plan. The plan will run from 2026 to 2030 and serves as China's overarching blueprint.

The plan also doubles down on the government's long-term goal of becoming more self-sufficient by overcoming technological bottlenecks that prevent China from making key technologies, such as advanced semiconductor chips [446].

 | https://doi.org/10.1515/9783112247952-008

In the year 2022, the USA enacted the "Creating Helpful Incentives to Produce Semiconductors" (CHIPS) Act and the Science Act, which are intended to enhance the nation's competitiveness, drive innovation, and strengthen national security. The legislation focuses on encouraging financial investment in semiconductor production capacities. Furthermore, it aims to promote research and development along with the commercialization of advanced technologies like quantum computing, AI, renewable energy, and nanotechnology. It provides between $240 billion and $280 billion in spending over the next 10 years. The majority of the funds, $200 billion, are dedicated to scientific R&D and commercialization. In the frame of the US CHIPS Act, $52.7 billion is allocated to semiconductor production ($39 billion), R&D ($11 billion), and workforce development, including an additional $24 billion in tax credits for the purchase of production and R&D equipment (Figure 8.1).

In addition to these grants, US fiscal support could reach up to $68 billion between 2022 and 2026, bringing total American state support close to €110 billion. But tax incentives are not the whole story. In addition, initiatives such as Pax Silica, along with the investment pillars embedded in trade agreements concluded with Japan, South Korea, and Taiwan, leverage US power to construct a multilayered international strategy that expands hardware manufacturing at home, reduces strategic import dependencies and ultimately sustains American leadership in AI. Large incentives, protectionist measures, and a strong state support for AI development are leading to the recreation of a complete server supply chain on American soil [444].

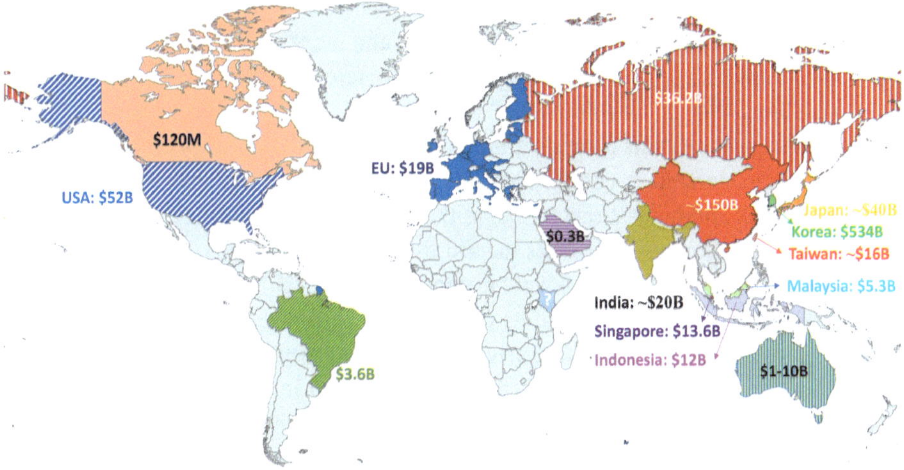

Figure 8.1: Government's incentives across the globe for securing parts of the semiconductor supply chain – updated from [1] – The Korean amount corresponds to private-public commitments – *These numbers are subject to changes due to political decisions.*

European Union announced a CHIPS Act in 2022 to reduce the EU's reliance on foreign states for supplies of critical components. The EU Commission, the Member States (MS) wish to provide incentives at a level close to $46 billion to domestic companies and to non-EU partners that will deploy semiconductor manufacturing capacities in EU (Figure 8.1). These incentives initially planned to mobilize around $90 billion in public and private investments. However, the European Commission directly controls only around 5% of these funds. The resulting regulation came into force in 2023 and targets two main objectives:

- Double its share of the global electronic components production market from 10% in 2022 to 20% by 2030. Given the global expansion of the semiconductor industry, leading to a projected doubling of global production from 2022 to 2030, this would mean that the EU would have to quadruple its current semiconductor production to reach 20% of global production.
- Produce logic components at the most advanced technology nodes in 2030, i.e., 2 nm or below, on its own soil to guarantee its sovereignty.

At the time this book is written, the effective EU and MS public funding amount is estimated around $19-20 billion (Figure 8.1).

Since 2022, Japan has earmarked about ~$40 billion for revitalizing its semiconductor punching power (Figure 8.1). Among the goals is a tripling of domestically produced chip sales by 2030. Japan's new chip strategy has two main targets. First, the country is seeking to reestablish itself as a prime location for manufacturing legacy chips by luring to Japan the biggest foreign names in the industry with generous subsidies of up to half of the setup costs. The second and more ambitious part of the strategy is the "Rapidus" project that aims at restoring Japan's place as an important and indispensable player at the forefront of silicon-chip by producing 2 nm chips in 2027.

Rapidus Corporation is a consortium of eight Japanese companies aiming to revive the country's leadership in cutting-edge semiconductors. This initiative is described as "the highest-risk and highest-reward bet of the bunch." It has justified the huge government investments and contributions toward the approximately $34.5 billion needed to start mass production of cutting-edge semiconductors by 2027 [447].

In 2024, South Korea published its plan to build the largest semiconductor cluster by 2047, corresponding to a public-private investment around $534 billion, investing in both its traditionally dominant sector of memory and in AI chips [448]. The incentive funding of the Korean government is currently estimated around $74 billion, essentially in tax incentives (Figure 8.1).

In response to the various initiatives, Taiwan is defending its central role in the chip world with its own CHIPS Act. The Taiwanese government aims to support semiconductor companies in developing innovative technologies and consolidating their crucial position in the global supply chain. One of its highlights is granting a tax credit of up to 25% for R&D expenditures in "future technologies" and up to 5% for investments in production equipment in advanced technologies. These incentives represent an aid amount about $16 billion (Figure 8.1).

India is working to become one of the key partners in global semiconductor value chains, by extending its current competencies in chip design to chip manufacturing. This initiative proposes an aggressive roadmap that aims to position India as a global powerhouse in the semiconductor value chain. This includes the domestic production of critical inputs such as specialized chemicals, gases, and the machinery required for chip production to create a self-reliant semiconductor ecosystem. With the Indian semiconductor market projected to exceed $100 billion by the end of the decade, the Indian government has set up very ambitious goals: be in the top six semiconductor nations by 2032 and in the top three by 2047 [449].

Furthermore, increasingly, ASEAN economies (Malaysia, Singapore, Thailand, Indonesia, and Vietnam, notably) want to become key players in the semiconductor industry. They wish to benefit from the US-China rivalry by attracting parts of the supply chain, which are currently localized in China, on their soils. The expected influx of investments represents a generational economic opportunity, bringing millions of jobs and technology transfer.

Building on their long-standing traditions of industrial policy and state intervention, ASEAN governments are pursuing new strategies to advance beyond assembly and testing into higher-value semiconductor segments (Figure 8.1). Until recently, these economies were viewed primarily as sites for foreign-owned manufacturing. Today, policymakers and industry leaders are working to establish more self-sufficient semiconductor ecosystems, anchored in domestic innovation and supported by interconnected regional supply chains.

Countries such as Australia, Canada, Saudi Arabia, United Arab Emirates, and Kenya are also willing to welcome chip manufacturing facilities on their territories. While Russia has invested in chip manufacturing since 2010 and targets to produce 28 and 14 nm chips in 2027–2028 and 2030, respectively (Figure 8.1).

8.1 Main beneficiaries of those government incentives

Government initiatives related to the CHIPS acts favored greenfield foreign direct investments (FDIs). FDIs have drastically transformed the semiconductor supply chain since 2022, and shaped the worldwide economy, accordingly. However, only technologically advanced and capital-intensive global firms have the capabilities to compete.

Until very recently, production of leading-edge chips was heavily concentrated in Taiwan and South Korea. In 2022, these economies were home to about 65% and 25%, respectively, of global leading-edge capacity. Hence, amid escalating geopolitical tension, companies and countries are seeking to diversify production away from Taiwan and South Korea and build capability closer to home. Correspondingly, associated greenfield FDI announcements have been reconfiguring the semiconductor ecosystem.

Since 2022, FDI announcements into semiconductors have reached $115 billion per year, a fivefold increase from the 2015 to 2019 period (Figure 8.2). Strikingly, half of this amount was focused on building new leading-edge facilities, which did not feature in the previous round of FDIs. However, building capacity for the production of leading-edge semiconductors at scale is very expansive (Figure 4.5). They are incredibly complex and precise; less than a handful of companies globally have the capabilities to develop them [450].

Against this backdrop, FDI appeared as an engine that could expand capacity globally. If all announced FDI projects were realized, they could add the equivalent of almost 60% of total global leading-edge capacity in 2022. FDI could add five times the existing capacity outside Taiwan and South Korea. Japan and the USA's combined share of global leading-edge semiconductor production capacity could increase from about ~10% in 2022 to more than 30% in 2030, with more than half of that increase driven by FDI projects (Figure 8.2) [450].

The USA could become the second-largest producer of leading-edge chips by the early 2030s, home to more than 20% of global capacity, thanks to FDIs from TSMC and Samsung, the leading players from Taiwan and South Korea (Figure 8.2b).

Europe's (primarily Ireland) and Japan's shares in leading-edge chips capacity may grow too but would likely remain in the single digits, as only a few projects have been announced.

As FDIs announced in advanced economies increased, the value of announced semiconductor-related investment in China dropped about 80% during the period since 2022, compared with 2015–2019, when China was the destination of most flows (Figure 8.2a).

In addition to leading-edge semiconductors, FDI also plays a role in the production of mature and advanced chips. In Japan and Germany, FDI has helped scale their production. Construction is also underway on megadeal-driven projects in India; one of these could be first to develop semiconductor production capacity in the country [450].

Despite the drop of FDI, the expected number of production lines (fabs) to be built in China between 2025 and 2030 is the highest one (+21 fabs). While the number of expected fabs to be built between 2025 and 2030 is 11, 7, and 6, in the USA, Europe, and South Korea, respectively (Figure 8.3).

Therefore, the manufacturing semiconductor capacity of Chinese companies will strongly increase in the near future (Table 8.1). It should reach around 32% of the worldwide capacity around 2030.

The success of FDI-driven projects in building leading-edge semiconductor capabilities is not guaranteed. Developing new and competitive leading-edge manufacturing centers outside Taiwan and South Korea is a formidable endeavor, especially without the same ecosystem of value chain partners and talent. Moreover, projects in newly seeded hubs may face different regulatory or environmental requirements, as well as potential challenges in accessing utilities and infrastructure.

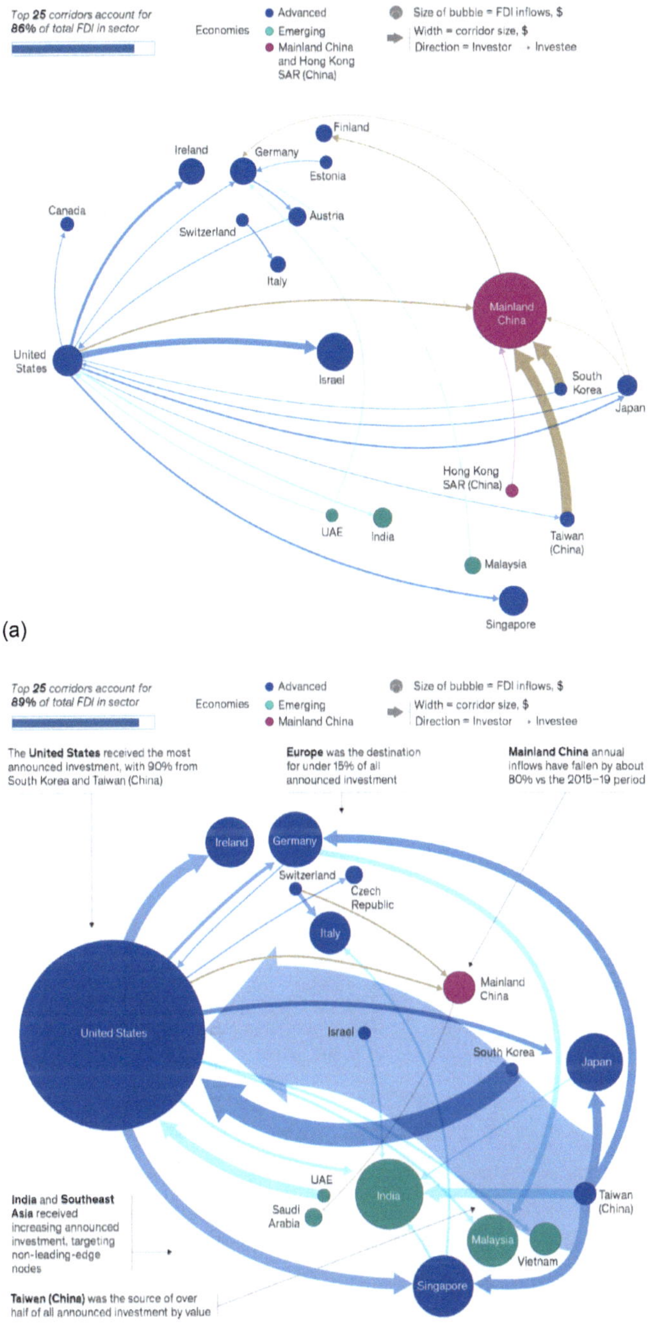

Figure 8.2: Semiconductors – top 25 FDI corridors between (a) 2015–2019 and (b) 2022–2025 [450].

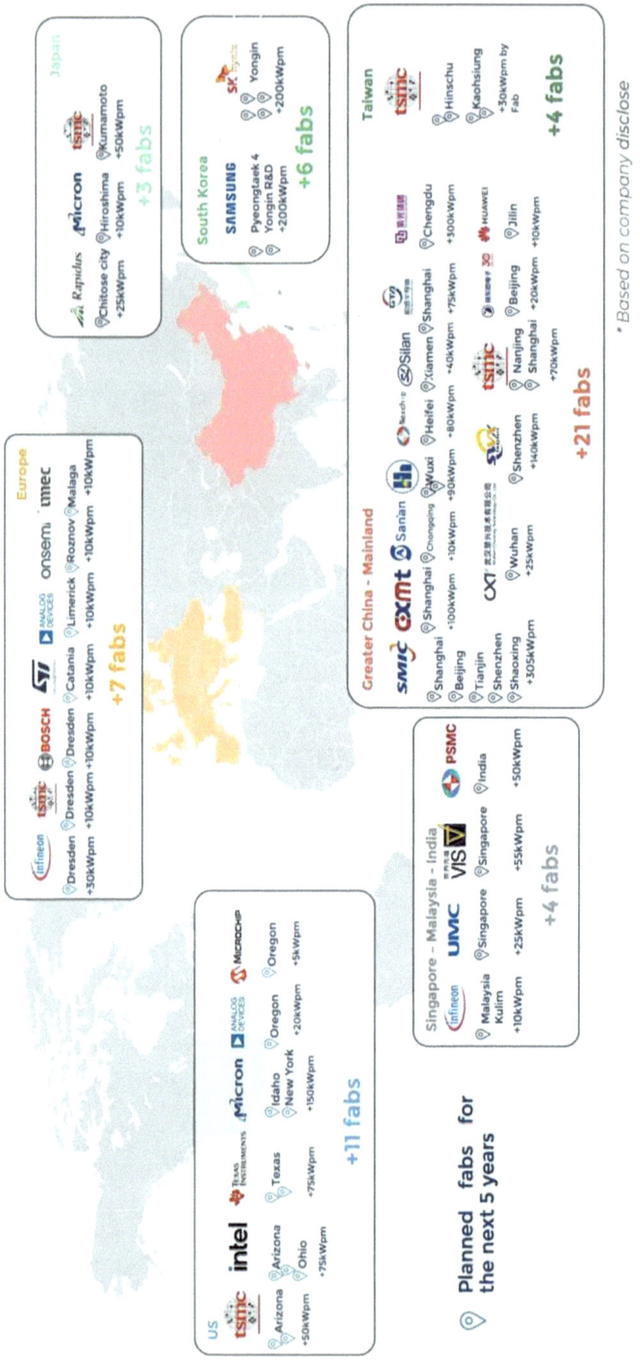

Figure 8.3: Expected number of fabs to be built between 2025 and 2030 – courtesy of Yole Development – Status of the Semiconductor Foundry Industry 2025.

Table 8.1: 2024 semiconductor manufacturing capacity and its expected increase between 2025 and 2030.

Country/region	USA	Europe	Taiwan	Japan	Korea	China	Others
2024 semiconductor capacity (%)	10	8	23	13	19	21	6
Increase in capacity 2025–2030 (kwspm)	355	80	60	85	400	1,265	140

Accessing low-cost electricity may be difficult, especially in Europe. Indeed, economic concerns loom large: Operating expenses of semiconductor manufacturing facilities may be 30% higher in Europe and the USA than in Taiwan [450].

9 The European semiconductor supply chain

Europe is a collateral victim of the current technological and economic rivalry between the USA and China. Despite Europe being the world's largest economic market, its low competitiveness in advanced chip design and manufacturing (Figure 7.3a), especially for AI chips, the lack of critical raw materials (CRMs) mining and refining, and the scientific skill shortage risk negatively impacting the entire European economy in the coming years. Europe must therefore react urgently to enhance its competitiveness in the overall semiconductor supply chain.

On the one hand, in order to stay or become more competitive, the European semiconductor industry must reinforce its strengths such as its equipment suppliers, its subsystems producers, its material manufacturers, and its R&D ecosystem. Europe is also leading the development of sustainable manufacturing and eco-design of semiconductor chips, which will become of utmost importance in the future due to its deleterious impact on climate change.

On the other hand, Europe must reduce some of its strategic vulnerabilities within the semiconductor supply chain.

9.1 European strengths in the semiconductor supply chain

The European (EU) semiconductor market was valued between $53.5 and €87 billion in 2023. The total revenue is expected to grow at a compound annual growth rate (CAGR) of 8.9% from 2024 to 2030 reaching between $97.30 and €150 billion by 2030 (Figure 9.1) [451, 452].

The EU semiconductor market aligns with the consumption patterns of end-user industries within the EU. In comparison to various countries and regions worldwide, the EU stands out for its distinct disproportion between sizable "embedded electronics industries" (automation, robotics, aerospace-defense, healthcare, energies, and telecommunications infrastructures), and relatively smaller "consumer electronics industries" (phones, consumer PC, and consumer electronics such as TVs, speakers, and headphones), as illustrated in Figure 9.2.

Embedded electronics industries account for 64% of the EU semiconductor consumption in 2022, while they account for 22% of the global semiconductor consumption.

The needs of the EU market in terms of products are aligned with the needs of its embedded electronic production: power and analog, optoelectronics and sensors, and microcontroller units (MCUs) account for 63% of the EU semiconductor consumption in 2022, while they account for 39% of the global semiconductor consumption. The European suppliers of these components are strong and supporting the needs of the EU industries for decades. As a result, the EU has currently little to no dependencies on these products [20].

 | https://doi.org/10.1515/9783112247952-009

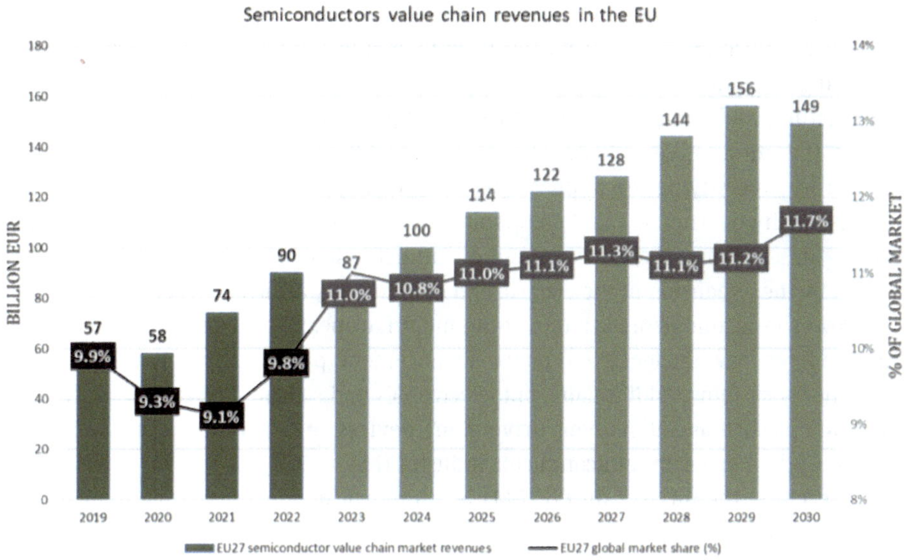

Figure 9.1: EU semiconductor value chain revenue and share versus the global market. Values are estimated as of 2023 (source: IDC) [453].

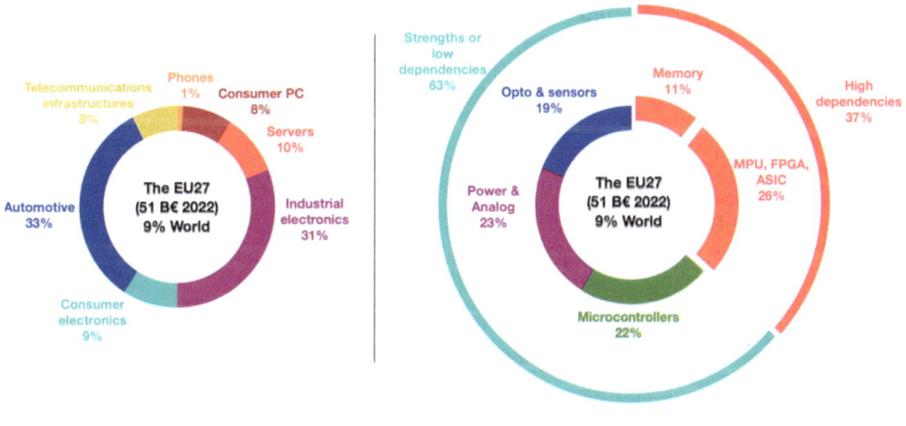

Figure 9.2: Spread of EU semiconductor demand by application in 2022 [20].

9.1.1 Automotive and industry chips

The automotive and industrial sectors are the biggest end markets for semiconductors in Europe. The automotive industry's shift toward electric vehicles (EVs) and autonomous driving is a major driver of the European semiconductor industry. EVs rely heavily on specialized chips for power management, battery efficiency, and advanced driver-assistance systems, making the automotive sector a critical demand driver. The European semiconductor industry has been pivotal in supporting this transformation, supplying chips that are integral to the operation of electric powertrains, charging infrastructure, and smart vehicle features. This growing reliance on semiconductors within the automotive industry accelerated the European semiconductor market growth by diversifying the applications of chips beyond traditional consumer electronics. With governments and companies investing in green technologies and automotive innovations, the European semiconductor market has poised for growth, driven by the expansion of the electric and autonomous vehicle markets.

Power semiconductors are one of the main European strengths due to their critical role in energy efficiency and power management, particularly in EVs and renewable energy systems. Additionally, optoelectronics and sensors are gaining traction in applications such as autonomous driving, IoT devices, and healthcare, are also major drivers of the European semiconductor industry (Figure 9.3).

Therefore, the three main EU IDMs lead the worldwide automotive semiconductor market with more than 35% market shares (Figure 9.4). However, these EU companies compete with each other for increasing their market shares.

The defined characteristics of chips for automotive are not only governed by the technology node used for their manufacturing but mainly by their reliability. Chips for automotive require zero-defect. Wafers used in the manufacturing of automotive chips subject to tighter in-line monitoring, more conservative process control, and failure thresholds that are orders of magnitude stricter than in consumer electronics. The "defect" threshold is roughly parts per million for consumer devices, but in automotive, this threshold is parts per billion.

The transition toward smarter and more autonomous vehicles will be crucial in changing them into ever-developing software systems. Gaining a competitive edge will rely more on embedding sophisticated AI-driven services, strong cybersecurity and connectivity features, unique user interactions, and effective collaboration between hardware and software codesign.

The automotive industry will continue to transition toward zonal vehicle architectures, which will open up new opportunities for a broader range of vendors (MCUs and SoCs) (Figure 9.5). Therefore, the automotive revenue is expected to increase in the forthcoming years but at a quite lower rate than the AI chips (Figure 9.5).

9.1.2 Research and development

Throughout the semiconductor value chain (Figure 7.1), chip design, semiconductor manufacturing equipment, electronic design automation (EDA), and core intellectual property (IP) are significantly characterized by high R&D intensity, especially chip design, which accounts for 53% of the industry's total R&D expenditure. Front-end manufacturing requires not only substantial R&D investment but also exhibits a nota-

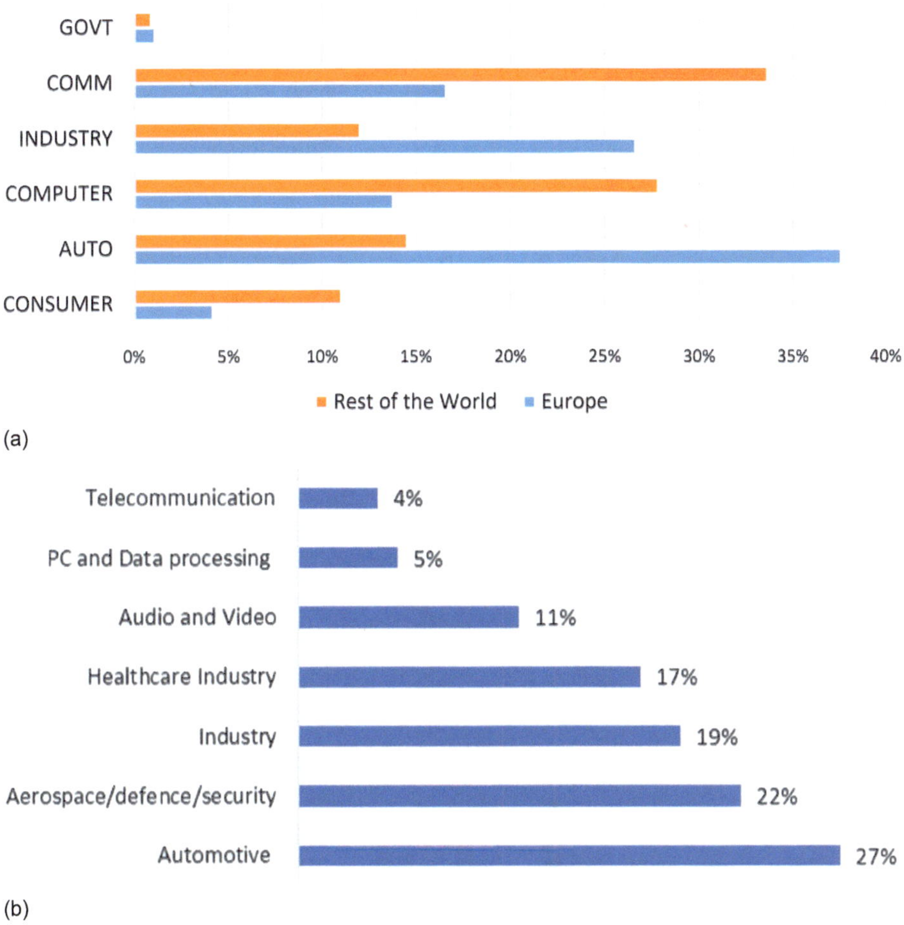

Figure 9.3: (a) Semiconductor demand by end-use sectors [454] and (b) Europe's market share in chip production for different sectors in 2023 [451].

ble high capital intensity, with capital investment accounting for 64% of the industry's total capital expenditure (CapEx).

Activities that require a lot of R&D, including early-stage research, design of chips, equipment for manufacturing, electronic design automation, and essential IP, are largely driven by the USA and Europe. In these areas, the investment in R&D compared to sales in the semiconductor sector greatly exceeds that of other parts of the world (Figure 9.6).

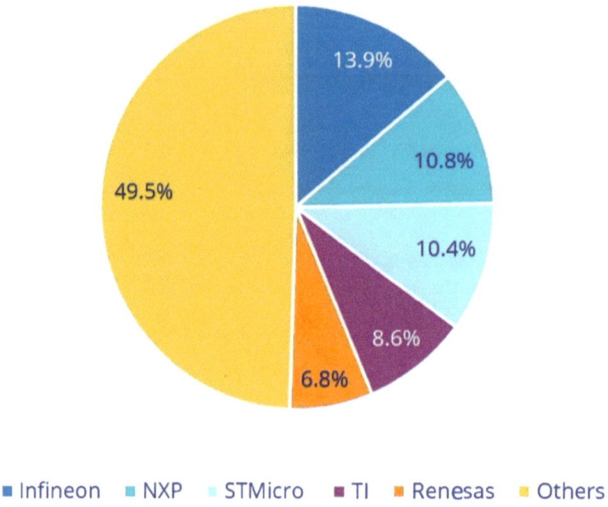

Figure 9.4: Worldwide automotive semiconductor market share 2023 – courtesy of IDC .

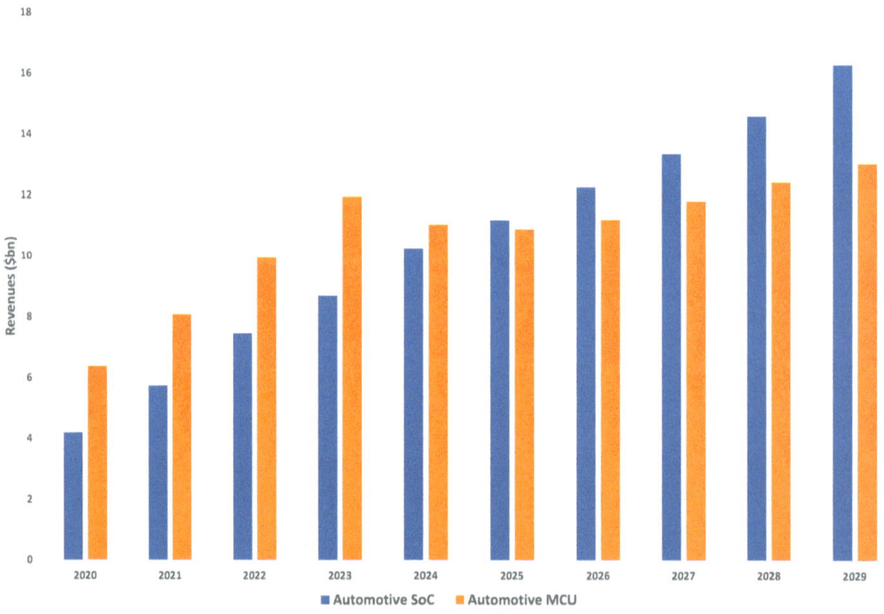

Figure 9.5: Automotive SoC and MCU revenue forecast – data from Omdia 2025.

In contrast, capital – and labor-intensive processes, including FEOL wafer fabrication and ATP, are predominantly concentrated in Asia where has lower operational costs (including labor, land, utilities, equipment, and materials) [455].

To remain competitive in the global chips race, the European semiconductor industry annually invests around 15% of its revenue into R&D. To ensure R&D investment expands in Europe, the EU Commission, in collaboration with the EU Member States (MS), launched the EU CHIPS Act, which allows funding:

- Pilot lines governed by several research and technology organizations (RTOs) for developing key enabling technologies
- A virtual design platform for giving access to support companies, particularly small and medium enterprises (SMEs) and start-ups, in designing advanced high-performance and AI chips.
- Competence centers for coordinating those activities and training or re- and up-skilling the workforce

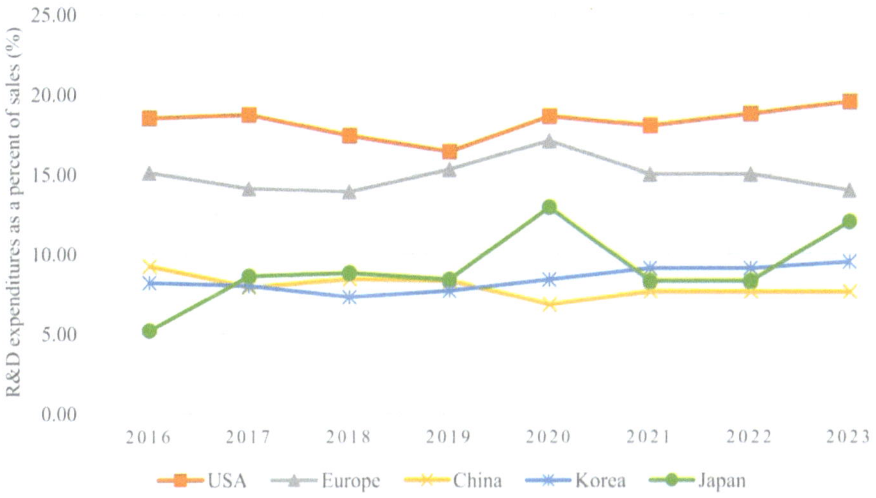

Figure 9.6: R&D expenditure as a percentage of sales in the semiconductor industry across major economies [455].

Europe is home to world leading innovative semiconductor research facilities or RTOs including Imec in Belgium, CEA-Leti in France, and Fraunhofer institutes in Germany. RTOs are nonprofit organizations that serve as intermediaries between research institutes and industrial players. RTOs work closely with pure research institutes to identify promising innovations and collaborate with industrial partners to develop prototypes and demonstrators that prepare those innovations for mass production. EU RTOs also cooperate with a wide array of public actors to help develop strategies for technological development within a given country.

Hence, thanks to funding important R&D works, the EU CHIPS Act represents a vital step toward ensuring the region's competitiveness in the global semiconductor market while also securing Europe's technological future.

9.1.3 Equipment suppliers

The wafer fabrication equipment sector is essential to the semiconductor industry. Semiconductor equipment suppliers play a crucial role in working with all actors throughout the supply chain to develop the relevant equipment sets, which fulfill the industry needs to meet the ever-growing chip demand and technology challenges.

European and Israelian wafer manufacturing equipment suppliers generated 28–29% of the $30 billion market worldwide in 2023 (Figure 7.3a). European vendors generate considerable revenues from supplying lithography technologies, followed by deposition, metrology, inspection, and wafer bonding equipment.

European leaders have particular strengths in patterning, where advanced semiconductor materials lithography (ASML) owns 90% of the market; deposition, where ASM International and Aixtron are key players; and metrology and inspection, where there are many European suppliers, such as NOVA, Zeiss, and Semilab, as well as ASML. The region also leads in wafer bonding and packaging equipment with companies such as EV Group, BESI, and SUSS MicroTEc.

Figure 9.7 presents the market shares of the wafer front-end/fabrication equipment suppliers from 2017 to 2023. The EU market shares are close to the North America (the USA + Canada) and Japanese market shares, and much higher than the Korean and Chinese ones. Even if the Chinese market shares is increasing for some years.

The market shares of the EU assembly (bonding and packaging) equipment suppliers are lower than their Japanese and Singaporean counterparts (Figure 9.8). However, such market shares guarantee a certain independency compared to other countries or regions like Taiwan, Korea, and North America.

Moreover, Europe is also leading in the subsystem market, such as vacuum valves and pumps, optical lenses, high precision optics, and sorters with 39% of the global subsystem revenue in 2023 (Figure 9.9). These subsystems are used to produce new wafer fabrication equipment or for their replacements and are compulsory for most equipment vendors operating worldwide.

9.1.4 Chemical and material suppliers

Producing semiconductors requires a multitude of chemicals and materials. If a single input is missing or postponed, the whole chip production falters. For example, fabs use hundreds of different chemicals, materials, and gases, including ultra-high-purity chemicals (parts per trillion). These chemicals are used to achieve the core processing

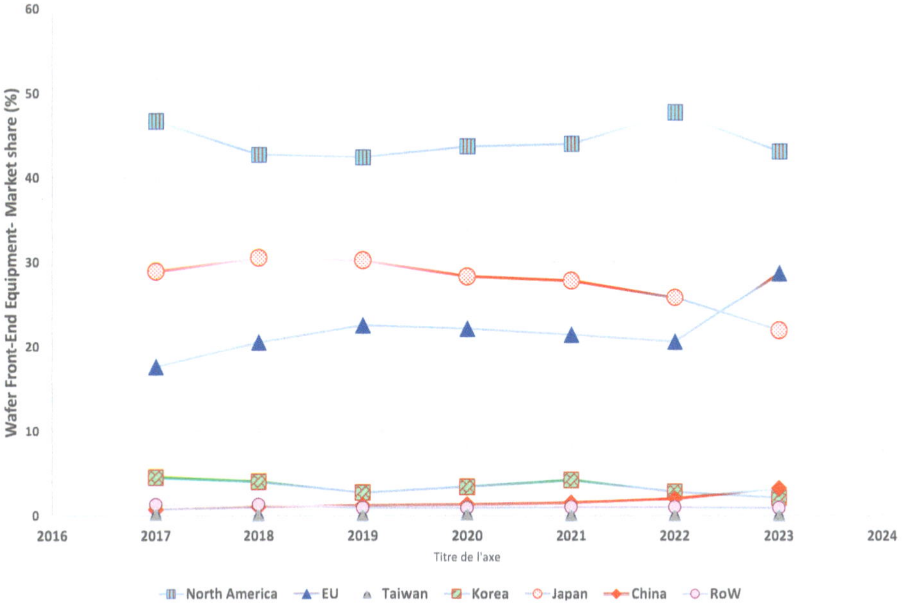

Figure 9.7: World sales of wafer fabrication equipment (WFE) for chip production: market share by jurisdiction – data from [454].

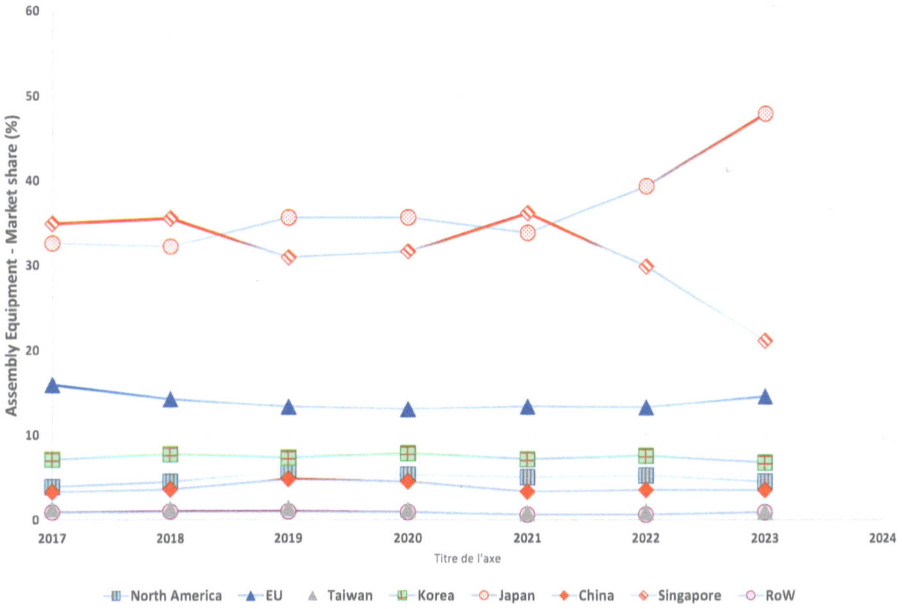

Figure 9.8: World sales of assembly equipment for chip production: market share by jurisdiction – data from [454].

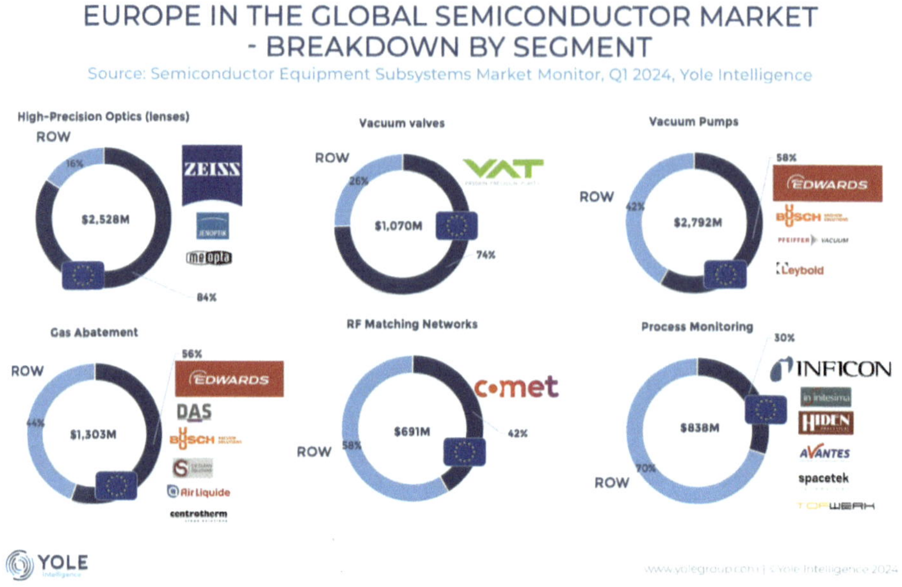

Figure 9.9: EU equipment subsystems suppliers – courtesy of Yole Development.

steps for semiconductor manufacturing: adding layers of metal and metal oxide, etching these layers, or cleaning the wafer surface.

The EU semiconductor supply chain has EU high-purity gas suppliers, such as Air Liquide, Merck, Linde, and BASF (Figure 9.10). But many gases such as He, Ne, Kr, and Xe are imported from various countries, such as Ukraine for Ne and Qatar for He.

The aforementioned EU suppliers also supply high-purity solutions and solvents for manufacturing chips. However, many of these solutions and solvents are generally imported. Ultra-high-purity hydrogen fluoride, which is used to etch away silicon dioxides, is imported from Asia, for instance.

The Si wafer market is dominated by Japanese suppliers with almost 50% of market shares in 2021, followed by a Taiwanese company and an EU firm Siltronic. In 2021, Siltronic has nearly 12% market shares of a Si wafer market estimated around $14 billion [460]. An EU company SOITEC leads the SOI wafer market. However, the Si metal, which is compulsory to produce the Si single crystals and, then, the wafers is highly dominated by Chinese suppliers (Figure 6.11).

The SiC wafer market is currently dominated by US suppliers, while Chinese suppliers gain market shares. But EU has, at least, two SiC wafer manufacturers to ensure its self-sufficiency in a market so important for power devices used in EVs. The EU CHIPS Act should increase the number of SiC wafer providers on its soil. Because, more than 10 Chinese companies have already begun small-scale production of 200 mm SiC substrates. It is highly likely that more than half of the world's SiC substrate production will be in China by 2026.

Figure 9.10: The EU microelectronics ecosystem – inspired by SEMI.

9.1.5 EU universities and engineer schools

Europe hosts some of the most ancient and esteemed universities globally, celebrated for their dedication to educational superiority. Universities like Oxford, Cambridge, Sorbonne, and ETH Zurich are internationally acknowledged for their strict academic criteria and significant research achievements. A focus on analytical thinking, creativity, and cross-disciplinary studies cultivates an atmosphere that encourages inquisitiveness and scholarly discipline.

The worldwide number of students with a first university degree in science and engineering in the leading regions in the semiconductor supply chain – the USA, China, Taiwan, Japan, South Korea, and Europe – grew at 4.5% annually between 2000 and 2015. The number of doctorates in science and engineering showed a very similar growth rate (Figure 9.11).

Europe presents the highest number of doctoral degrees compared to the other countries, which is a very good opportunity for the EU semiconductor ecosystem to hire high-skilled talents (Figure 9.11).

In 2002, EU reached 1.13 million science, technology, engineering, and mathematics degrees. Among them, 28% got a degree allowing them to be hired the semiconductor ecosystem. Unfortunately, only 1.6% of them, i.e., 17,800, were hired by the semiconductor industry.

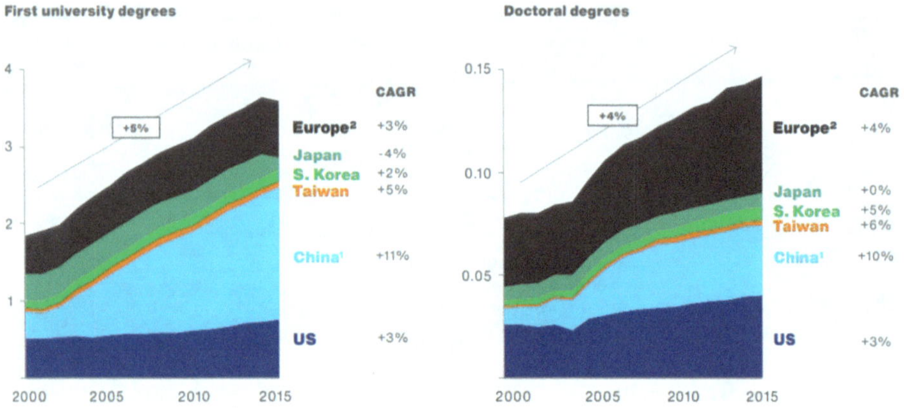

Figure 9.11: Annual graduates in science and engineering (millions) [454].

9.1.6 Answers to the sustainability challenges

Through the Green Deal initiative announced in 2019 with an objective that EU will no longer emit any GHG in 2050, the semiconductor industry has started to reduce its GHG emission for 15 years.

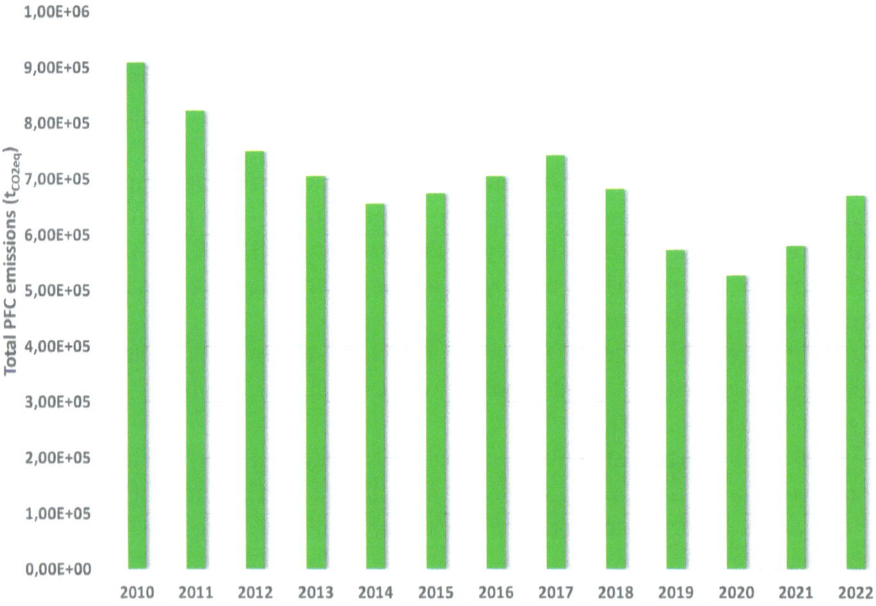

Figure 9.12: European semiconductor “perfluorocompound” (PFC) emissions, 2010–2022 - data from [457].

The EU semiconductor industry has achieved a 42% absolute emission reduction of "perfluorocompound" (PFC) gases from 2010 to 2020 (Figure 9.12). The PFC (scope 1) emission reductions have been achieved through the aggressive implementation of process optimization and more efficient alternative processes, use of alternative chemistries, and the installation of abatement equipment in Europe-based facilities. Accordingly, the EU IDMs are among the lowest emitters of scope 1 emission (Figure 6.3).

Thanks to its extensive use of nuclear and renewable energies (Figure 6.8), the EU semiconductor IDMs are also the ones with the lowest scope 2 GHG emission (Figure 6.7).

The EU semiconductor industry has demonstrated a substantial commitment to water sustainability through significant investments in advanced water recycling and reuse technologies, implementing sophisticated closed-loop water systems that enable multiple cycles of water reuse within production facilities, and mitigating the need for external water supply and generation of wastewater. EU chipmakers have also achieved measurable improvements in water efficiency through systematic process optimization and equipment upgrades, successfully reducing water-intensive operations while maintaining high production quality standards [458].

However, the water recycling rate of EU IDMs is lower than the ones of their counterparts, which suffered from severe droughts and must optimize their water recycling rates (Figure 9.13).

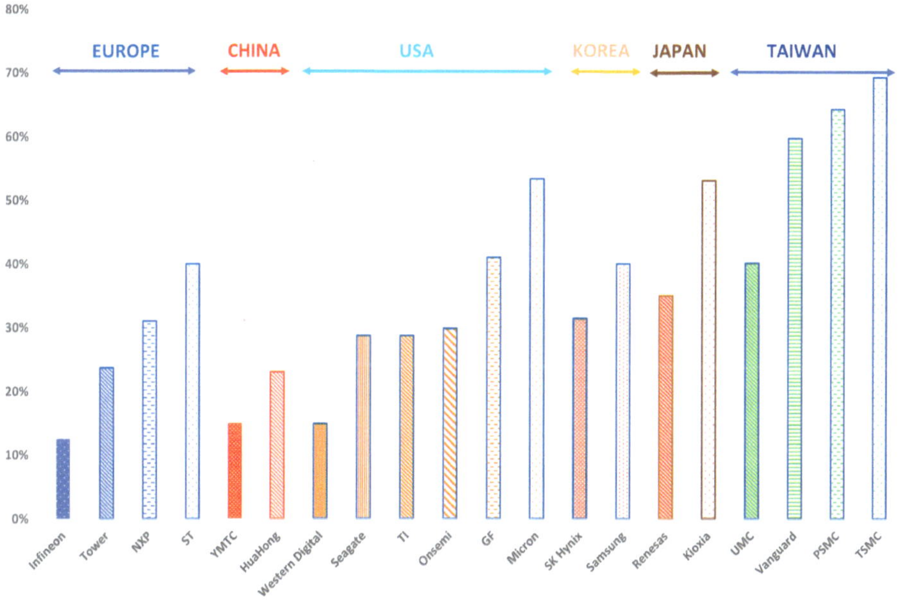

Figure 9.13: Water recycling rate in semiconductor manufacturing corporations [182] (water recycling rate = recycled water/total water use, total water use is the sum of recycled water use and water intake) – data from [420].

Since the customers of electronic systems and cars are more and more eager to purchase sustainable items, the electronic system and car suppliers demand for more sustainable chips and sustainable processes for their manufacturing. Therefore, since the EU semiconductor industry is involved in developing sustainable manufacturing for 15 years, while their counterparts are not so proactive in this domain, the EU chip ecosystem has a significant competitiveness advantage compared to their competitors, in the USA and in Asia.

Europe was the region that generated the most e-waste (17.6 kg per capita) but had the highest documented collection and recycling rate (7.5 kg per capita), recycling 42.8% of the e-waste generated. The recycling rate was only 30% and 11.8% for the Americas and Asia, respectively [425].

9.2 European weaknesses in the semiconductor supply chain

As previously described, the value chain involved in the creation and production of any semiconductor is extraordinarily complex and globalized. At a high level, it consists of four broad steps: precompetitive research, design, wafer fabrication, and ATP,

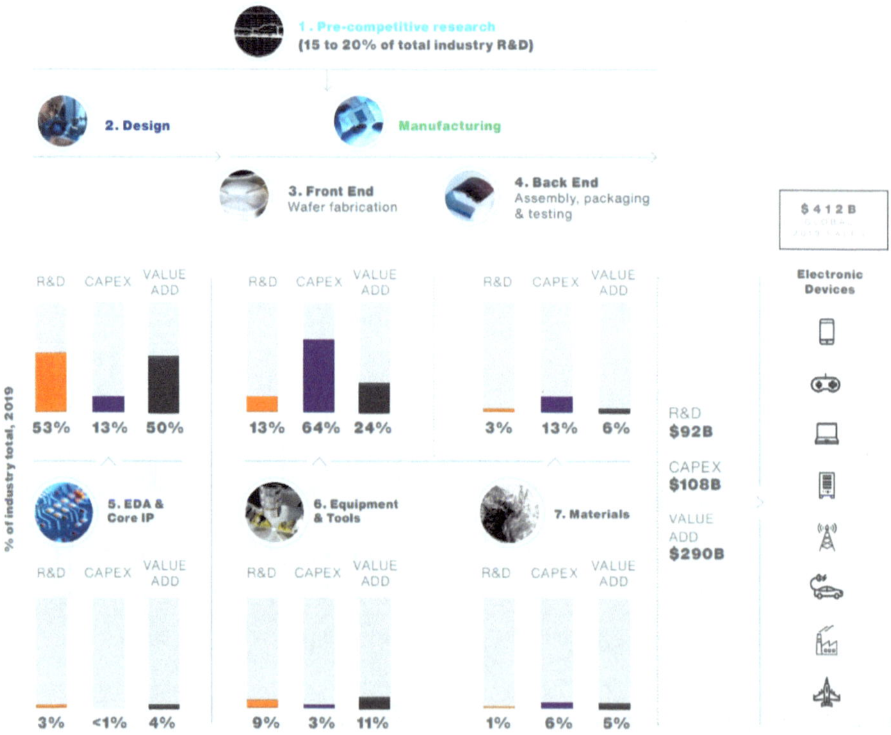

Figure 9.14: Distribution of the semiconductor value chain in various categories with an estimation of the R&D, CapEx, and value added per category in 2019 [459].

supported by a specialized ecosystem of materials, equipment, software design tools, and core IP suppliers (Figure 9.14).

As shown in Figure 7.3a, the main weaknesses of the EU semiconductor industry are the small number of design fabless, the absence of memorymakers and leading-edge and advanced foundries, the lack of lithography masks and resists suppliers, and its quasi-absence in BEOL technologies. To those EU weaknesses, its strong dependency in Critical Raw materials (CRMs), high cost of energy, and high wages must be added.

Figure 9.15 summarizes the strengths and the weaknesses of the EU semiconductor value chain.

In this chapter, leading-edge technologies arbitrary correspond to technology nodes ≤7 nm, advanced technology between 14 and 28 nm, legacy technologies between 45 and 90 nm nodes, and mature technologies ≥130 nm.

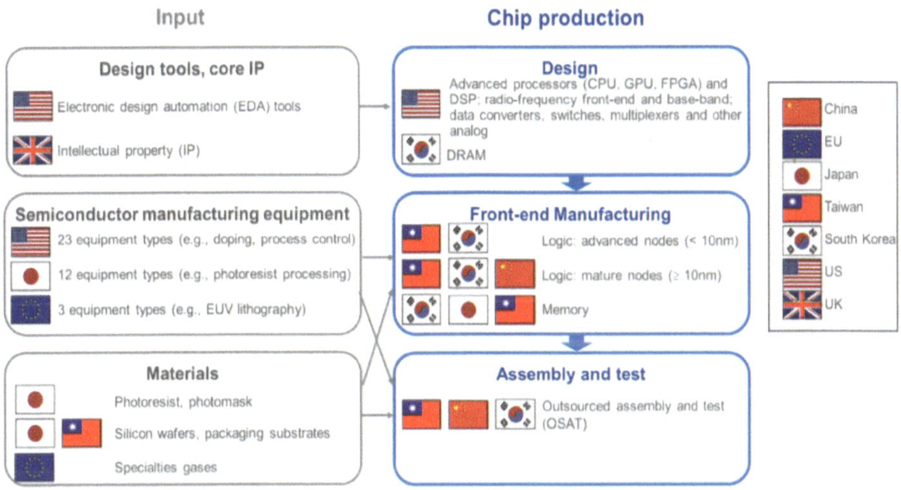

Figure 9.15: Geographical concentration of the semiconductor supply chain: value chain activities where one region accounts for about 65% or more of the global share measured in capacity for the manufacturing, assembly, test, and in revenues for the other activities [441].

9.2.1 Chip design

Chip design refers to all the steps required to design a chip. The first is to derive a schematic model of the integrated circuit meeting the technical requirements of the chip while accounting for the electronic component specifications and their operating conditions. The schematic model (logic or memory design) is then converted into a physical design, the integrated circuit layout, with interconnected electronic components (e.g., transistors). Finally, validation and verification via software simulations will ensure that the physical design will operate as expected (Figure 9.16).

According to Figure 9.14, chip design is the step in the semiconductor value chain that involves the highest value added (50%), and therefore, the greatest proportionate revenue capture, compared with wafer fabrication (24%) and ATP (6%). So-called "fabless" chip design companies typically spent more than 25% of their revenue in R&D (Figure 9.14), while IDMs spend around 15% of their revenue in R&D [460].

Chip design is also talent-intensive and relies on close business relationships with semiconductor fabrication firms that manufacture the chips.

Fabless revenues represented only 30% of the semiconductor industry revenues in 2019. In 2024, they represented around 50% of the semiconductor industry chip sales, and they should reach nearly 62% of the semiconductor industry revenues in 2028 (Figure 9.17).

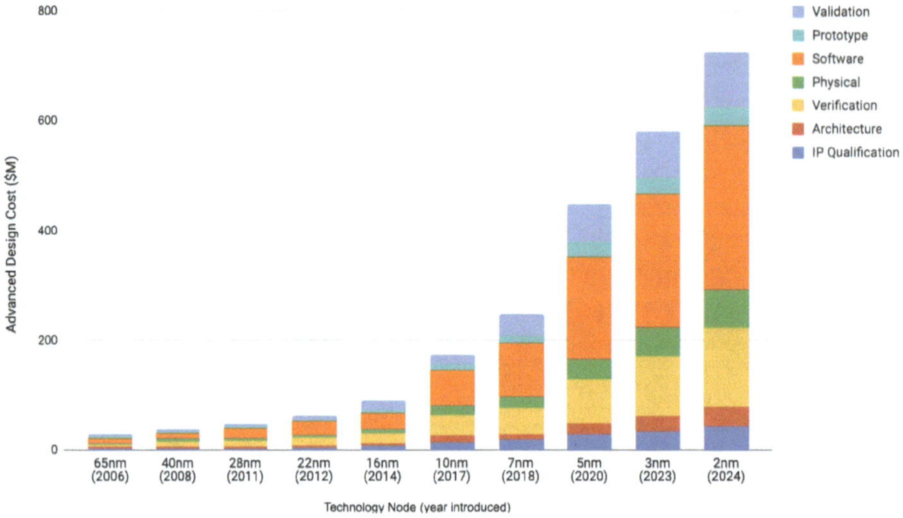

Figure 9.16: Cost of advanced designs [461].

This trend explains why the US fabless companies like Nvidia, Broadcom, AMD, and Qualcomm present the highest market capitalization and the 2024 revenues within the semiconductor industry (Figure 9.18).

Furthermore, the semiconductor industry's record of steady technological improvement has created a winner-take-all dynamic that makes leading-edge capabilities vital within several segments, such as AI nowadays. If a company's product or service is even slightly better than a competitor's, it typically captures an outsize portion – or even the vast majority – of industry revenue. This phenomenon is apparent along the entire value chain, from equipment production to chip manufacture. Com-

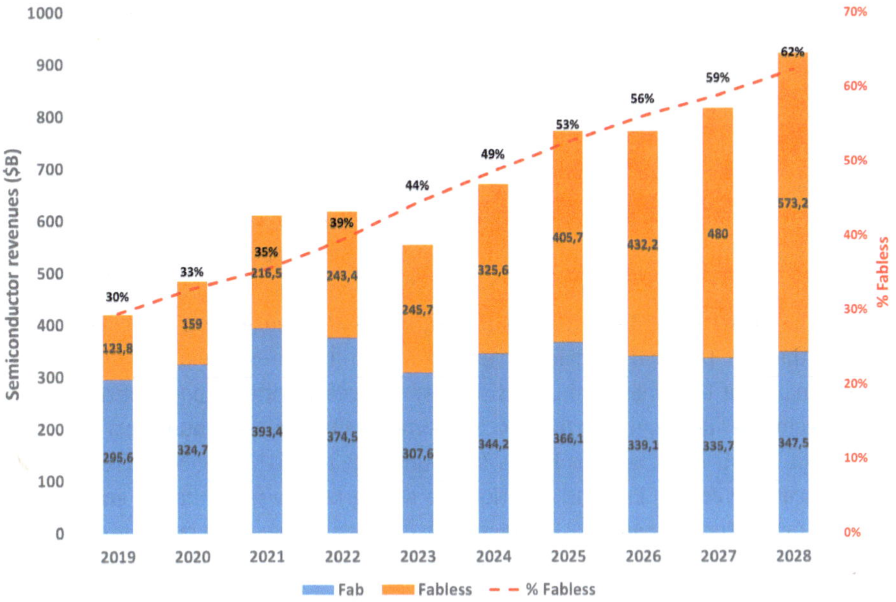

Figure 9.17: IDM versus fabless sales – data from IDC, SEMICON West, July 2024.

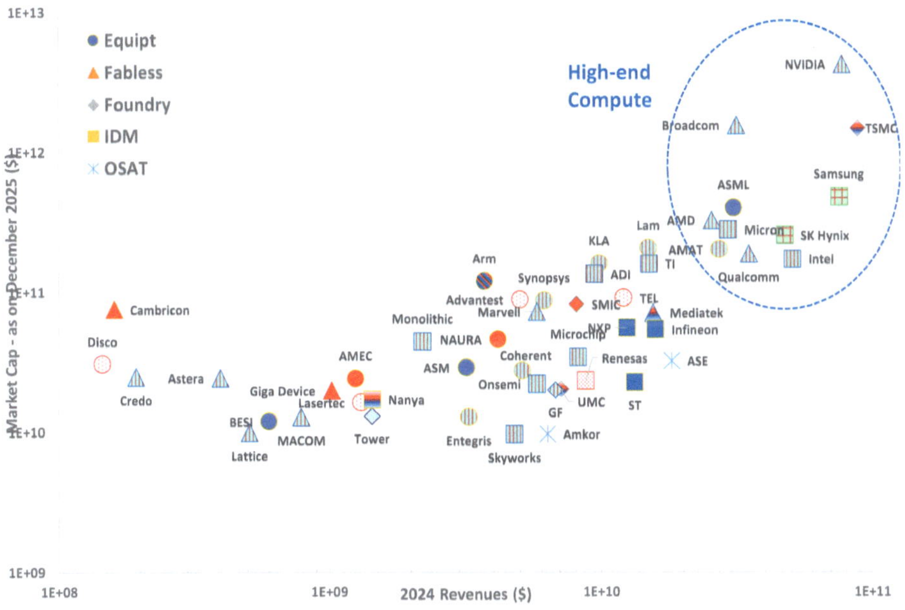

Figure 9.18: Market capitalization (as in December 2025) versus 2024 revenues per company present in the top 50 market capitalization ranking – data from annual company reports (*The company's headquarters location is depicted with the same colors as in Figures 9.7 and 9.8*).

panies that want to challenge the winner may find it difficult to catch up, since the leading players are often several years ahead in technology development [462].

Only a few companies are capable of designing the leading-edge and advanced chips with node sizes below 5 and 14 nm because of the skills and large investment required for design such chips (Figures 4.5 and 9.16). Meanwhile, demand for these chips is soaring [392]. In some major market segments, including high-performance computing (HPC), AI and ML, and smartphone processors, chips produced at a technology node ≤5 nm are critical because they combine strong performance with lower power consumption (Figure 4.2). Only the fabless companies in the top of 2024 revenues can design such high-end computing chips (Figure 9.18). Therefore, design of leading-edge chips is highly concentrated in the USA. Due to their market capitalization and their expected annual revenues, the dominance of US fabless companies will even increase with the boom of AI, deepening the technical and financial gap between them and their smaller counterparts.

Except for EU IDMs, such as Bosch, Infineon, NXP, and STMicroelectronics, and telecommunication vendors Ericsson and Nokia, a strong Europe's chip design ecosystem is lacking. European car manufacturers were slow to invest in their own chip design capabilities, and the EU lacks large consumer electronics and mobile technology companies with the resources and incentives to develop such capabilities, or that can provide the market demand for other companies to them [463]. Therefore, no EU fabless company is present in Figure 9.18, which gathers the top 50 market capitalized semiconductor companies.

Hence, Europe has a minimal share of the semiconductor design market, which is one of the most important and valuable parts of the semiconductor industry. The share of the EU fabless companies even decreased from 4% to 1% during the last decade [464].

One reason for this low market share in chip design could arise from the lower funding rate provided by European and national public authorities for design R&D, compared to the other regions (Figure 9.19). Only the US public authorities provide less incentives than EU for funding design R&D. While Chinese and Taiwanese public authorities fund the design R&D at a 45% and 28% rate, respectively (Figure 9.19).

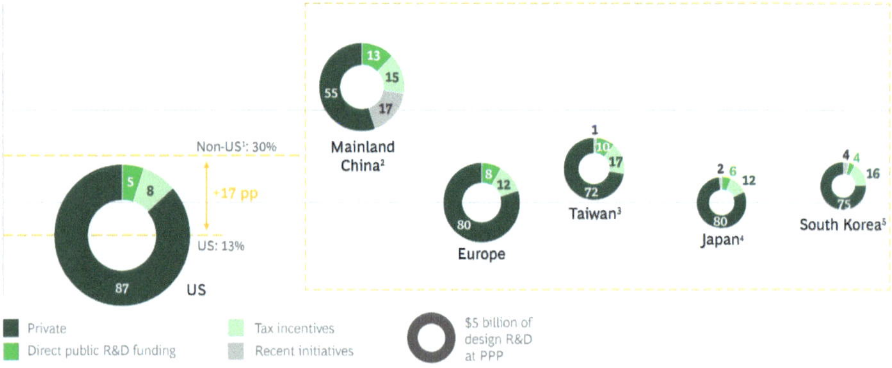

Figure 9.19: Estimated share of semiconductor-specific design R&D funded public investment (%) [465] – pp = percentage points; ppp = purchasing power parity.

Such discrepancies can explain why Chinese and Taiwanese companies and start-ups (see later) are so active in chip design.

Furthermore, the EU market shares are mainly concentrated on discrete, analog, and optoelectronics (DAO) devices. A few EU market shares are represented by its activities on logic devices, and EU is not deeply involved in the memory chip design (Figure 9.19). These observations mean that the EU chip design community, mainly IDMs, focus its activities on the main markets of the European industry: automotive, industry, and healthcare.

Since the EU chip design community is almost absent in memory design and only slightly involved in the logic device design (Figure 9.20), its activities do not currently concern any AI chip design, which require both logic devices and memories.

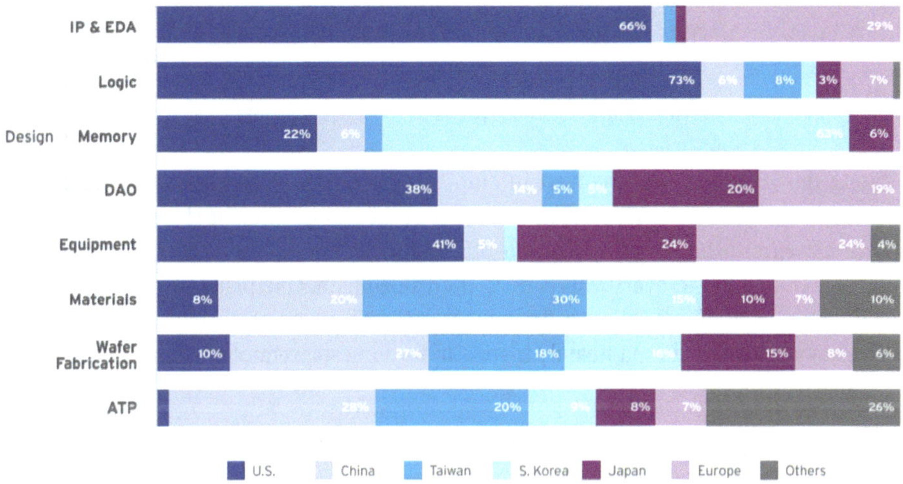

Source: IPnest; Wolfe Research; Gartner; SEMI; BCG analysis

Figure 9.20: Semiconductor industry value added by activity and region 2024 [442].

Chip design is the process step in which the majority of start-ups are active worldwide during the last 3 years (2020–2023) (Figure 9.21). Chip design is indeed an attractive field for investors because it has the highest added value in the semiconductor supply chain (Figure 9.14) and relatively low upfront investment costs compared to semiconductor manufacturing. Chip design has lowered entry barriers than other process steps in the semiconductor ecosystem, and no expensive manufacturing equipment or clean rooms are needed.

9.2.2 Funding of start-ups and scale-ups

The number of start-ups active in chip design is the highest (750), while those active in IP and wafers is particularly small: 18 and 15, respectively (Figure 9.21).

Figure 9.21: Number of semiconductor start-ups per value chain position [466].

Among the 750 start-ups involved in chip design, China is the most active country (Figure 9.22). China is present in every market of the semiconductor start-up ecosystem. There appears to be no prioritization of specific segments; the activities of Chinese start-ups can be found everywhere along the value chain. One plausible reason for this is China's pursuit of self-reliance. To become less dependent on foreign (especially US) technology providers, China heavily invests in its domestic semiconductor ecosystem.

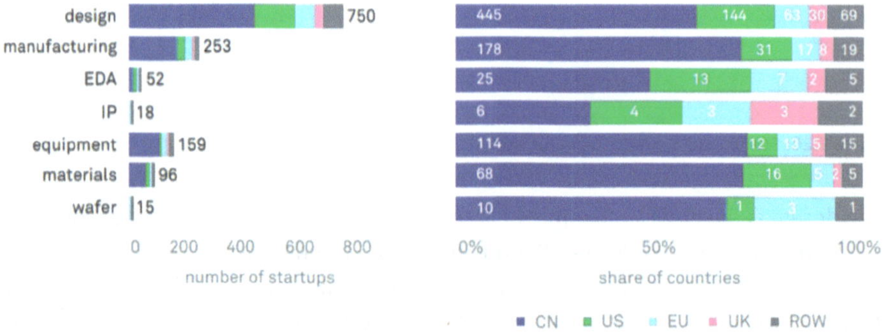

Figure 9.22: Country shares of start-ups in each value chain position [466].

The US fabless start-ups are the second most important behind China (Figure 9.22). Their activity mirrors their strong suits in the transnational semiconductor value chain. US companies have a high market share in chip design, due to US fabless companies (e.g., Qualcomm, Nvidia, and AMD) and system companies (e.g., Apple, Tesla, Amazon, and Google).

One of the current shortcomings of Europe is its inability to nurture start-ups for becoming unicorns, especially in a capital-intensive market of design chips using leading-edge technologies. Even if Europe has recently gained experience in venture capital (VC) funding, its highly fragmented national equity markets unlike in the USA, and its general preference for short-term returns continue to hinder large-scale investments in fabless start-ups, which can seriously disincentivize their long-term growth and innovation, the so-called "valley of death."

Comparing how start-ups progress through funding milestones reveals one of the clearest illustrations of Europe's scale gap. Among companies founded in 2016, US start-ups are twice as likely to have raised $50 million or more than their European counterparts, and more likely to have raised at every level of funding. This divergence doesn't reflect weaker early-stage performance since European founders are nearly as likely to raise any form of initial investment. But what follows is a systemic drop-off in capital availability as companies mature, limiting Europe's ability to consistently scale its most promising start-ups into global category leaders on equal terms [467].

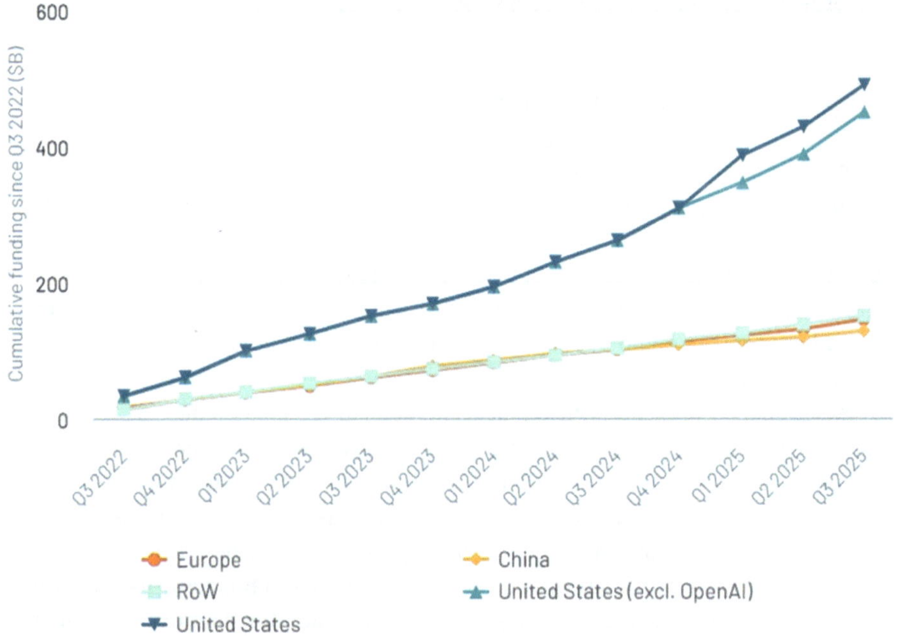

Figure 9.23: Cumulative tech start-ups funding raised Q3 2022–Q3 2025 [467] (excluding biotech, debt, capital, and grants).

This trend is highlighted in Figure 9.23, which gathers the cumulative tech start-ups funding raised between Q3 2022 and Q3 2025. US funding of start-ups is far higher than the EU one, which is comparable to the Chinese one.

In the first three quarters of 2025, US funding corresponds to 0.61% of its gross domestic product (GDP), excluding the exceptional funding of Open AI ($40 billion), i.e., $177 billion. During the same period, EU funding reached $33 billion (0.17% of its GDP), while China and the rest of the world (RoW) funding were $44 billion and $9 billion, respectively [467].

Because of the lack of deep-tech financing for long-term investments in high-risk, high-return ventures, innovative fabless start-ups in the EU struggle to get funding to scale up and overcome the so-called "valley of death" to become unicorns with a market capitalization >€1 billion. Accordingly, EU promising small fabless companies often become acquisition targets for foreign players, further eroding Europe's innovation ecosystem.

Scaling start-ups to create large European design houses, which are able to support the design costs of advanced chips with the critical human resources for performing these designs, is currently very challenging. The EU commission and the MS have set-up the European Innovative Council for funding start-ups and scale-ups. However, owing to the huge cost for designing advanced AI chips (Figures 4.5 and 9.16), a specific European initiative should be set-up quickly.

9.2.3 Leading-edge semiconductor technologies

The emergence of the fabless-foundry model occurred concomitantly at a time when European mobile phone leaders gradually lost their prime position to US and Asian actors. Europe's remaining chip demand was mainly focused on automobile and industrial applications, which do not require a similar innovation level and advanced technology nodes than mobile phones and AI. Empirically, a 5-year gap is usually observed between the deployment of a logic chip in mobile phones and in computers compared to automotive.

Accordingly, European semiconductor manufacturers focuses their manufacturing capacities to produce embedded systems, power electronics, and sensors given their close ties to the automotive industry. As IDMs, European semiconductor manufacturers continues to invest in both design and manufacturing but were not involved in the worldwide race for the production of chips using leading-edge technology nodes, such as AI chips.

One obvious reason is the CapEx required to build a manufacturing line of logic leading-edge chips, which is quite higher than the CapEx for manufacturing mature nodes (Figure 4.5). European suppliers cannot currently afford to invest such a large amount of CapEx. In contrast, their competitors have invested more than $90 billion in CapEx in 2022 and $80 billion in 2023 for advanced logic and foundry (Figure 9.24a).

Figure 9.24b shows the evolution of the CapEx spent by the three major semiconductor suppliers from 2010 to 2025. Due to the increased demand for AI chips and memories,

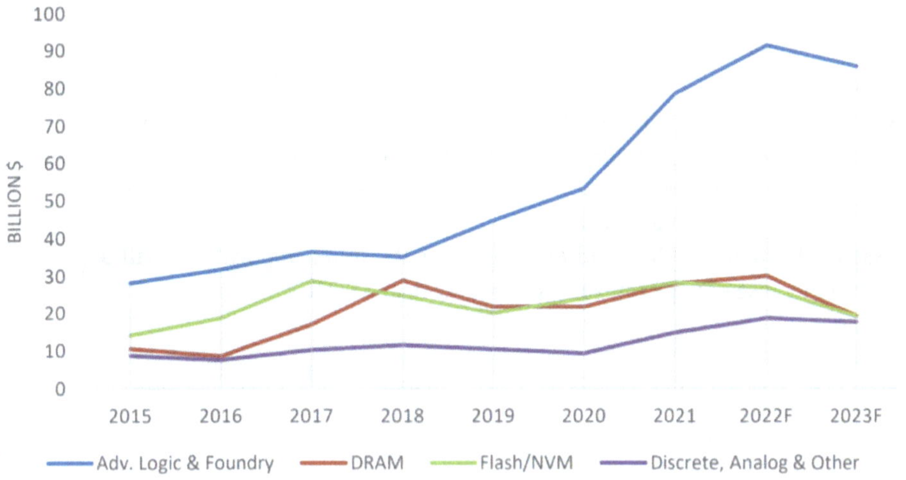

(a)

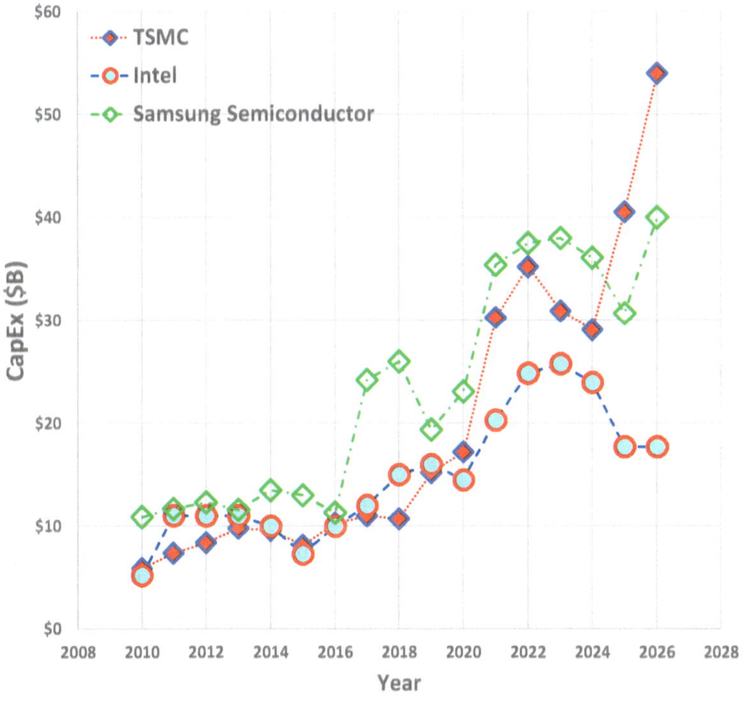

(b)

Figure 9.24: (a) Evolution of CapEx by type of products – courtesy of TechInsights CAPEX Forecast Q4-2025 and (b) evolution of the CapEx spent by the three major semiconductor suppliers – data from company annual reports.

the CapEx values globally increased during this period. Even if the erratic average selling price of memories tended to slow down the CapEx spent by Samsung from 2022 to 2025, the huge demand for high bandwidth memory (HBM) for AI applications and secure leadership in the AI semiconductor era will drive the CapEx toward summits in 2026 and beyond. For instance, Samsung would spend around $73 billion to expand manufacturing capacity and accelerate R&D on next-generation AI chips, especially HBM, in 2026.

The huge demand for AI chips – logic devices and memories – associated with the risks of shortage for this kind of devices for the other applications will lead to an overall increase of CapEx in 2026 (Figures 9.24b and 9.25).

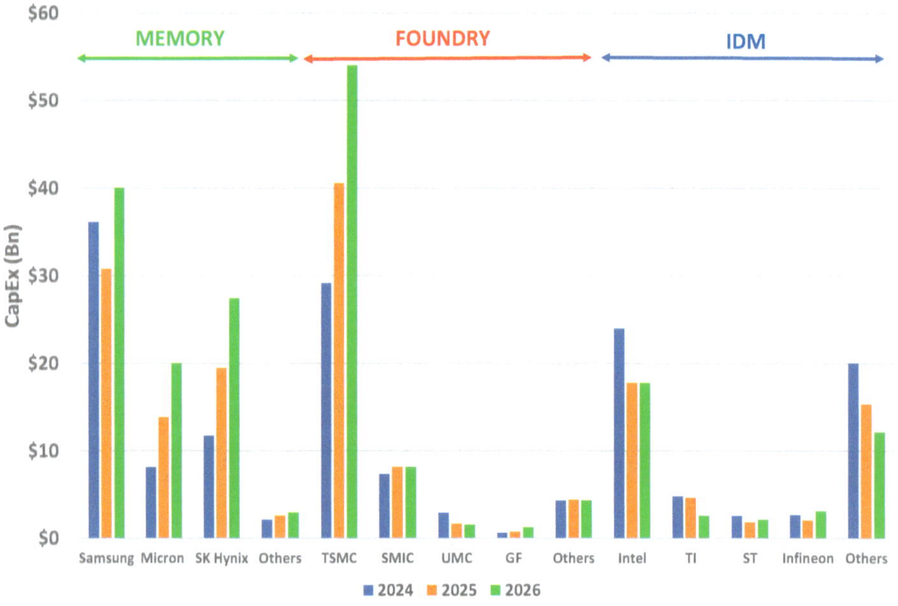

Figure 9.25: Evolution of the CapEx spent by the top chipmakers in 2024 and 2025, and expected in 2026 – source companies.

Driven by the requirements of AI, all memory manufacturers are expected to substantially raise their CapEx in 2026. There has been a marked rise in the need for memory attributed to AI training and inference, which leads to a memory shortage at the time this book is written. The training of AI has particularly resulted in an increased need for HBM, whereas model inference has generated a significant demand for storage capabilities, thereby enhancing the expansion of NAND flash applications within data centers. This robust demand has resulted in consistently elevated investment levels in the memory supply chain in both the short and long terms, providing a buffer against possible declines from conventional memory cycle variations.

In the foundry sector, TSMC will be the sole company to strongly increase its CapEx, widening the gap between its capacity and the one of the other foundries (Figure 9.25).

The CapEx of the major IDMs remains globally constant in the 2024–2026 period (Figure 9.25). Figure 9.25 also highlights that the chipmakers deeply involved in the manufacturing of AI chips present the highest CapEx, while those mainly producing semiconductors for analog and power electronics, such as TI, ST, and Infineon spend less CapEx in the 2024–2026 period.

Furthermore, the comparison between the CapEx spent in 2024 by the three main chip suppliers (Figure 9.24b) and the 2024 revenues of the main European chipmakers (Figure 9.18) shows that the CapEx spent by the leaders of the semiconductor manufacturing is much higher than the revenues of the main EU IDMs. This further widens the gap between the European chipmakers and their Asian and US counterparts.

However, one can assume that autonomous vehicle driving will change the automotive industry, and that high-performance chips manufacturing using more advanced technologies will be compulsory in the coming years.

Before the EU CHIPS Act was operational, the most advanced logic chips were only produced in Europe by Intel-Ireland using a 14 nm FinFET technology. The EU CHIPS Act and/or national initiatives now allow producing 3–5 nm FinFET-based chips, referred as Intel 3 and 4 technologies [468] at the Irish Intel premises (Figure 5.23). However, Irish Intel production lines are mainly devoted to manufacture Intel products rather than an open foundry. Furthermore, the US government, which recently became an Intel stake-

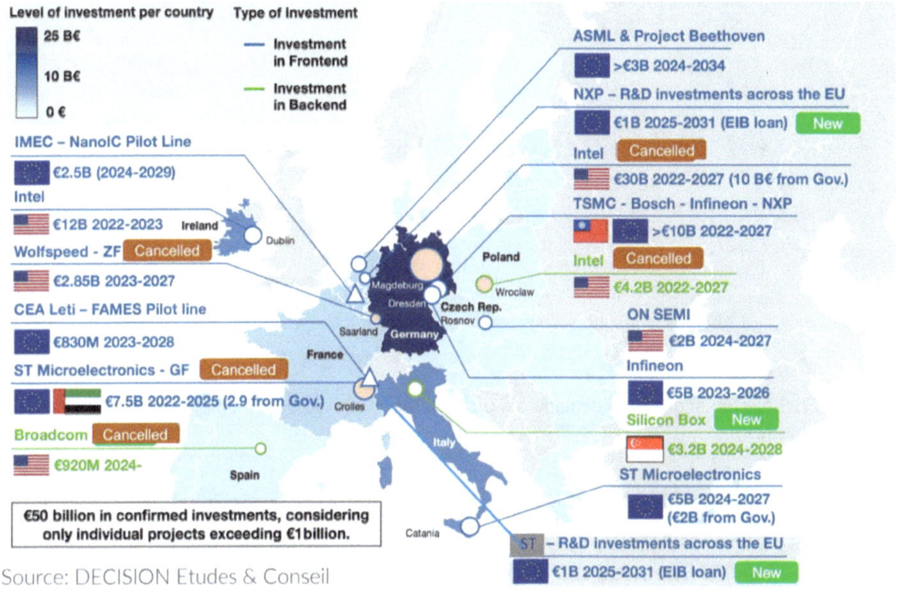

Figure 9.26: Evolution of semiconductor investment plans in the EU (over €1 billion), January 2024–December 2025 – updated from [469].

holder, could develop some "export controls" on the most advanced chips produced in Ireland, restricting the access to the Irish Intel foundry to its European customers.

However, despite expected large subsidies from the German and Polish public authorities to provide leading edge chip manufacturing and advanced packaging capability in Europe, Intel recently announced that it canceled plans to construct these new semiconductor factories (Figure 9.26). These cancelations resulted from (i) the current issues faced by Intel, (ii) the weak demand within the European electronics industry for leading-edge-technology chips, and (iii) overambitious market expectations.

According to the projections of the EU industry for the year 2022, it was anticipated that the need for advanced chips in the near to medium future would be minimal, with the majority of the demand being focused on microchips in the range of 65–90 nm (Figure 9.27). This forecast predominantly aligns with the requirements of the automotive sector. This diminished appetite for advanced chips is also partially due to the absence of key fabless EU firms capable of designing chips intended for HPC, broadband technologies (5G and 6G), and AI which necessitate the utilization of chips manufactured with cutting-edge technologies.

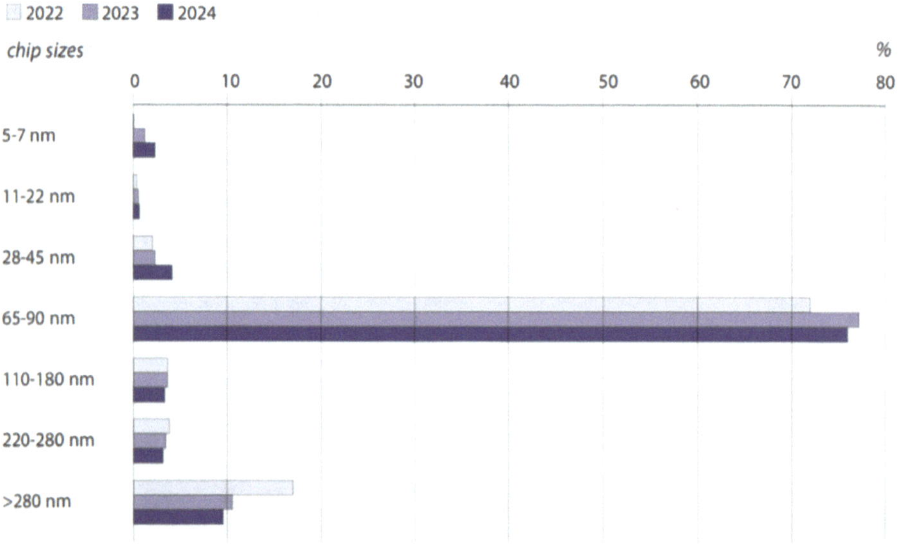

Figure 9.27: Projected aggregated demand of EU companies surveyed by technology nodes between 2022 and 2024 [470].

Creating advanced technologies holds no value without a target market, along with a purpose to propel it forward for generating significant income to fund further R&D for future innovations. The need for chip production in Europe using advanced technologies would likely surge with the rapid implementation driven by AI. The requirement for large-scale AI production facilities in the EU, and thus for AI chips, could transform

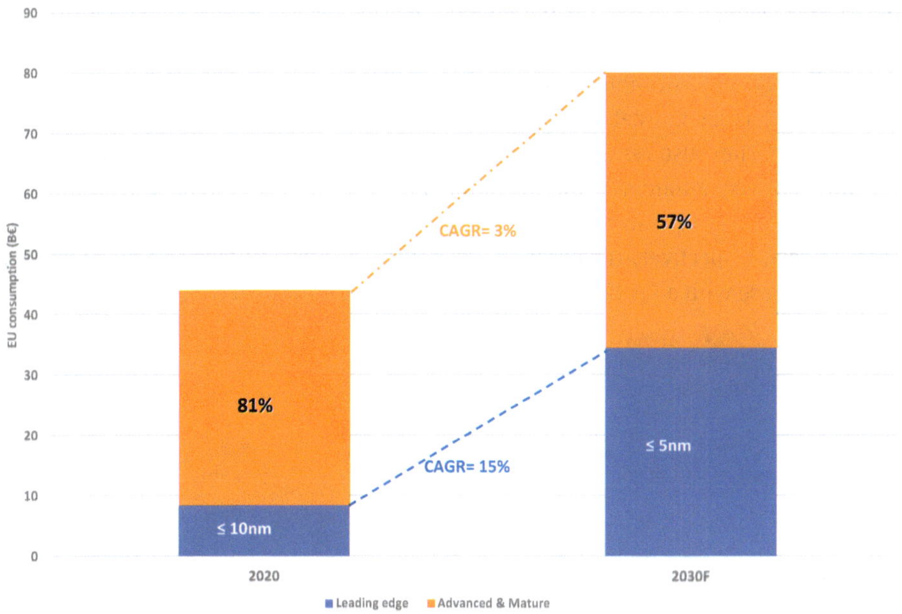

Figure 9.28: EU non-memory IC consumption by technology node – data from [471].

the European digital sector, opening up new markets and potentially replacing existing products, which would contribute to both European wealth and national security (Figure 9.28). This rising need could likely encourage the growth of major European fabless companies and the establishment of state-of-the-art foundries within Europe.

9.2.4 Stand-alone memories

The European semiconductor industry is almost absent in the design and manufacturing of stand-alone memories such as dynamic random-access memory (DRAM), HBM, and NAND flash memories since the bankruptcy of Qimonda, a subsidiary of Infineon, in 2009 [472], and the purchase of Numonyx, a joint venture between STMicroelectronics and Intel, by Micron in 2010 [473].

DRAM and NAND make up ~25% of the total semiconductor industry in 2025 (Figure 9.29) DRAM and NAND combined is a $170 billion industry in 2024 [474]. But DRAM and NAND markets are very cyclical and driven by supply and demand.

The overall memory market evolves very quickly as a function of the market demand. Such rapid changes drastically influence the revenues of the memory chipmakers (Figure 9.29), and their future investments. Accordingly, due to the memory demand related to the boom of AI, DRAM, and flash NAND shortages are forecasted in the coming years. Consequently, the memory market driven by HBM would represent nearly 50% of the semiconductor market by 2026 or 2027.

The Qimonda bankruptcy and the purchase of Numonyx partially resulted from the erratic memory market, which never exhibits a steady-state average selling price (Figure 9.30). They also resulted from the very low demand for memories in the main EU chip markets: automotive and industry. The automotive and industrial sectors are not big customers of stand-alone memories as they only represent 6% and 10% of these sectors, respectively (Figure 9.31). Even if the memory (DRAM and flash) market should increase with a CAGR of 21% between 2024 and 2030 [475].

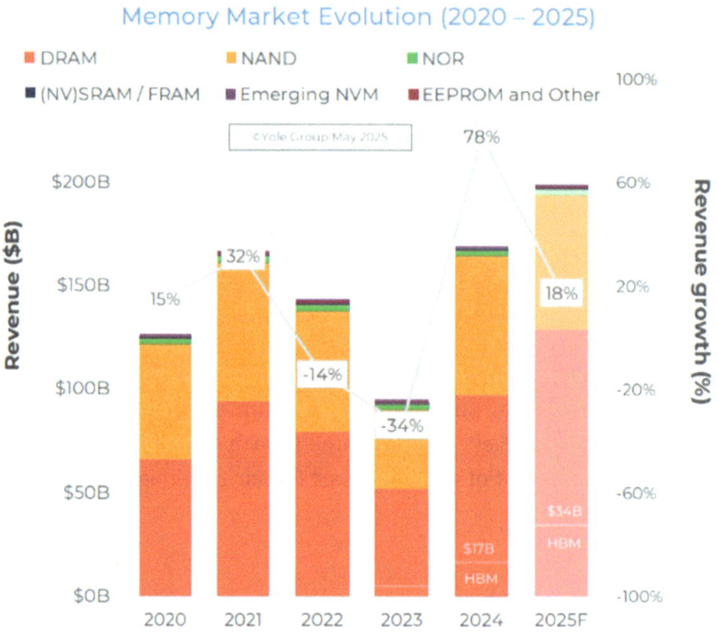

Figure 9.29: Memory market evolution (2020–2025) [474, 475].

STMicroelectronics still produces electrically erasable programmable read-only memory (EEPROM), which only represents less than 0.6% of the memory market (Figure 9.29) [475]. Flash memory is a distinct type of EEPROM, which is programmed and erased in large blocks. Flash memory is preferred for large volumes of data, while EEPROM is used for small volumes. In 2020, Infineon acquired Cypress that produces NOR flash memories. However, EEPROM and NOR flash memories are generally used for small data volume storage, which do not match AI accelerator requests, for instance, while NAND flash memories are used for AI data storage.

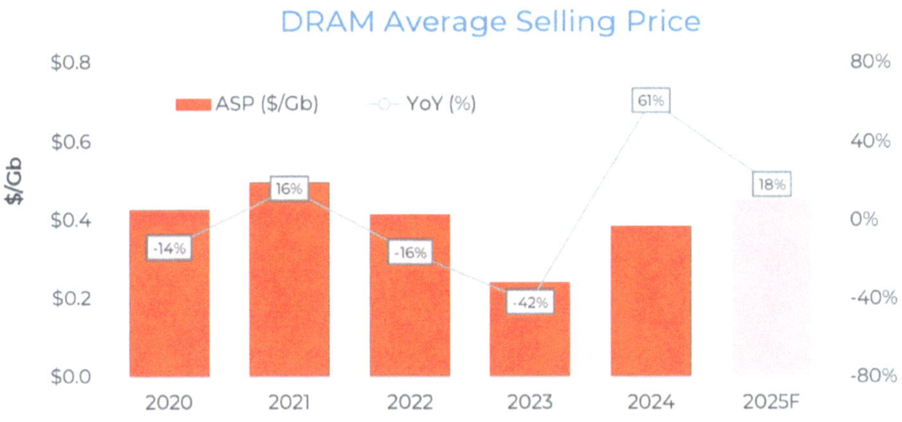

Figure 9.30: DRAM average selling price – courtesy of Yole Intelligence.

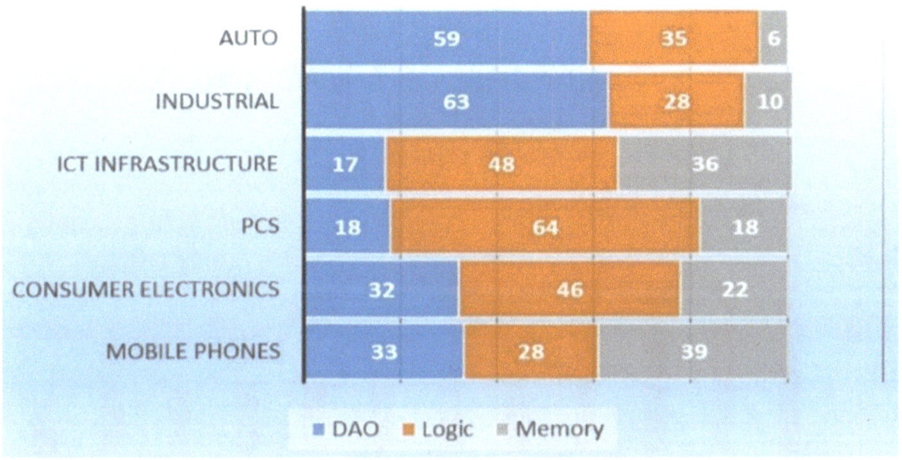

Figure 9.31: Global semiconductors sales by allocation market, 2019 (%) [441].

9.2.5 Advanced, legacy, and mature chips

As it has been aforementioned, the European semiconductor industry heavily focuses on serving the automotive and industry supply chains, with strengths in MCUs, analog products, power devices, sensors, and other mature semiconductors.

One of the main competitiveness issues faced by the European semiconductor industry is the higher cost for building and running a manufacturing fab.

Building a new fab in Europe is roughly similar to in Taiwan but significantly higher than in China. In contrast building a new fab in the USA is significantly more

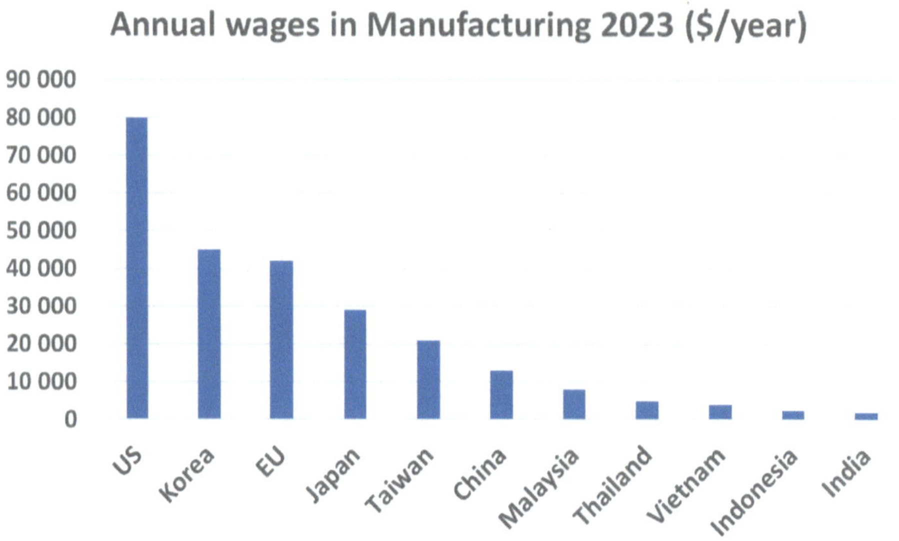

(a)

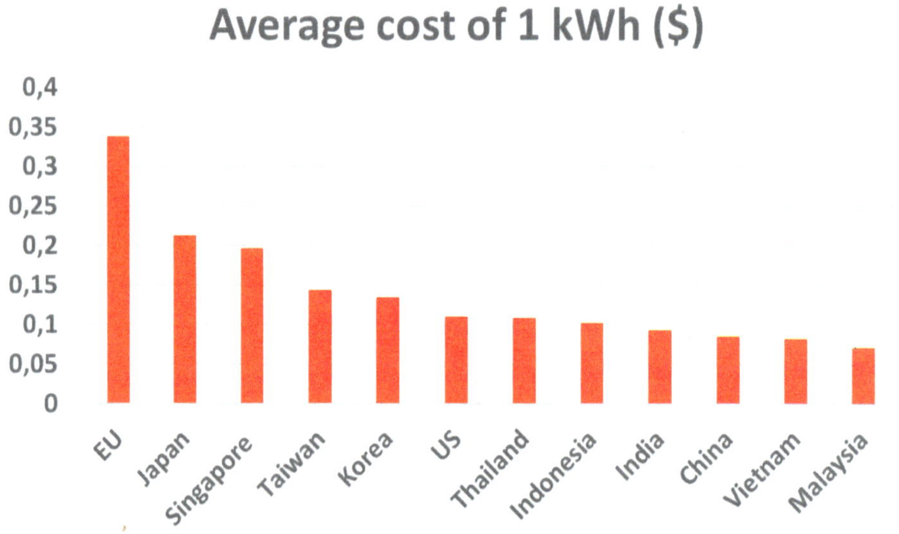

(b)

Figure 9.32: (a) Average annual wages in manufacturing in 2023 [478] and data from OECD, and average cost of 1 kWh [479] per region – (€1 = $1.12).

expensive than building a fab in Asia, mostly because of the cost of construction. One of the major factors in the construction cost difference is variations in labor cost. Constructing a fab is extremely labor-intensive, requiring input from a wide variety of white- and blue-collar trades, and labor contributes a large portion of the up-front cost of building a fab. In Europe, labor can be more than two to three times more expensive than in Asia (Figure 9.32a). Labor shortages across many industrial sectors also heighten labor costs [477].

The primary reasons for variations in operational expenses for semiconductor fabs situated in Europe, the USA, Taiwan, and China are labor (including both direct and maintenance labor) (Figure 9.32.a) and utility costs (Figure 9.33). In Europe, utility expenses are significantly elevated because of the high energy prices on average (Figure 9.32.b). The expenses associated with acquiring tools for the fabs and logistics are quite similar across different regions, owing to the globalized and unified structure of the semiconductor supply chain. As a result, the yearly operational costs for direct and indirect materials, including initial equipment, tend to remain equal across the examined regions, assuming subsidies are not taken into account [477].

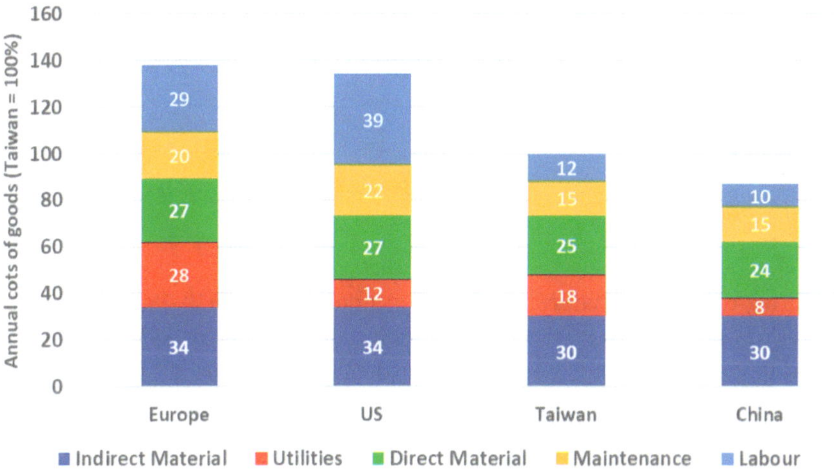

Figure 9.33: Annual good costs of 28 nm semiconductor fabs with an 85% utilization rate – data from [476, 477].

In summary, without considering subsidies, running a mature semiconductor fab in Europe costs 3%, 38%, and 59% more than in the USA, Taiwan,, and China, respectively. The difference between running costs between Europe and China is even higher, considering some subsidies provided by the Chinese government. The Chinese chipmakers benefit significantly from government-backed equipment leasing programs. The government purchases equipment and leases it to fabs at no cost. The higher costs for running a fab in Europe should naturally reduce the number of fabs owned by non-European chipmakers.

Due to the advantageous subsidies provided by the Chinese public authorities, and the export control of extreme ultraviolet lithography (EUV) lithography and advanced tools by US, which considerably slow the development of leading-edge technologies in China, Chinese chipmakers will strongly increase their capacity in advanced nodes (20–45nm), which are key for MCUs (Figure 9.34). They will also increase their

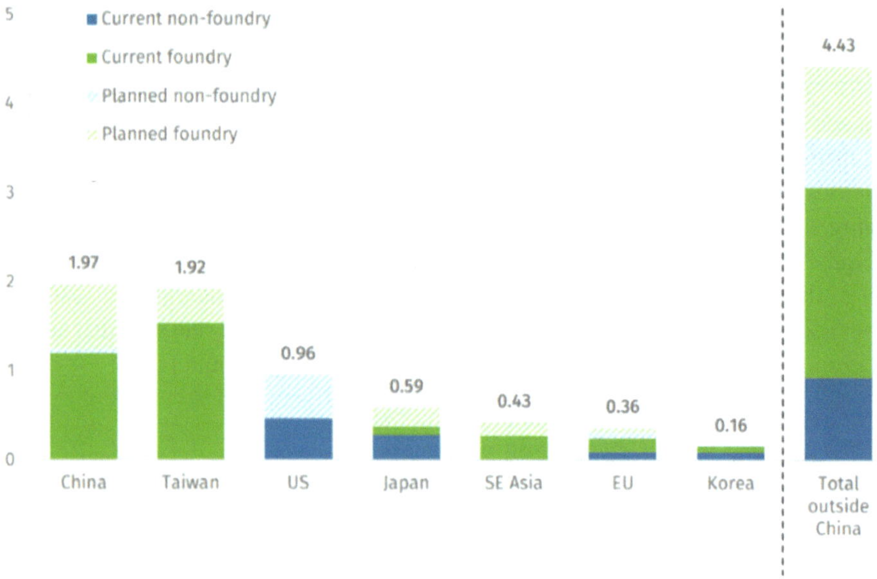

Figure 9.34: Current and planned wafer capacity (excluding memory) at 20–45 nm as of March 14, 2023, in million wafer starts per month [479].

manufacturing capacity in mature and legacy technology nodes (50–180 nm, key for power devices and sensors) due to the huge domestic demand for EVs. Figure 9.34 shows that China had already 27% of the worldwide wafer capacity for advanced nodes in 2023. This capacity could increase up to 31% in the 5 coming years, if the scheduled fabs are really built and fully operational.

In the near future, coupling China's growing manufacturing capacity (Figure 9.34) and cost competitiveness (Figures 9.32 and 9.33) will represent a very serious threat for the European MCU analog and power devices producers, which currently lead the market.

This picture looks quite similar for the mature nodes; in 2023, China controlled around 30% of the worldwide capacity of the worldwide 50–180 nm manufacturing capacity, and could reach 35% over the next 5 years. As mature nodes represent the largest European chip demand (Figure 9.7), Europe will be quickly swamped with Chinese chips at the expenses of the European semiconductor industry.

Moreover, with the current slowing down of EV sales in Europe and the tough competition between European and Chinese carmakers, the market share of European automotive companies will be reduced and, by extension, will limit semiconductor demand within the European automotive sector for mature semiconductor vendors that are provided by European semiconductor manufacturers. Therefore, the market shares of the European semiconductor industry could even eventually decrease despite the EU CHIPS Act objectives, in a sector in which EU chipmakers are strong.

9.2.6 R&D expenditures

As aforementioned, the European R&D is rather strong. However, translating R&D results into commercial products and then in revenues for the EU semiconductor industry remains rather difficult.

This can be due to the lowest revenues and market capitalization of the main EU chipmakers compared to their US and Taiwanese counterparts. Figure 9.35 presents the top semiconductor companies in terms of their 2023 R&D expenditures. Without any surprise, this ranking looks like the one depicted in Figure 9.18, which ranks the semiconductor companies in terms of capitalization and 2024 annual revenues. The R&D expenditures allow enhancing the innovation activities of the top-ranking firms, thereby increasing their competitiveness compared to their counterparts.

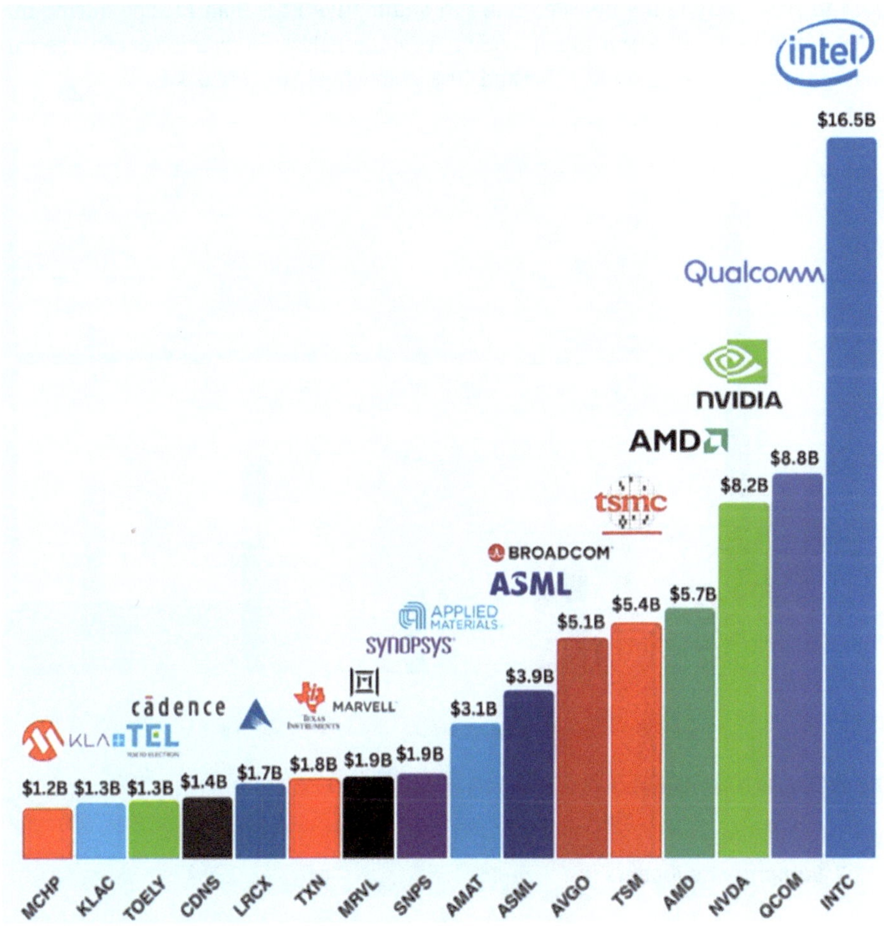

Figure 9.35: Semiconductor companies ranking by their 2023 R&D expenditures [480].

Accordingly, the European chipmakers, with lower capitalization and annual revenues (Figure 9.18), tend to limit the risks for investing in emerging technologies. Only one European company appears in terms of R&D expenditures: ASML (Figure 9.35).

In the case of an industrial transfer of R&D results to start-ups and spin-offs, this transfer can be hindered by the lack of VC investments in these companies as previously described for the EU fabless start-ups.

A third reason can be the low EU funding rate for the R&D works dedicated for designing and manufacturing chips (Figure 9.19).

Accordingly, while the R&D expenditures in the information, communications, and technologies (ICTs) sector increased from 2014 to 2024 by 417%, 105%, and 93% in China, the RoW, and the USA, respectively, the European R&D ICT expenditures only increase by 61% during the same period (Figure 9.36).

In contrast the EU R&D expenditures in the automotive increase by 82% from 2014 to 2024. Only China invested more in automotive R&D than Europe during this period. While the USA and Japan, which are nonetheless the countries of large carmakers, invested less than EU; 52% and 49%, respectively (Figure 9.36).

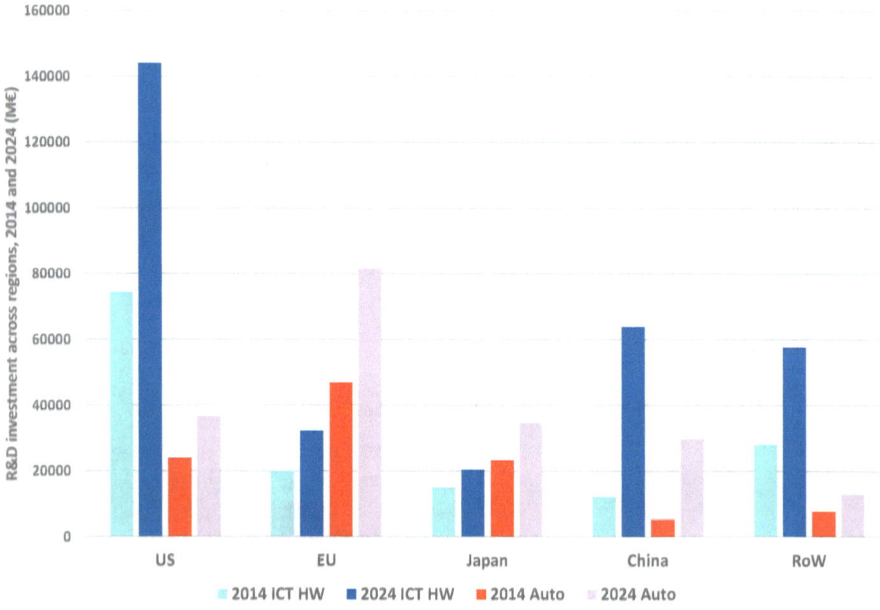

Figure 9.36: R&D investment across regions, 2014 and 2024 – data from [481].

9.2.7 Equipment suppliers

Even if the semiconductor equipment sector is one of the main strengths of the EU semiconductor value chain, Europe has no equipment manufacturer involved in the thinning and chemical mechanical planarization, ion implant, and etching/cleaning processes (Figure 9.37).

Furthermore, if ASML, the current EU semiconductor leader in terms of revenues strongly dominates the market of the lithography equipment (Figure 9.38); the threats arising from China and the USA are numerous.

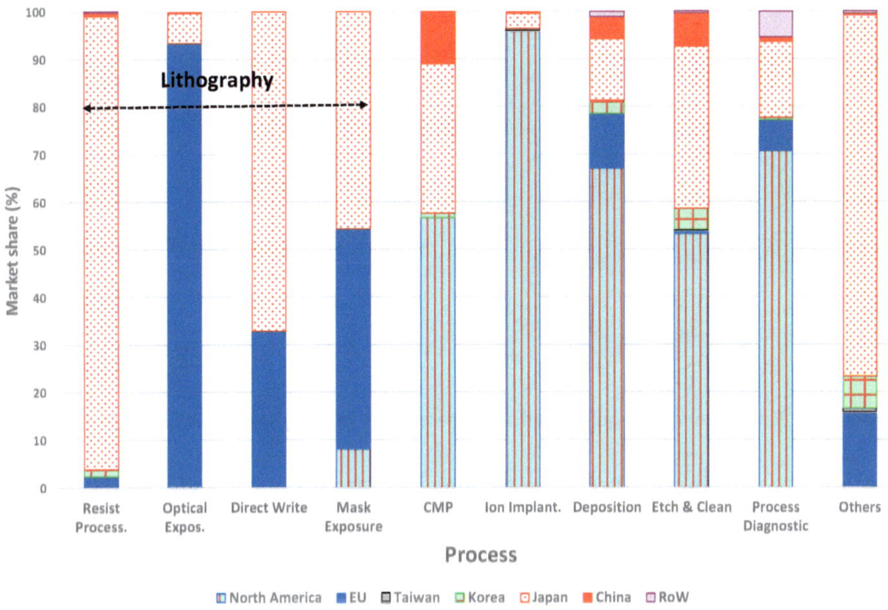

Figure 9.37: Wafer fabrication equipment for chip production market share 2023 – data from [454] (*The company's headquarters location is depicted with the same colors as in Figures 9.7 and 9.8*).

Owing to US export restrictions and seeking for its own technological independence, China is spending roughly $40 billion to develop domestic technologies aimed at breaking ASML's monopoly in EUV lithography [483]. China could have also assembled a prototype EUV machine using components sourced from older ASML systems that aims to produce functional chips using the prototype by 2028 or 2030 [484].

A Chinese start-up even emerged, creating equipment using plasma as a light source to bypass ASML's laser-driven plasma method. China is simultaneously upgrading technologies required for cutting-edge semiconductor manufacturing, such as light source technology, optical equipment manufacturing, and precision mechanical engineering. A Chinese ASML could emerge within a decade [485].

The USA is very strong for designing chips (Figure 7.3). However, the US government will bring everything to be self-sufficient in the capacity of manufacturing semiconductors. In this context, the USA pointed out that EUV lithography technology has become the key technology for mass – producing transistors above 7 nm. US compa-

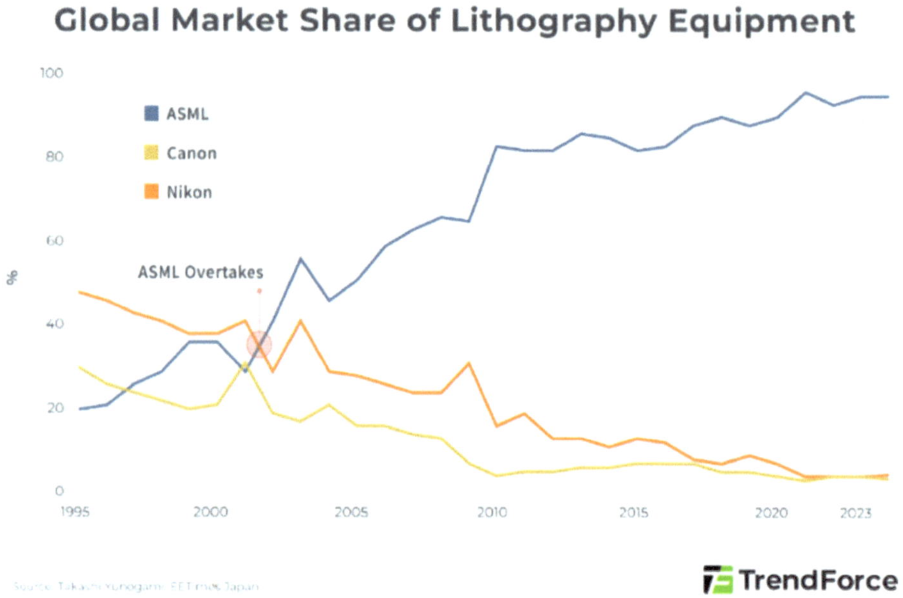

Figure 9.38: Global lithography equipment market share: ASML versus Nikon versus Canon [482].

nies are also working on alternative technologies for EUV lithography. An US start-up recently announced that it hopes to use a particle accelerator to generate light for lithography machines and claims that it can produce such a light source by 2028 while maintaining compatibility with existing tools. This start-up also pointed out that the company is building the world's most powerful laser to revolutionize semiconductor lithography, metrology technology, and other key economic and national security applications. This brand – new EUV free – electron laser FEL light source for the semiconductor market would tend to replace the current laser – produced plasma light source, which is approaching its physical limit. The FEL system would significantly enhance ASML's technology roadmap, reduce capital and operating costs, improve the production capacity of semiconductor fabs, and help the USA regain its leading position in the field of advanced semiconductors.

Another US company aims to use a "desktop" particle accelerator to generate the required high-power light. This accelerator can accelerate electrons to extremely high energies within a centimeter-scale range, rather than requiring a kilometer-scale facility like the large accelerators at CERN. They hope to achieve this goal by using plasma waves driven by high-power lasers. The target wavelength for the next generation is 6.7 nm. In addition, the company claims that it can double the transistor density with the same numerical aperture while achieving three times the throughput of existing machines. Moreover, the critical dimension uniformity can be improved by 25%, significantly improving the manufacturing of high-aspect-ratio features of new transistor architectures and computing paradigms, including quantum [486].

If Chinese equipment suppliers wish to compete with ASML on a highly complex tool such as a EUV lithography tool, driven by the strong demand in the China market, the Chinese semiconductor equipment continuously gains market shares in less complex semiconductor tools. Prior to the tightening of US and allied export controls, Chinese-made semiconductor manufacturing equipment accounted for only about 10–15% of China's domestic market. Despite sustained state support beginning in 2014, domestic penetration remained limited through 2021 due to low customer demand and adoption. Conversely, in the years following the imposition of export controls, that share has surged from 25% to 35% of the market between 2024 and 2025, exceeding the Made in China 2025 target of 30%. Progress has been particularly dramatic in etching and thin film deposition, where local firms now supply 40% of the domestic market [483] (Figure 9.37).

Three Chinese semiconductor equipment suppliers already appeared in the 2023 top 20 ranking, and one even appeared seven in the 3Q2025 equipment ranking, which represents a serious threat for US, Japanese, and European semiconductor equipment producers, such as ASML, ASM-I, Aixtron, and Besi.

9.2.8 Lithography masks and resists

Photoresist, a seemingly insignificant chemical material, plays a crucial role in semiconductor manufacturing. It is an indispensable part of the chip manufacturing process, using photochemical reactions to precisely transfer complex circuit patterns from the mask plate to the silicon wafer. The technological accumulation and innovation capability of Japanese companies in the photoresist field has enabled them to occupy an absolutely dominant position in the global market.

Besides the American firm Dupont, which holds the third position, the remaining share is dominated by Japanese companies, whose total market share surpasses 70% (Figure 9.39). There are some European firms, such as Merck, manufacturing photoresists, but this is solely for mature and legacy technology nodes.

A photomask contains the pattern of an integrated circuit. As transistors have become smaller and smaller, photomasks have become more complex to accurately transfer the pattern to silicon wafers. The process of creating photomasks has become correspondingly more advanced; even slight defects in a photomask can impact silicon device performance.

Some IDMs and foundries, such as GlobalFoundries and TSMC, are fully or partially involved in the production of semiconductor photomasks. They are represented by the captive category in Figure 9.40. The semiconductor photomask market is dominated by Japanese and US suppliers. A Scottish company had 1% of market shares in 2020 (Figure 9.40).

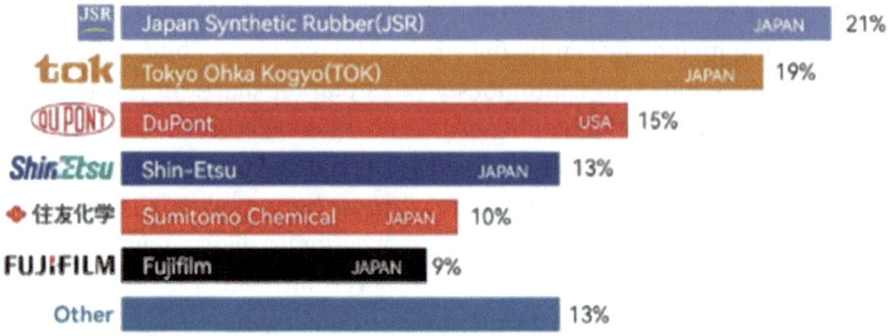

Figure 9.39: Market shares of the leading photoresist suppliers.

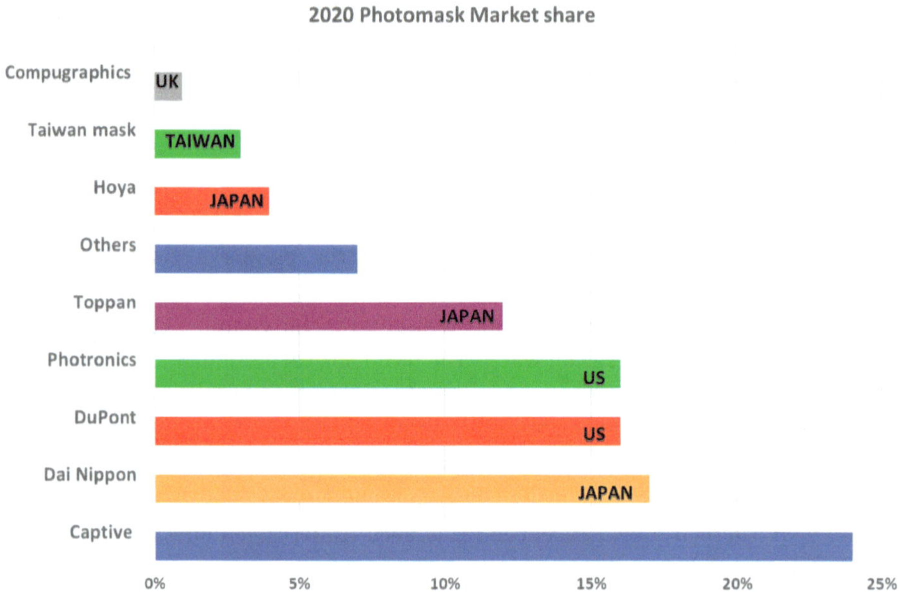

Figure 9.40: Market shares of the leading semiconductor photomask suppliers – data from [487].

9.2.9 Conventional and advanced packaging

Figure 9.41 depicts the market shares of OSAT companies in 2023. The conventional chip packaging is dominated by the Taiwanese and Chinese companies. An US company ranks at the second place in terms of market shares. Neither European nor Japanese company is present in the top 20 ranking of packaging company.

Despite (i) its substantial presence in the assembly equipment domain (Figures 9.8 and 9.42), (ii) its strengths in BEOL material suppliers (Figure 9.42), and (iii) the high number of small European OSATs (Figure 9.42), no European OSAT is present in the top ranking of the worldwide OSATs (Figure 9.41).

Since conventional packaging still requires a lot of manual process steps, the high European wages (Figure 9.32a) lead to higher packaging costs for the customers of European OSATs.

So, the European semiconductor industry strongly depends on Asian packaging suppliers. Due to the lower annual wages in Asia, the European semiconductor industry outsources the packaging of the chips it produces in Asia for lowering the overall cost of the electronic devices it sells.

European IDMs develop and manufacture various conventional packages, but only for their own products. When the packaging activities of European IDMs are included in the European conventional chip packaging activities, the European ATP capacity reaches 3% of the 2022 revenues, even if these companies generally outsource ATP activities outside Europe. This EU ATP capacity is not expected to increase between 2022 and 2032 (Figure 9.43).

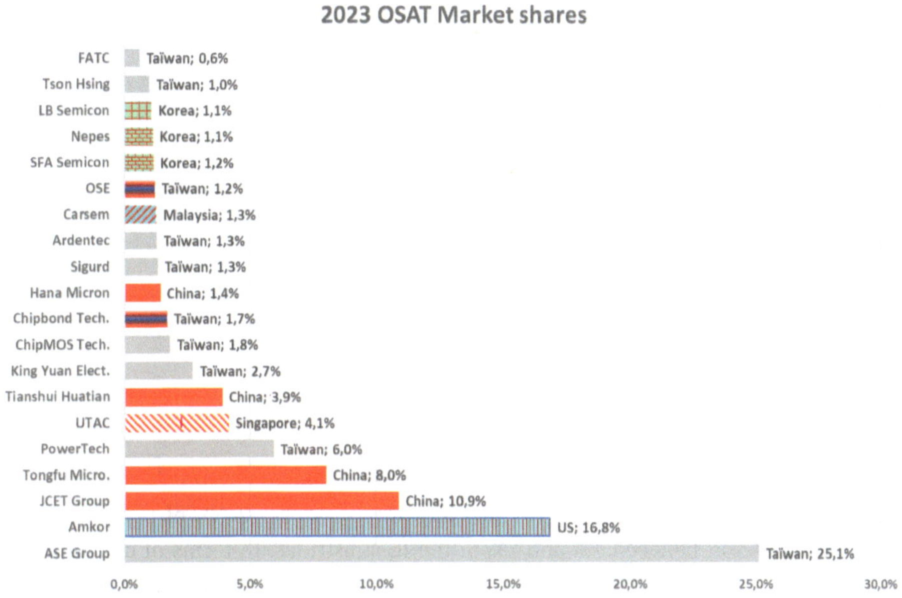

Figure 9.41: Market shares of OSATs in 2023 – data from [488, 489] (*The company's headquarters location is depicted with the same colors as in Figures 9.7 and 9.8*).

Figure 9.43 also shows that, in 2032, the conventional packaging will be even more dominated by China at the expenses of Taiwan in terms of capacity. It also highlights that, between 2022 and 2032, the conventional packaging capacity will spread to Asian countries where the annual wage is lower than in China (Figure 9.32a), such as Vietnam, Thailand, and Malaysia.

Figure 9.42: European conventional packaging ecosystem – courtesy of Yole.

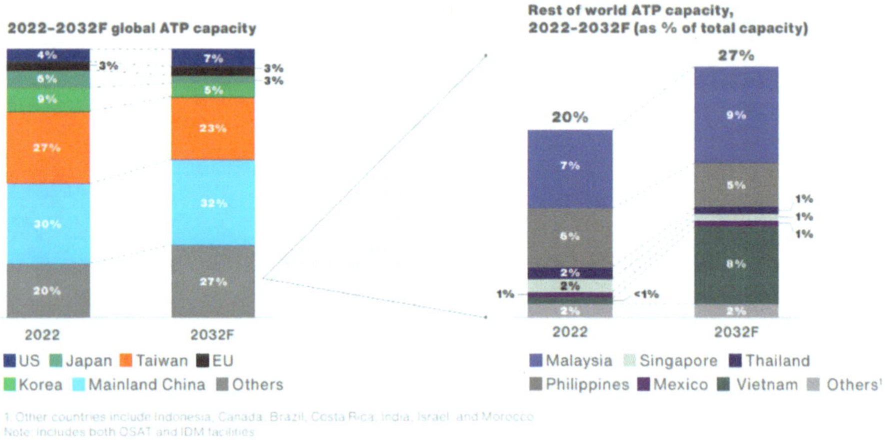

Figure 9.43: ATP capacity distribution by region [489].

The European electronic packaging industry only produces 2–3% of printed circuit boards (PCBs) and between 1% and 2% of the IC substrates. In terms of conventional packaging, the decline of PCB and electronic manufacturing services in Europe could induce serious issues to the continent's security and sovereignty.

The rising demand for compact, quicker, cheaper, increased performance, and more energy-efficient electronic devices drives the expansion of advanced (chiplets and 2.5D/3D integration) semiconductor packaging. Advanced packaging has become an essential development rather than a mere choice; it serves as the cornerstone for the expansion of semiconductors. Ranging from mobile phones to data hubs, and from AI processors to military technologies, advanced packaging is establishing itself as the driving factor in the revolution of the semiconductor sector.

Advanced packaging help reduce pressure on leading-edge nodes; it shifts a significant portion of demand to mature-node manufacturing. It is now becoming a key enabler for producing AI accelerators, autonomous driving, 5G connectivity, and next-generation consumer electronics that are further driving market growth (Figure 9.44).

Due to the huge demand for advanced packaging, traditional FEOL foundries alongside logic and memory IDMs have moved into the advanced packaging business as a natural continuation of Moore's law. As a result, the advanced packing activities of the top FEOL players lead to significant revenues (Figure 9.44).

Although advanced packaging and 3D integration applications are currently concentrated within HPC, AI data centers, and telecommunication infrastructures, European chipmakers have already used advanced packaging and 3D integration for manufacturing some devices, like CMOS imagers. Advanced packaging is also slowly propagating from HPC and AI down to automotive (Figure 9.44) and medical applications, which perfectly fits with the strengths of the European value chain.

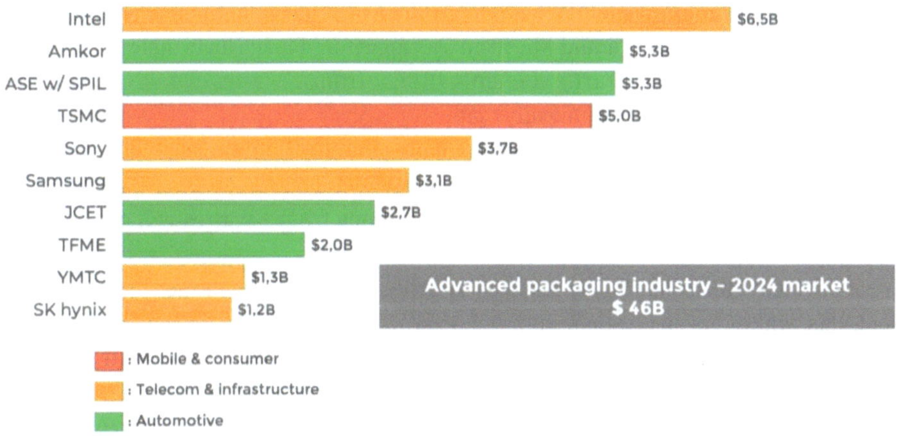

Figure 9.44: Top advanced packaging revenues in 2024 – data from [490].

Moreover, the European semiconductor industry has developed and patented numerous technologies allowing advanced packaging and 3D integration such as hybrid bonding and micro-transfer printing. Therefore, the current technological gap of the European semiconductor could be quickly filled.

9.2.10 Critical raw materials

The European semiconductor industry has only three ways to get access to CRMs for manufacturing chips: (i) recycling the e-wastes (WEEE), (ii) fostering the development of the mining and refining of those materials, and (iii) setting up partnerships with like-minded countries.

The recycling of e-wastes is witnessing significant growth due to the escalating volume of electronic waste generated globally. Indeed, WEEE generation reached a volume of 65 million tons in 2024 worldwide, compared to 57 million tons in 2021, i.e., an increase of 15% in 3 years. Europe is the biggest producer of e-waste with 17.6 kg per capita [425, 491]; therefore, WEEE recycling could represent an important source of CRMs for the European semiconductor industry. The most common e-wastes are PCBs, but they contain toxic and hazardous compounds like heavy metals, brominated flame retardants, and epoxy resins, which are harmful to humans and the environment. The development of new, more automated, and environment-friendly recycling technologies of its own e-wastes could thus represent a good opportunity for Europe.

The current e-waste recycling procedure is only economically viable for precious metals (Ag, Au, and Pd) and for metals that are not too difficult to extract from PCBs such as Cu, W, and Co (Figures 6.13 and 6.14). Research works are currently developing the extraction of other metal elements from electronic components at the surface of PCBs such as recycling Ta and Mn from Ta capacitors [427]. Nevertheless, the recycling of Ga, Ge, and rare earth elements (REEs) from e-wastes would remain extremely difficult and not economically viable.

Therefore, mining and refining will remain compulsory for ensuring the needs of the European semiconductor industry of those elements, as well as for guaranteeing the delivery of CRMs for which their recycling volume from e-wastes will not be sufficient to fulfill the semiconductor industry.

As depicted in Figure 6.9, European countries are absent from the countries mining CRMs, while they act in their refining or processing process. Figure 9.45a presents the worldwide market share of European countries for some critical materials, for which these countries exhibit mining market shares higher or close to 1%. It shows that, except for the silica sand, which can be used for producing high-purity silicon ingots, Europe is far being self-sufficient for the other CRMs.

In terms of critical material processing or refining, Europe is rather well positioned for some of those metals (Figure 9.45b). However, the critical elements that are subject to China export controls such as Ga, Ge, and REEs are not well processed in Europe. Therefore, the refining of those critical metals must be encouraged across Europe in a long-term strategy as proposed in the Critical Raw Materials Act, which has been adopted in March 2024 [491].

The Critical Raw Materials Act represents the most comprehensive attempt so far to reverse this trajectory. It sets ambitious 2030 benchmarks: 10% domestic extraction, 40% domestic processing, and 25% domestic recycling, while seeking to cap single-country dependencies at 65% across the value chain. At the time, this book is written these ambitious targets largely fell short of translating ambition into reality [492].

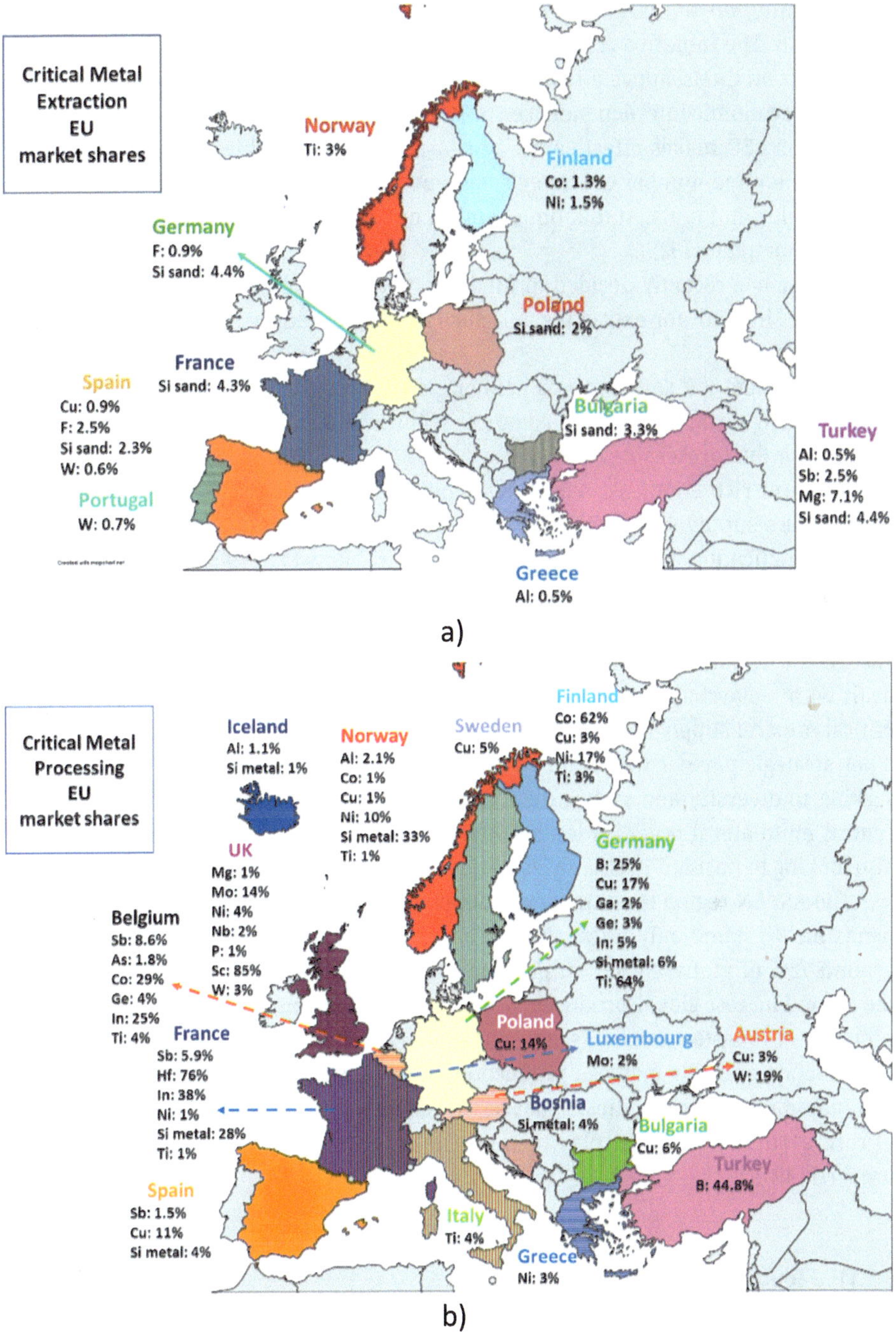

Figure 9.45: Critical material market shares of European countries (a) extraction and (b) refining – data from [421].

Consequently, on December 3, 2025, the EU Commission unveiled the ReSourceEU Action Plan. The initiative reflects the EU's attempt to reduce highly concentrated dependencies on CRMs, above all on China, after a year marked by the growing weaponization of economic interdependence.

ReSourceEU makes effective use of constrained financial leeway and can viably stabilize a limited number of strategic bottlenecks: permanent magnets, batteries, and defense-relevant inputs. At that time, semiconductors were not considered as strategic in this action plan [492].

Europe has recently decided to fund the modernization of a Greek bauxite mine and processing unit for extracting and purifying Ga, to reduce its dependency on Chinese Ga imports.

As complete self-sufficiency is neither geologically feasible nor achievable in the near and long term, hence, third countries will continue to be the EU's main suppliers of CRMs for the foreseeable future.

Resource-rich Southeast Asia holds significant potential as a partner in addressing Europe's supply chain vulnerabilities. Yet, it has so far remained largely absent from diversification efforts through the EU's Critical Raw Materials Act, as well as Germany's own raw materials strategy. This is a missed opportunity. Others have already recognized the region's potential: at the October Summit of the Association of Southeast Asian Nations (ASEAN), for example, US President Donald Trump inked new deals with Malaysia, Thailand, Cambodia, and Vietnam to strengthen cooperation on critical mineral supply and investment. Indeed, Southeast Asia is rapidly emerging as a key strategic partner for countries like South Korea, Australia, the USA, and others seeking to diversify and secure their critical minerals supplies. ASEAN, the region's central multilateral organization and the EU's long-standing institutional partner, is also looking to position itself as a strategic actor in this space [493].

The ASEAN region is an important supplier of several critical minerals the EU depends on. To name only a few, Indonesia and the Philippines together account for around 72% of global nickel output and 14% of global cobalt production. Burma is the second-largest global producer of REE after China. Vietnam is the second-largest producer of tungsten and holds approximately 17% of global REE reserves. In addition to extraction and raw material production, the region is home to strategic processing and manufacturing capacities. Malaysia, for example, is the world's largest processor of rare earths after China, playing a crucial role in global REE supply chains (Figure 6.11) [493].

9.2.11 Geopolitical shifts

The technology rivalry between the USA and China has drastically changed the semiconductor value chain. Taken together, US and Chinese policies amount to a quasi-death warrant for the traditional model of market-driven semiconductor globalization, even if elements of interdependence and cost-based trade are likely to endure. The strategies adopted in Washington and Beijing to effectively nationalize segments of the global semiconductor value chain send a clear signal: semiconductors must be treated as critical infrastructure, indispensable to strategic resilience and national security. More specifically, US policies demonstrate that reshoring can be rendered commercially profitable through the right mix of leverage and incentives [444].

Furthermore, the Chinese government continues to orient its policy toward industrial upgrading. At the heart of this strategy, detailed in the 15th Five-Year Plan (2026–2030) and the "AI+" initiative, innovation, and AI are emerging as the true drivers of the country's future development.

China views AI from two major perspectives. A domestic economic driver: AI is identified as the catalyst for "new, high-quality productive forces" aimed at boosting productivity gains and supporting growth in the face of deflationary pressures. The country benefits from an ideal environment for the widespread adoption of these technologies thanks to highly developed digital infrastructure and a high-performing education system.

A key geopolitical weapon: In its rivalry with the USA, AI is the pillar that must guarantee China's technological autonomy, national security, and global industrial leadership. On the international stage, China is already leading the race alongside the USA. It is present across almost the entire AI value chain and is currently developing computing models (including its own LLMs) that are among the most powerful in the world. This technological advantage is reinforced by overwhelming logistical dominance: China supplies 21% of global exports of AI-related goods and, decisively, controls the supply of critical materials needed for these technologies. Ultimately, driven by a proactive state policy, China possesses all the strategic, financial, material, and energy resources necessary to pursue accelerated AI development over the next 5 years and solidify its technological supremacy.

Hence, the global semiconductor industry is facing rapid technological changes, trade restrictions, and geopolitical shifts that Europe is not prepared for. The main driver of rising demand for chips is AI, an area in which Europe is comparatively weak. European firms are also being squeezed out of China's market, which used to be a primary source of growth for them. And the USA, which used to be a close partner, is threatening tariffs that would limit European sales to the US semiconductor market. Merely setting targets for domestic chip production will not solve these challenges.

US policies have also created tension and chaos in the global chip supply chain, leading to rising chip prices and supply shortages. For European semiconductor manufacturers, this means facing higher costs for raw materials and equipment procurement while exploring new markets; at the same time, they contend with price competition and market share disputes from Chinese and other regional manufacturers, as the US restrictions were not designed for European firms.

10 Call for a European long-term vision

10.1 Main results of the EU Chips Act at the end of 2025

The proposed European Chips Act, presented by the Commission in February 2022, aims to mobilize €43 billion in "policy-driven investment" for the European semiconductor sector by 2030. The Commission expects long-term private investment to exceed this, and aims an overall investment amount around €90 billion (~$100 billion) by coupling the European Union (EU) Chips Act and the microelectrics-related Important Projects on Common Economic Interest (IPCEI), which are only funded by the Member States (MS) (Figure 10.1). The Chips Act serves to enable immediate European coordination of responses against supply disruptions, strengthen, and scale-up production and innovation throughout the European semiconductor value chain, and enhance Europe's technological leadership, practical applications, and digital sovereignty in this field [494, 496].

The EU Chips Act is structured in three main pillars [470, 496]:

- The Chips for Europe Initiative (C4EI), targeting research to reinforce technological and innovation capabilities and addressing gaps in the ecosystem. A co-funding between EU and MS allowed to set up:
  - pilot lines, for developing advanced semiconductor and quantum technologies;
  - a virtual design platform for lowering the access barrier to chip design for start-ups and SMEs;
  - competence centers for facilitating the access to pilot lines and the design platform for start-ups and SMEs, mainly.

- Security of supply, targeting enterprise: to improve the security of supply by attracting investment in advanced production capabilities and related innovations thanks to:
- a €2 billion Chips Fund that would facilitate access to finance for start-ups to propel innovations and attract investors;
- support for first-of-a-kind facilities (FOAK) categorized as "Open EU Foundries", which design and produce semiconductor components mainly for other industrial players, and "Integrated Production Facilities," which design and produce components for the European market.
- Monitoring and crisis response, targeting coordination: a coordination mechanism between MS and the EU Commission will monitor the supply and value chains of semiconductors. The EU industrial Alliance of semiconductors, created in 2025, provide advices to this coordination team.

 | https://doi.org/10.1515/9783112247952-010

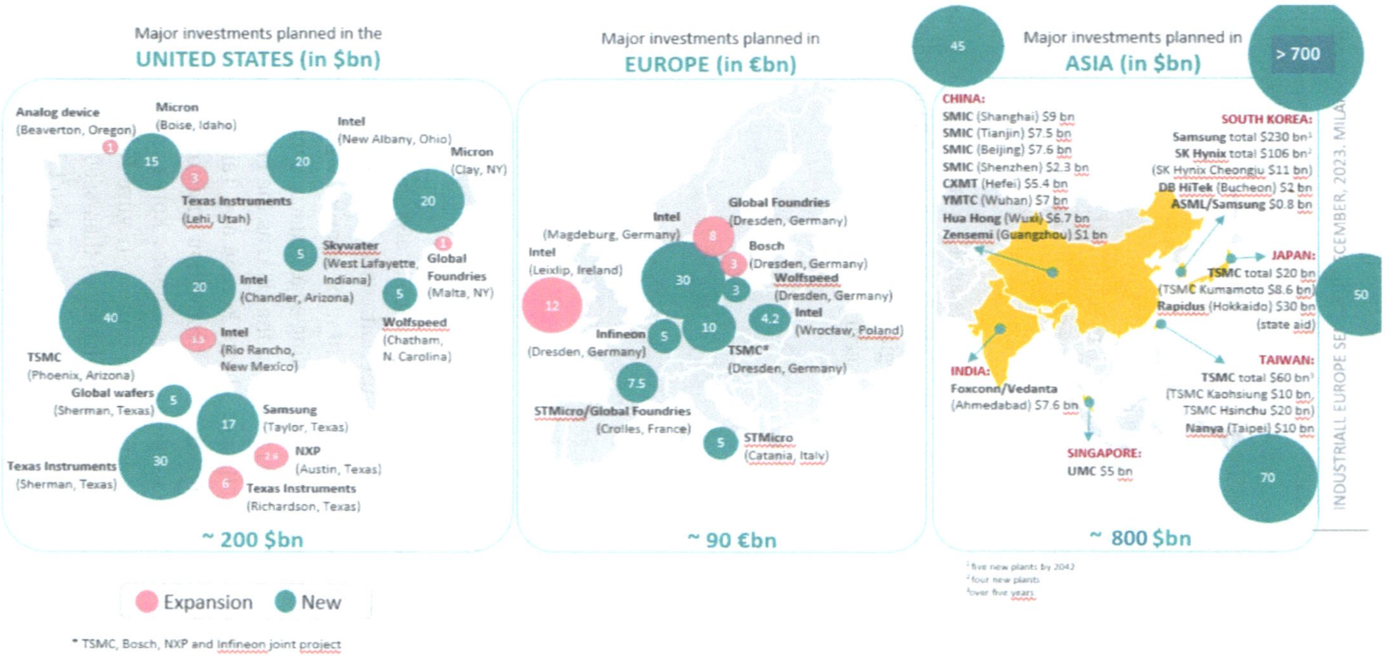

Figure 10.1: Major investments planned in the USA, Europe, and Asia, thanks to the Chips Acts – updated from [495].

In its report [470], the European Court of Auditors stated that (i) the C4EI is progressing well but suffering some delays, and (ii) the coordination and crisis-monitoring mechanisms, which were expected to be available in the short term are still in the very early stages. It also noticed that the uptake of FOAK investments is slow and unlikely to be sufficient for the overall digital target of 20% to be met by 2030.

Hence, the EU Chips Act ambitions targeting 20% of the worldwide semiconductor manufacturing capacity and the implementation of one fab producing chips at a ≤2 nm technology node on its soil will not be reached by 2030.

Considering the chip shipment, the US-based firms maintain the largest market with 50.4% of chip shipments in 2024. The Korean-based companies rank second with 21.1% of worldwide chip shipments, while the European chipmakers have shipped 9.2% of the worldwide shipments in 2024 (Figure 10.2).

The first reason for not reaching these initial targets arises from the fact that, as the semiconductor industry is global, the EU faces intense international competition. Countries around the world have their own strategies for attracting investment, increasing their market share and strengthening the security of their supply (Figure 8.1).

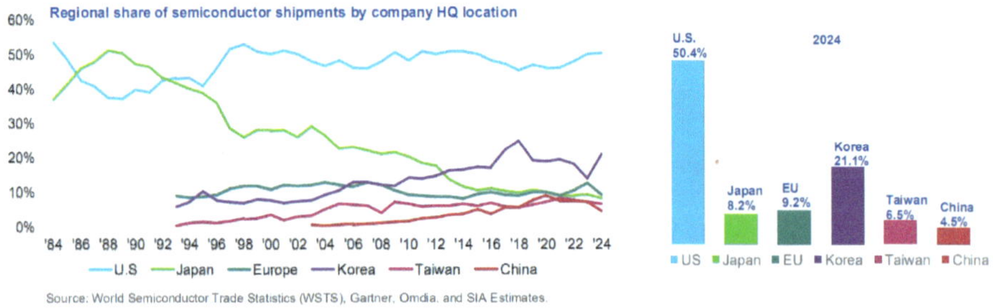

Figure 10.2: Regional share of semiconductor shipments by company headquarter location [497].

This fierce competition is highlighted in Figure 10.3, which shows the forecasted evolution of the capital expenditures (CapEx) of different countries between 2024 and 2032. Due to the huge demand for AI chips, Taiwanese companies will continue to massively invest in new production lines using leading-edge semiconductor technologies. As it has been already depicted in Figure 8.2b, the USA, as the great winner of the various Chips Acts, succeeded in attracting large Korean and Taiwanese firms, and some European chip makers, to locate some production lines on its territory (Figure 10.1). Thanks to internal private-public investments, Korea is expected to strongly increase its semiconductor manufacturing capacity (Figure 10.1) reaching a forecasted CapEX of $300 billion (Figure 10.3), to which more than $500 billion should be added, according to very recent announcements. Japan has also taken significant steps to support fab construction (Figure 10.1). The Chinese government is pursuing its whole-of-nation approach to promote its semiconductor industry, even if the planned CapEx between 2024 and 2032 is lower than in the previous decade.

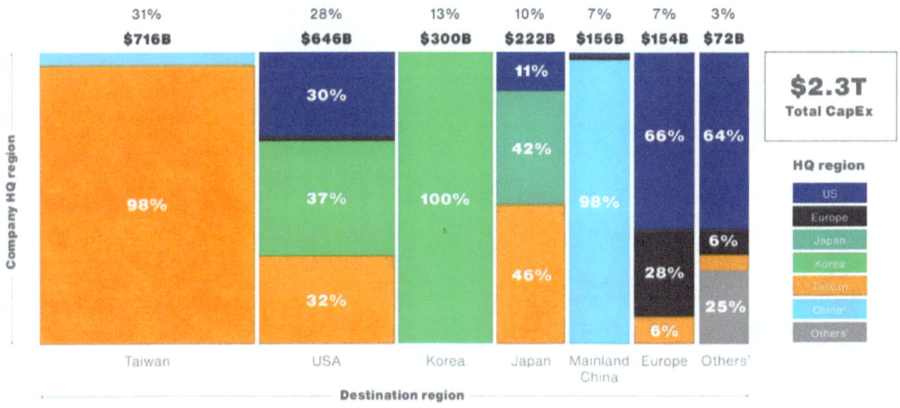

Figure 10.3: Future flows of CapEx from company HQ region to destination region, 2024–2032F [245].

Since the EU Chips Act was prepared in urgency, its ambitions were lower than the ones of these competitors (Figure 10.1). Europe was expected to increase its manufacturing capacity thanks to CapEx expenditures from various semiconductor companies around $150 billion (Figure 10.3).

But even with such a significant CapEx, in 2032, the European semiconductor capacity is forecasted to be at the same level by 2032 as in 2022, i.e., 8% (Figure 10.4) or even lower (6% in 2025), due to the fierce worldwide competition. This means that one of the main EU Chips Act objectives (20% of the worldwide semiconductor fab capacity in EU) would even not be reached by 2032.

Furthermore, the overall EU semiconductor manufacturing capacity has been affected by the cancellation of some important FOAK investments (Figure 9.24). Indeed, the biggest FOAK investment with the building of Intel's fabs in Germany and Poland (Figure 10.1) has been canceled. One of the main root causes for this cancellation is the low current European demand for leading-edge technologies (Figure 9.27).

The second reason for not reaching the initial EU Chips Act targets effectively relies on the big focus of the EU semiconductor industry for mainly answering to the chip demands for the domestic automotive, industry, and healthcare markets. Those EU markets do not currently need leading-edge technologies. That is the reason why TSMC is currently building a fab in Germany, using advanced technology nodes between 40 and 12 nm for the EU automotive industry, and not fab-producing high-performance chips.

So, only Intel-Ireland currently produces leading-edge technologies but mainly for its own products and not so much for the European electronic components community.

The EU manufacturing capacity of Discrete, Analog, Optoelectronics and sensors (DAO) would increase between 2022 and 2032, reaching 19% of the worldwide capacity by 2032 (Figure 10.4). Those components perfectly match the needs of the automotive, industry and health care sectors, the strengths of the European industry (Figure 9.10).

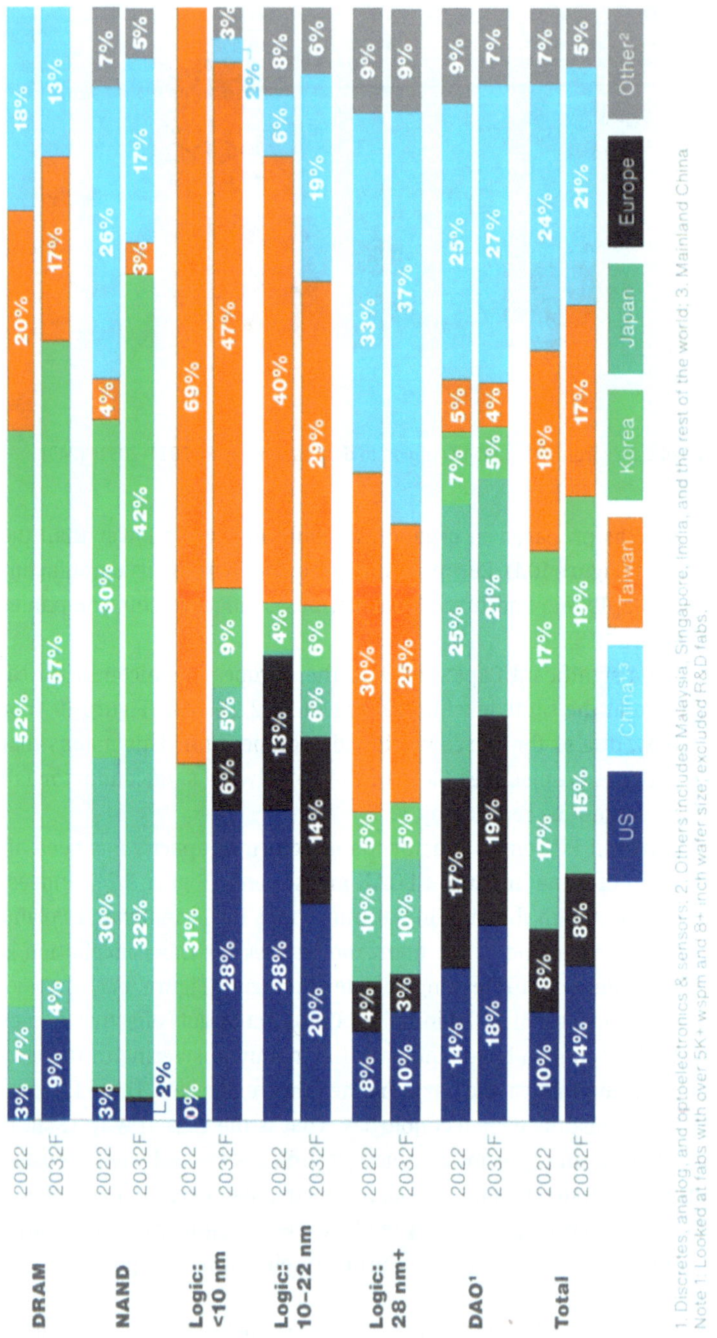

Figure 10.4: Global wafer fabrication capacity by technology category by region, 2022 (top) and 2032 forecast (bottom) (%) [498].

Accordingly, the EU expected ambitions, the forecasted CapEX, and the overall manufacturing, respectively, shown in Figures 10.1, 10.2, and 10.3, would be even significantly reduced through the cancellations of some FOAK projects (Figure 9.26).

The third reason for explaining that the translation of the EU Chips Act into industrial output has limited success so far is the elaboration of the EU chips Act. It was established in urgency as a response to the chip shortage induced by the COVID-19 pandemic. Hence, the EU Chips Act lacks a coherent long-term strategy, leaving MS to define and develop their own national policies without any cooperation between them. This fragmentation is further reflected in the Chips Act's financing model, where EU funds depend on national and private contributions, often fueling competition among MS over the scale of state support rather than fostering collective capacity [470].

The fourth reason is that the EU Chips Act related funding may seem like a lot of money on paper, but it pales in comparison with the sums being recently invested by the big semiconductor companies such as TSMC or Samsung (Figure 9.24b).

10.2 Toward an EU Chips Act 2.0

The first EU Chips Act provided new impetus, although without an impact assessment and clearly defined targets. But it is likely to be insufficient to stimulate the level of investment needed, with success also depending on global competition and other crucial factors [470].

Due to the worldwide fast-moving semiconductor industry, Europe must thus quickly react by defining a long-term (≥10 years) cohesive semiconductor policy for guaranteeing its digital and economic sovereignty.

The EU Chips Act was preceded by a 2013 strategy that aimed to strengthen the micro- and nano-electronic sectors [499]. This strategy initially aimed to reverse the decline in the EU's share of the global semiconductor supply, and reach a level of production closer to the EU's share of world GDP within a decade. This would have required doubling the economic value of semiconductor-component production in Europe from 2020 to 2025. This initiative was named "Airbus of Chips" by the European Commission. The Airbus company was deliberately assembled from national defense and aerospace companies across Europe to provide strategic defensive independence at a continental level. Airbus now produces about half of the world's jet airliners.

Unfortunately, some European countries outright rejected the Airbus for chips proposal that could have positioned the continent more competitively today. However, this proposal pinpoints that indigenous semiconductor manufacturing is fundamental for the European prosperity and competitiveness, which has been later highlighted by the chip shortage occurring during the COVID-19 pandemic.

A similar industrial policy must be rapidly proposed and coordinated by the European Commission, whatever its future name: "Airbus of Chips 2.0" or EU Chips Act

2.0. This long-term (≥10 years) strategy should be setup in collaboration with all MS and relevant stakeholders of the European electronic component and systems community. It could take, as an example, the development of the EUV lithography tools by ASML. ASML used a customer-oriented approach and spent more than 15 years to build an ecosystem that enables the fabrication of this highly complex equipment, gathering optics, electronics, robotics, etc.

10.2.1 EU Chips Act 2.0: preparation and governance

The EU Chips Act 2.0 could consist in a "precompetitive cooperation" or "precoopetition" through strategic alliances between industrial stakeholders involving in similar or complementary sectors, including European and MS public authorities, for designing agenda, milestones, and objectives of this very ambitious program. Those strategic alliances would be favorably driven by top European industry leaders. They will aim to enhance the European supply chain and the European demand for advanced chips by clearly understanding the technical requests of future electronic systems in strategic and critical sectors, as well as the current strengths and weaknesses of the European semiconductor value chain.

So, it should involve some major end-user companies for providing their major inputs about their long-term roadmaps. It will finally target to manufacturing chips using leading-edge and more advanced technology nodes, and/or designing new chip architectures, and implemented new technologies. Furthermore, incentivizing end-user industries to jointly develop products with European chip designers and makers must be a key priority of this strategy.

It should also focus on making the global semiconductor supply chains more dependent on European technology. It will be quite worthwhile to learn from the Japanese economic security strategy based on the notion of "maintaining, boosting and obtaining strategic indispensability" in the semiconductor supply chain.

It can also benefit from the Taiwan's example, even if some aspects of Taiwan's experience are not readily transferable to the EU, and EU should not attempt to replicate or mimic Taiwan. Taiwan's economic success cannot be reduced to chip manufacturing alone. Rather, the Taiwanese semiconductor sector has been embedded within a much broader electronics and industrial ecosystem. In addition to leading-edge fabrication, Taiwanese firms have also emerged as global leaders in semiconductor packaging and testing, while maintaining a meaningful presence in chip design and upstream materials such as wafer substrates and other critical inputs. This becomes particularly consequential in the context of AI. Modern AI accelerators require not only advanced logic chips but also sophisticated packaging to integrate HBM. Even when the memory itself is supplied by Korean or US firms, much of the final integration, advanced packaging, and testing takes place in Taiwan [500].

Three very important lessons Europe must learn from the Taiwanese experience are (i) industrial policy must focus on sustaining a long-run innovation pipeline, ensuring continuity from research through experimental development, and building institutions that systematically reduce the frictions that prevent scaling, and (ii) that competitiveness is not bought through subsidies alone – it is built through ecosystems that consistently convert technological potential into globally scalable production. [500], and (iii) in industries like semiconductors, investments that have provided the most enduring benefits are those which have been backed by long-term government support, often over decades.

The European strategic sectors concern the economic domains in which Europe is already strong, but needs to be reinforced: the automotive, industry, communications, and healthcare domains (Figure 9.10). Due to the worldwide competition in those sectors, European actors will need some serious structural and technological changes to remain competitive. The transition to more energy-efficient efficient and more secure systems, and the implementation of AI hardware and software in these electronic systems are ambitious technical challenges, which will be very positive for the European semiconductor industry.

The strategic sectors also concern critical sectors in which Europe is relatively weak, such as AI and HPC, as well as parts of the semiconductor supply chain.

As AI is the backbone of the modern economy, and owing to its unprecedented and accelerated growth in AI and edge-AI, Europe cannot afford to not be an important actor in this domain. Such an involvement would necessitate massive investment in the semiconductor industry for designing and producing AI chips including GPU- and HBM-like devices using leading edge technologies coupled with advanced packaging, but also highly efficient power electronic components, and photonic integrated circuits (PICs).

The accessibility and risk associated with critical materials could derail plans to reshore and develop semiconductor manufacturing. Minerals needed for chip fabrication are increasingly difficult to obtain, with the location of mines and purification processes mainly localized in China. Hence, Europe is strongly dependent on China for the supply of critical materials. This weakness must be strongly improved either by strongly enhancing the recycling efficiency of electronic wastes, and even by reopening mines or discovering new critical ore reserves in its subsurface. Reopening mines or opening new mines will take between 10 and 15 years, and could induce social and ethical issues, but this long-term strategy must address those critical topics.

Due to the utmost importance of AI for all the application sectors, and due to the expected growth of AI-related devices (AI chips, memories, power semiconductors, PICs, etc.) in the semiconductor market (Figure 10.5), this long-term strategy must absolutely include the design and manufacturing of AI-chips in Europe. Building a strong European position in chip design is another leverage for semiconductor demand will be particularly important for creating a leading-edge technology ecosystem in Europe.

Indeed, IDMs and foundries need a “first and best” customer with large demand and deep pockets to fund their continued advance. Apple has been that customer for the last decades at TSMC. Such a juggernaut partnership has propelled both companies to new heights, put competitors in the dust, and fueled the chipmaking industry. Hence, Europe urgently needs a company or a group of companies technically and economically able to design advanced chips that will further require high-volume manufacturing using leading-edge technology nodes. This company or this group of companies would probably request some European and/or national incentives for designing advanced chips and covering parts of their expenses.

Close technical collaboration between the European company(ies) and the European foundry(ies) and/or IDMs will be of foremost importance for reaching their economical viability in order to become competitive, and for guaranteeing the European sovereignty in AI chips.

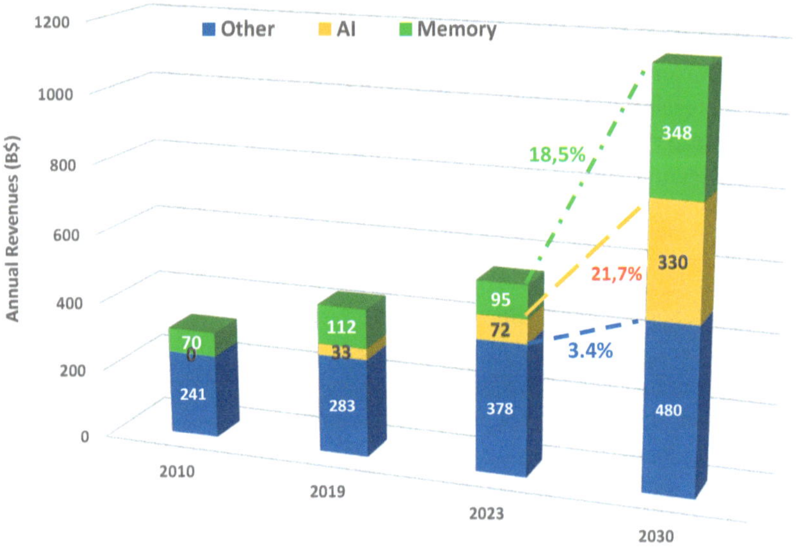

Figure 10.5: Evolution of the semiconductor by technology typology. $CAGR_{2023-2030}$ = 3.4% (other), 21.7% (AI), and 18.5% (memory) – data from [501].

The European semiconductor industry currently mainly focuses on the automotive and industry sectors, it only addresses less than 30% of the overall semiconductor market (Figure 10.6). Even if the CAGR of those sectors from 2024 to 2020 is good and expected as 11.6% and 10.7% for the automotive and industrial sectors [388], respectively, it is quite lower than the CAGR forecasted for AI chips: 21.7% from 2023 to 2030 (Figure 10.5), and 27.88% between 2026 and 2035 [502].

Digital health is an application segment in which the European semiconductor competencies, like those in sensors, imagers, analog technology, and low-power and RF devices, are strong. This sector will drastically grow in the coming years with population ageing, and could represent a fast-growing sector for the European supply chain.

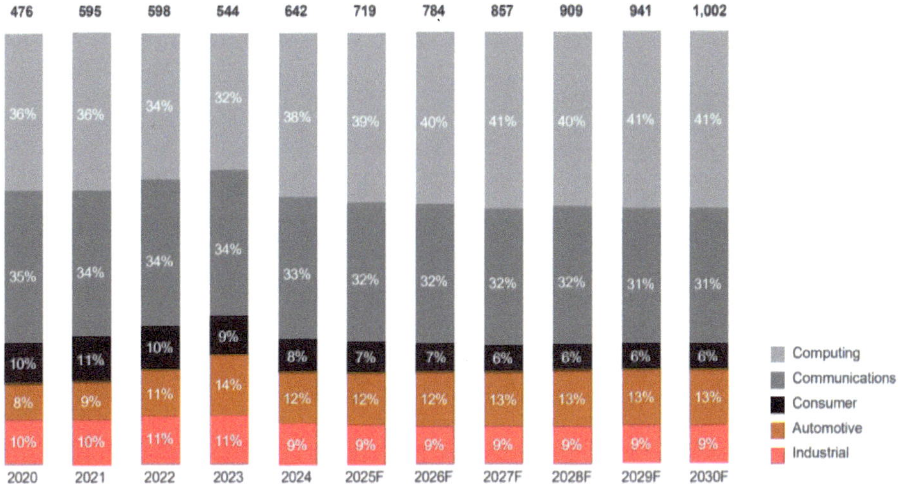

Figure 10.6: Global semiconductor market by application share, 2020–2030 ($ billion) [503].

Due to the higher costs for running a leading-edge fab in Europe (Figure 5.5), the European semiconductor industry currently faces serious constraints to participate in high-volume segments of the semiconductor market such as smartphones, personal computers, and consumer electronics.

The European sectors, such as Defense, Energy and Aeronautics are so critical that their electronic systems require a large number of chips "made by Europe" to guarantee their security and sovereignty. Defense applications led to the birth of the worldwide semiconductor industry. Even if the defense needs have decreased with time, they will unfortunately tend to increase a lot, especially in Europe. Therefore, the long-term European semiconductor strategy must take into account the future needs of these critical sectors. European and/or national procurements look like relevant tools for fulfilling the future needs of these critical sectors.

10.2.2 Potential EU cloud-AI chip prospects

As a backbone and key enabler of AI, data centers, autonomous vehicles, and smartphones, the global semiconductor market is projected to almost double within 5 years from ~$790 billion in 2025 to more than ~$1 trillion in 2026, $1.3–1.6 trillion in 2030 and $1.5–2 trillion in 2036. The AI chips (including GPUs, FPGAs, ASICs, AI accelerators, and microcomponents such as CPUs and MCUs in Figure 10.5) deployed to support AI workloads, including edge and data centers, should represent 34%, 68%, and 78% of the semiconductor market in 2025, 2030, and 2036, respectively (Figure 10.7), with the main market shares relying on edge-AI chips within 10 years [504].

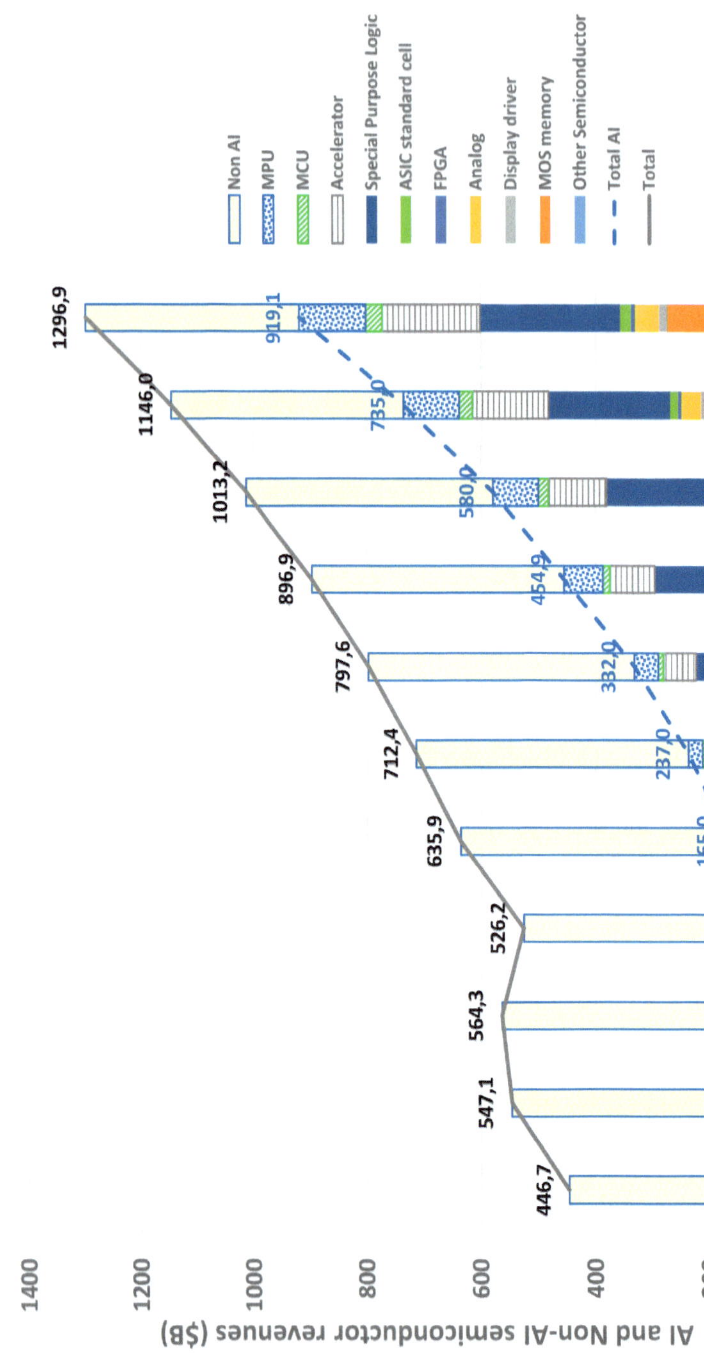

Figure 10.7: Forecasted evolution of the semiconductor market - data from IBS and [505].

Due to the huge CapEx spent by the big hyperscaler giants, some doubts recently emerged on the return on investment of AI-related data centers due to the amazing spending, which is forecasted in the coming years. However, huge investments in designing and manufacturing AI-related chips "made in/by EU" look compulsory.

Europe lags behind the USA and China in AI computing power. Nowadays, Europe welcomes 5% of global AI data-center-installed capacity, compared with 74% in the United States while European and US GDPs are roughly comparable and 14% in China [506]. And, in most cases, these European AI data centers are owned by US companies (Figure 10.8).

AI is one of the most important technologies in human history, and it has influenced human life significantly due to the progress in AI in the last decade. It is partially built on European science and scientists. Europe's AI scientists are at the heart of practically all US technology companies and research groups that are advancing this scientific frontier and developing today's remarkable technologies. However, Europe's technological capacity and expertise have declined over many years. The AI systems Europe relies on today are developed elsewhere. Europe has become a technological vassal of a few major US technology companies and now faces a critical challenge. This is a challenge that must be addressed to protect both the EU economy and strategic sovereignty.

Now, data centers and/or cloud infrastructures have become a holistic enabler of the broader economy. European cloud service providers have experienced a steady decline in market share, dropping from 26% in 2017 to just 15% in 2022 and down to 10% in 2025 – a trend expected to reach a point of no return within 3 to 5 years [507, 508]. Such a weak presence of European cloud providers represents a serious threat for data sovereignty and privacy, the new oil of the economy.

AI adoption could help economies accelerate labor productivity, driving growth, and countering potentially negative economic, political, and societal consequences. But it also has the potential to worsen Europe's competitive position if the continent misses the opportunity to take a more central role in AI's development.

Europe's window of opportunity to accelerate its embrace of AI is very narrow but clear. The biggest impediments to AI adoption across the continent are concerns over trust, security, and national dependency. Addressing these will likely accelerate adoption and, as a result, help reinvigorate the world's third-largest economic region while building economic resilience. Sovereign AI capabilities potentially provide such a path if Europe can develop and control critical AI capabilities, enabling greater flexibility in the form of national, economic, operational, and technical independence [509].

Therefore, Europe and some MS decided to speed-up massive investment (€20 billion) in sovereign data infrastructures throughout the building of AI gigafactories [510]. In 2025, the Commission proposed the Cloud and AI Development Act, with the aim to at least triple the EU's data center capacity within the next 5 to 7 years and fully meet the needs of EU businesses and public administrations by 2035. The Act will streamline the deployment of data centers by identifying suitable sites and simplifying permitting processes for projects meeting sustainability and innovation criteria. Simultaneously, it will address the growing energy demands by promoting energy efficiency, the adoption of innovative cooling and power management technologies, and the integration of data centers within the broader energy system [338].

The evolution of CapEx for building data centers, mainly for training AI models and the further inference process is impressive (Figure 10.8). It would reach more than $300 billion in 2025 (Figure 10.8) and would range from $400 to $450 billion in 2026. Moreover, the global data center CapEx will continue to grow for fulfilling the demand for performing better AI models (Figure 10.9). It would reach $1 trillion by 2030 with a major CapEx devoted to IT equipment including AI chips. Accordingly, the demand for AI chips produced using leading-edge technology nodes would increase at the same CAGR.

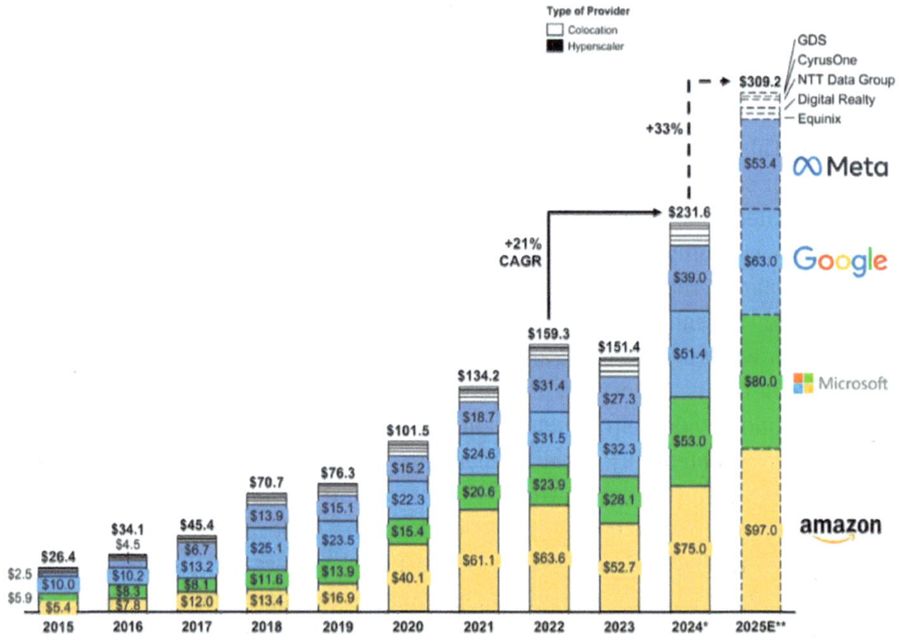

Figure 10.8: Global data center industry-leading provider CapEx (in $ billion) [511].

AI data centers consist of powerful data processing hardware, AI chips like CPUs, GPUs, NPUs, TPUs, MCUs, and memories such as HBMs, and high-speed networking equipment for allowing AI models to get access to massive datasets and to run across thousands of servers working together. They also include robust storage systems to handle vast datasets, advanced cooling systems for thermal management, software components like AI frameworks and security infrastructure ensuring data safety.

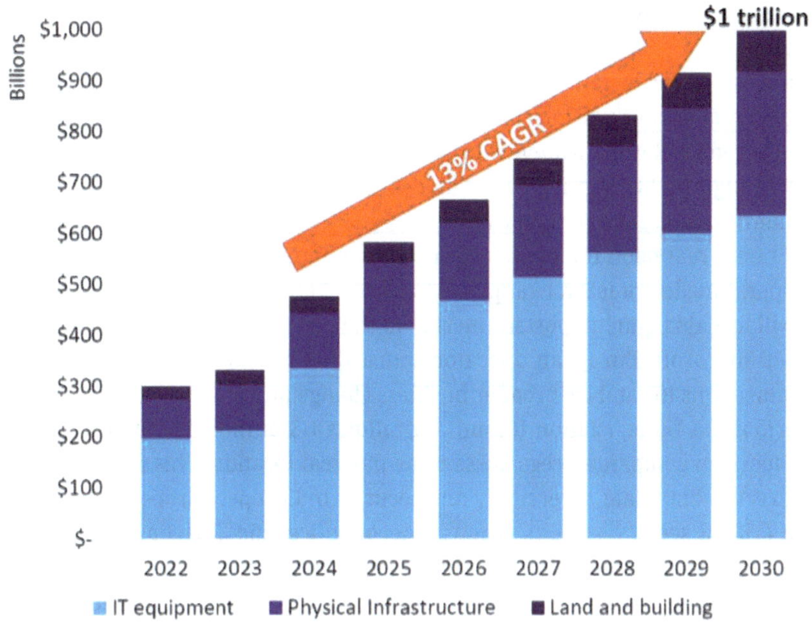

Figure 10.9: Evolution of the global data center CapEx – courtesy of Omdia.

They also rely on energy-efficient power systems and optical networks to optimize performance and sustainability.

Among the AI data center capital expenditure for 2026, $400–$450 billion globally, over half of that spending is for the chips inside devices ($250–$300 billion) and the rest is for everything else (land, construction, power, permitting, and more). It is further predicted that AI data center CapEx will rise to $1 trillion in 2028, with AI chips being over $400 billion in that year.

The European semiconductor industry is currently only present in few of these domains. Therefore, when Europe and its MS wish to massively invest in AI giga-factories, the benefits in terms of semiconductor devices and their corresponding revenues will mainly go to the US, Taiwanese, and South Korean chipmakers, and a very few to the EU semiconductor industry.

Despite the expected future deployment of AI in numerous industrial and defense sectors (Chapter 2), investing in the manufacturing of leading-edge chips for Cloud-AI applications in Europe is still under discussion since the domestic demand for these applications remains currently markedly low.

Such a low demand essentially arises from the lack of European fabless actors or big European OEMs, which are involved in designing AI or advanced chips for driving the demand for leading-edge and very advanced technologies. As an example, Chinese EV suppliers design their own chips for their vehicles. Why European car makers could not design their own AI and AI-related chips using less than leading-edge technology nodes?

Nevertheless, if, for instance, Europe needs to enhance its data sovereignty and reduce its dependency to US data centers, EU will need to deeply invest in data infrastructures. Therefore, Europe will need to favor the growth of relevant design-houses in order to "design by EU" cutting-edge Cloud-AI chips for data-communications, 5G and 6G communications, and AI applications.

AI could be a real boon for the European telecommunication sector. However, if European operators do not learn to apply AI to internal efficiency and speed of execution, they will lose this quite important market to the benefits of hyperscalers such as Amazon and Microsoft. European operators must thus consider themselves as the first AI organizations to catalyze broader business change and enhance their markets.

Furthermore, a US legislation (Cloud Act) allows US authorities to access data stored by American companies, regardless of its physical location. This Act raises significant concerns about data sovereignty and security in Europe. European industries involved in critical sectors, such as Defense, Aeronautics, and Telecommunications, will never use AI data centers owned by US companies for security and privacy reasons. Therefore, large European stakeholders involved in these critical sectors can also play an important role by defining a long-term strategy for developing some "design and/or made by Europe" devices for AI data centers and data infrastructures, such as next-generation accelerators, for instance. This initiative should be actively supported by Europe and the MS, even if defense activities are currently governed by each European country and not by the European commission.

Current and potential future European semiconductor manufacturers should also develop or gain access to "design and made by Europe" AI Chips that need to support activities for European AI data centers (Cloud-AI chips). Since the AI chip market will exhibit the most significant growth of the semiconductor sector in the forthcoming years and becomes a strategic geopolitical market (Figure 10.5), the European semiconductor industry must be deeply involved in this high-growing market.

Indeed, due to the unexpected geopolitical shifts between China and the USA, as well as the current and future conflicts across the world, it is extremely difficult to claim that the market availability of AI chips (CPUs, GPUs, memories, etc.), produced using leading-edge technologies, will remain as it is today. Hence, for guaranteeing its own technological sovereignty, Europe must have access to "design and made by EU" AI chips manufactured on its soil.

The growth of some important European fabless companies and the involvement of OEMs for designing chips and the mandatory European sovereignty in data centers, communications and defense would increase the demand for "made by EU" Cloud-AI chips. However, since manufacturing of leading-edge chips requires massive investments and huge operational costs (Figure 5.5), the European Commission and the MS must commit to a long-term strategy (≥10 years) for such a new initiative, and provide some substantial incentives to start the design and manufacturing of "made by EU" AI chips.

European designers could design AI chips from the ground up for the demands of AI using new computing architectures and/or paradigms, which could achieve orders-of-magnitude improvements in energy efficiency, performance, cost, and size.

Hence, Europe could gain strategic autonomy by creating an alternative supply chain: design, manufacture, and deploy AI systems tailored to sovereignty applications, small footprints, high security constraints, and low-power budgets. Areas such as government, defense, corporate, and financial systems are all waiting for such solutions [512].

Europe could also copy the Japanese "Rapidus" initiative, for instance. But this will require a tremendous number of subsidies that would be comparable to the USA, Japanese, Korean, and Chinese ones. An alternative to this strategy would consist in setting-up friendship agreements with trustable Asian Cloud-AI chip suppliers.

During the duration of this strategy, one can expect that (i) more energy-frugal AI accelerators using new computing architectures will replace GPUs, (ii) that the compulsory needs to embed AI in the next data communication networks will allow the European communication suppliers to become again world supplier leaders, and (iii) the high-volume manufacturing of "made by EU" non-volatile memories will start to replace NAND flash and DRAM memories.

Without massively investing in the manufacturing of AI chips, the EU economy would be in a very difficult position in 2035, depending on the USA, Taiwanese, and Korean AI chips and their leading-edge technologies on the one hand, and will become more and more dependent on Chinese chips for more mature technologies like the ones required for automotive and industrial applications, on the other hand.

Even if the European semiconductor industry would not be currently able to massively invest for manufacturing semiconductor devices using leading-edge technologies (≤7 nm) that are mandatory in AI data centers, it can diversify its activities by benefiting the market opportunity offered by AI in many growing market segments (Figure 10.6). The European semiconductor industry can lead the development and production of innovative power systems since the needs of AI data centers are similar to those of electric vehicles (EVs). It is also well positioned for providing network systems, which uses silicon photonics-based optical interconnects.

In a near-future, the European Commission could also set up regulations for requiring that new AI giga-factories acquired under European and MS program contain a very significant number of components (AI chips – including logic devices, memories and advanced packaging, power devices, Si photonics, etc.) manufactured by the European semiconductor industry.

Whatever the Europe commitment about the manufacturing of Cloud-AI chips using leading-edge technology nodes on its territory will finally be, the European semiconductor industry should pursue the development and manufacturing of power electronics, optical interconnects, advanced sensors, PICs, 3D integration and advanced packaging for enhancing the energy efficiency of the future data centers and their data infrastructures.

10.2.3 Edge-AI chips: an EU opportunity

As it has been aforementioned, data processing using AI models is performed in two main steps: training and inference. Training models is currently essentially performed in AI-dedicated data centers due to its drastic needs for large computing abilities, while data inference using trained models can be achieved both in AI data centers and at the edge (edge-AI).

Edge-AI can be defined by a cloud-edge-IoT continuum, in which devices of varying capability – from ultra-lightweight sensors to powerful edge gateways – are used to run increasingly sophisticated AI models. Due to its expertise in designing and manufacturing embedded electronic systems, the European semiconductor industry is very well positioned for developing energy-efficient edge-AI devices. The European semiconductor industry needs to reinforce its current strengths by pursuing the development of advanced technology nodes, and new and disruptive technologies. They could include in-memory (Figure 4.11) and neuromorphic computing (Figure 4.24) for AI accelerators using nonvolatile memories, energy-efficient logic devices like FPGA, ASIC, and MCU, which could also include NPUs, and cybersecurity blocks to protect data privacy and confidentiality.

Future growth of the semiconductor market is anticipated to be propelled by AI functions at the edge. The existing edge-AI sector is valued at roughly $24 billion, with a significant portion allocated to edge devices and edge gateways that primarily consists of nanoelectronic chips (Figure 10.10). By the year 2035, this market is projected to exceed $350 billion, growing at a compound annual growth rate (CAGR) of over 27%.

As a result, the market for edge-AI chips is expected to expand at a similar pace.

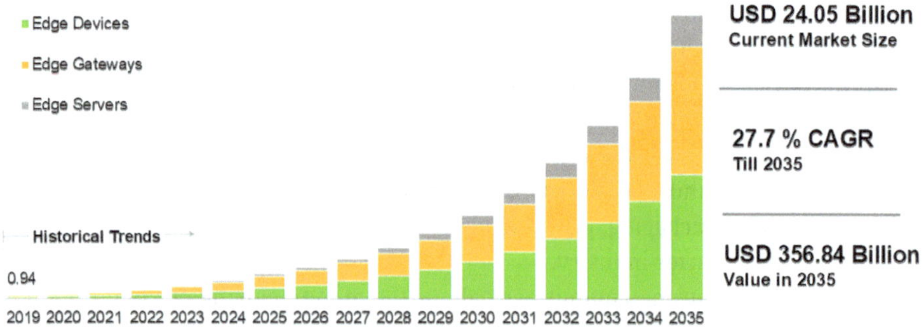

Figure 10.10: Evolution of the edge-AI market [513].

Moreover, the main applications of edge-AI perfectly match the strengths of the EU industry, i.e., automotive, industry, IoT, and healthcare (Figure 10.11).

Therefore, in the long-term, the European edge-AI chip suppliers would not only serve the European markets but the worldwide market for such devices, targeting an ambitious global leadership in this market.

Edge-AI is also key for the European Defense. The vast majority of foundational AI models in the EU are American or Chinese. This leaves the EU vulnerable as EU defense companies may have to wait for updates from abroad that may not be optimized for their specific needs. The statement is similar for AI chips; AI chips are schematically designed in the USA, produced in Taiwan and packaged in China. Some AI chips could be less efficient than those designed in the USA or could contain any spy chip embedded in the package [463].

Edge-AI chips do not usually require leading-edge technology nodes (≤7 nm) but 10–22 nm nodes typically. Therefore, the European semiconductor industry must strongly invest now or risk falling behind permanently in this fast-moving field. This investment would be favorably achieved thanks to close collaborations within the full European semiconductor ecosystem by defining a shared technology roadmap, common goals, and milestones before differentiating through competition in a further step.

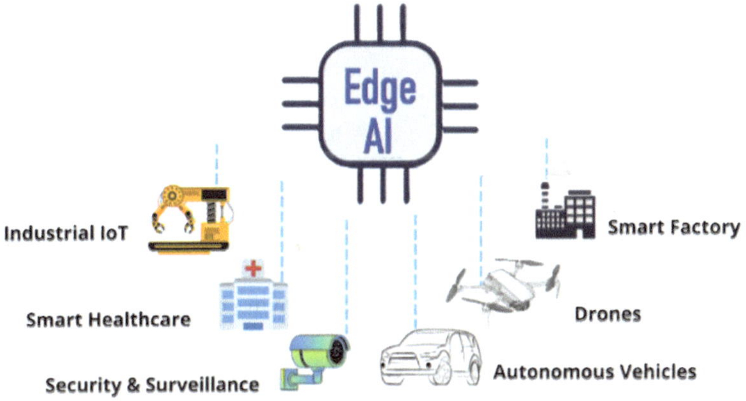

Figure 10.11: Main applications of edge-AI [514].

Europe, the MS and the European semiconductor industry must urgently define a long-term strategy for producing "design and made by EU" edge-AI chips that cope with the current strengths of the EU industry. This initiative will generate the high-volume production of edge-AI chips. If they do not commit on such a strategy, the manufacturing of chips in Europe – estimated between 6% to 8% of the worldwide manufacturing capacity and will not be drastically change with the current EU chips Act (Figure 10.4) – will progressively decline to reach less than 5% of the worldwide manufacturing capacity in 2040 (Figure 10.11). The future huge demand for edge-AI chips is an excellent opportunity for the current European semiconductor industry to recover some market shares in the future. Europe cannot afford to miss this edge-AI chip design and manufacturing opportunity (Figure 10.12).

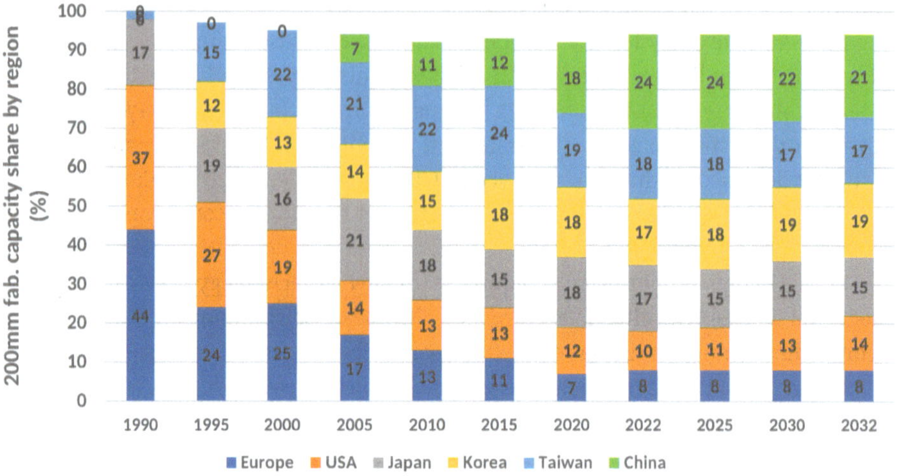

Figure 10.12: Evolution of the regional wafer manufacturing capacity – data from [489].

10.2.4 Chip design

In addition to its low number of fabless companies, the European Original Equipment Manufacturers (OEM), companies that produce parts and equipment that may be marketed by another company, do not often design the chips they need. US and Chinese OEMs are responsible for more than 60% of semiconductor demand, while the European OEMs are not so much involved in designing their own chips, with only 9% of the OEM semiconductor demand (Figure 10.13).

Europe must increase its sovereignty in chip design as a basis for long-term value growth in Europe. A strong position in design is a precondition for a thriving semiconductor ecosystem including manufacturing. Design capabilities must be strengthened across the ecosystem, from start- and scale-ups to IDMs and critical digital systems providers.

Europe lags behind China and the USA for funding chip design start-ups (Figure 9.22). EU should invest in EU's dwindling chip design ecosystem, focusing on improving conditions for start-ups, small and medium-sized enterprises (SMEs) and spin-offs from research institutions. This should include lowering entry barriers to chip design, investing in chip design infrastructure, and improving access to funding, private and public equity.

China already has a substantially stronger chip design ecosystem than EU in several areas and can scale faster due to heavy investment in chip design by China's hy-

Figure 10.13: Global semiconductor demand by region of OEM headquarters in 2020 (%)[1] [465].

perscalers, IoT, and mobile technology companies. Under these conditions, Europe will increasingly depend on Chinese designed chips.

With newly established European initiatives, such as the virtual design platform in the Chips Act to lower the entry barrier for start-ups and SMEs, and the European Innovation Council's (EIC) "Accelerator Challenge," which focuses on chip design [464], EU has understood the utmost importance of chip design. Moreover, promoting Europe's chip design ecosystem could be a cost-efficient way to climb the semiconductor value chain, diversify EU economy and earn a strong position at the technological frontier.

However, drastic and rapid efforts are still needed for decreasing the EU dependency in chip design to the USA, Taiwan, and China. These efforts must focus on designing AI chips using leading-edge technologies for AI data servers, data-communications, and other strategic sectors in order to create a demand for high-volume manufacturing of AI chips in Europe, for guaranteeing EU technological sovereignty and prosperity. Europe urgent needs to massively invest and deeply support R&D in advanced chip design to create an efficient European leading-edge semiconductor ecosystem.

10.2.5 Leading-edge semiconductor technologies

Smallest nodes (≤2 nm) are the cutting-edge technology that is driving applications such as AI and HPC. They enable future technological advancements, along with strategic applications. Especially for such critical applications, Europe needs to be able to provide for its own essential demand. Establishing new factory(ies) will therefore be crucial for Europe to secure a relevant production share and digital sovereignty. Otherwise, it will be highly dependent on production from other regions for these critical technologies.

Only Intel-Ireland currently produces leading-edge technologies (≥3 nm) in Europe but mainly for its own products and not for the needs of the European electronic components and systems community (Figure 10.14).

The manufacturing of leading-edge technologies is mainly concentrated in Taiwan (Figure 10.14). Since China has claimed the “ownership” of the Taiwan island for several years, the geopolitical risks associated to a Chinese invasion of the island cannot be neglected. In this case, the supply chain for advanced logic nodes will be broken for European customers, and it will strongly impact the European technological sovereignty.

More than 80% of global semiconductor CapEx through 2026 will target leading-edge nodes below 10 nm. For Europe’s automotive and industrial sectors, however, the critical dependencies lie elsewhere:

- Automotive MCUs are still produced at 40–90 nm.
- Industrial controllers and sensors rely on 65–180 nm technologies (Figure 9.27).
- Power devices such as SiC and Si based transistors depend on mature, highly specialized process lines (Figure 10.15) [516, 517].

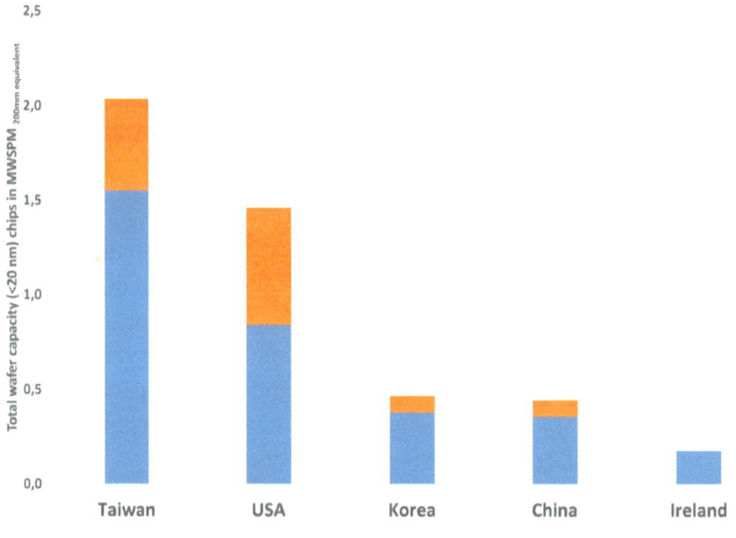

Figure 10.14: Total wafer capacity for advanced logic (<20 nm) chips, measured in million wspm in 200 mm equivalents. The blue shade corresponds to the capacity accounted for by the in-production fabs while the orange shade reflects upcoming fabs in each of the economies - data from [515].

As a consequence, currently there is no European end-user company that could drive the very expansive design of ≤2 nm chips (Figure 9.16) and benefit from a large market allowing high-volume manufacturing of such chips, which could justify the implementation of fab producing chips using leading-edge technologies in Europe.

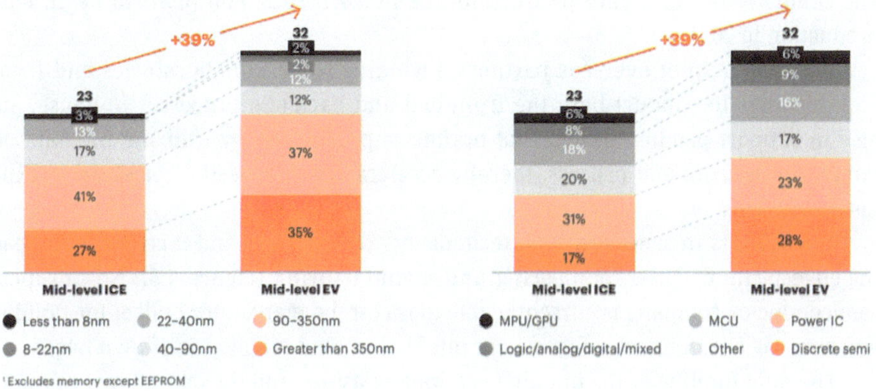

Figure 10.15: Semiconductor content in mid-level internal combustion engine (ICE) and EVs per technology node and types [517].

One of the major objectives of the proposed Chips Act 2.0 must be the creation of a broad European design community gathering OEMs, Communications, Automotive and Aeronautics suppliers and others that already need AI and will require more and more AI for their future products. This initiative is mandatory for increasing the European demand for AI and HPC chips and would enable the creation of at least one European fab to manufacture AI chips using leading-edge technologies.

Moreover, no EU semiconductor firm has currently the financial power to take the ambitious risks for moving from advanced or mature nodes to leading-edge technology nodes (Figure 9.18). Competing policies in other nations (the USA and several nations in Asia) also offer better prospects than in EU. Therefore, creating or attracting production sites in EU will require successful compensation of locational disadvantages when compared to Asia and the USA, which are mainly driven by high government subsidies. As a complement to this, policymakers should consider sourcing a minimum share of high-end chips from European manufacturing sites for critical applications, such as data centers.

Europe could copy the ambitious but highly risky Japanese “Rapidus” initiative. Rapidus Corporation was founded in 2022 to work on a project at the heart of Japan’s semiconductor strategy. Rapidus aims to develop a system for mass producing leading-edge 2 nm chips, which have yet to be realized anywhere. In 2022, the most advanced generation produced in Japan now was the 40 nm chip. Major Japanese companies – including Toyota Motor Corporation and the Sony Group – have invested a total of $47 billion in this massive public-private partnership, while the Japanese government has announced significant support, for an initiative totaling roughly $31 billion. A large-scale R&D facility and production plant have been built in Japan.

The company set up a pilot production line in April 2025, and plans to begin mass production in 2027.

With the help of overseas partners, including some EU laboratories and firms, Rapidus aims to conduct both the front-end and back-end processes in-house, and also to support product design that optimizes production by utilizing big data obtained in the front-end process, thereby accelerating the speed of production exponentially [518].

Investments in semiconductor technology are very high and increasing for leading-edge technology in engineering and manufacturing (Figure 4.5). No European semiconductor company is currently able to invest the mandatory budget for building and running a leading-edge fab, except Intel-Ireland but mainly for its own needs.

The total funding of the building, equipment assets, and the operational costs of a leading-edge technology fab will be very high. Due to the geopolitical shifts and current wars across the globe, such an investment looks nowadays strategic for the EU technological sovereignty, for reducing its strong dependencies versus the USA, Taiwan, and China. One alternative consists in setting up a close partnership between the European semiconductor ecosystem and a leading-edge technology supplier, who has the experience in running a leading-edge technology fab.

In order to ease the start of a European manufacturing line for AI Chips, the EU semiconductor industry could favorably partner with a leading-edge technology supplier, like Bosch, Infineon, and NXP which are partnering with TSMC for creating ESMC in Dresden.

Two parallel efforts are required to make progress against the EU's strategic ambition in the next 10 to 15 years that would allow for combining the EU's strengths with the partners' expertise and de-risk the realization of its ambitions. First, reinvigorate the EU semiconductor ecosystem from R&D, design, equipment, and chemicals to assembly process. Second, partner with one of the leading-edge semiconductor technology companies pledged to invest the mandatory CapEx and operational expenditures (OpEx) for running such a fab [471].

The main risk associated with a Rapidus-like and/or a partnership with a top semiconductor company approach is a possible low utilization rate of the European leading-edge fab. With high utilization, semiconductor companies can expect to see a positive cash flow from their new fabs within about 5 years, even with little to no government support (Figure 10.13). If the fabs are utilized below capacity, however, this time frame could increase significantly. In extreme cases, where utilization drops below about 55–65% of full capacity and remains at that level, companies could be in a dire situation. Even in a best-case scenario, where a semiconductor company receives high government subsidies, it would take almost 10 years to see a return on investment when operating at 55% capacity (Figure 10.16) [462].

The technological frameworks classified as "single-digit nanometer" that stem from advancements related to "Moore's law" are designed to handle digital computational tasks. However, they fall short for various applications demanding exceptional

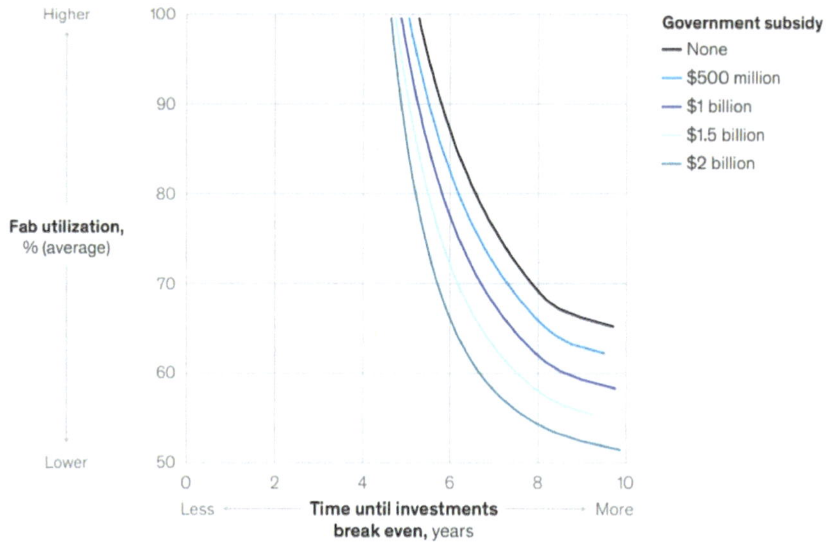

Figure 10.16: Evolution of the breakeven point of a 5-nm fab as a function of its utilization rate [462].

analog capabilities, high-frequency operation, and high-power efficiency, along with requirements for minimal noise levels, and extremely low power usage [519].

Nearly 80% of the non-memory semiconductor total available market is currently comprised of technology nodes >12 nm. The progress of innovation toward single-digit nanometer technology is often used as a bellweather for industry health. It should not be assumed that innovation in these disparate market spaces is disconnected; in fact the opposite is true. New materials, transistor geometries, and design tools, which are first appear in single-digit nanometer technology are rapidly adapted into less advanced technology nodes to realize the performance features, which are required for the balance-of-system components in computation and communication networks [519].

Some recent examples of solution adaptation include partially and fully depleted SOI (strained) material migration from high-performance compute (HPC) to high-performance RF, multi-work function modules from scaled CMOS to ultra-low-power solutions for physical AI, 3D "FinFET" transistor structures from HPC to highly efficient wide bandgap semiconductor devices, and advanced packaging modules from HPC into RF systems [519].

10.2.6 Mature and advanced chips

In terms of manufacturing volume, the mature and advanced chips (from 28 to 500 nm technology nodes) shipments represented roughly 86% of the overall chip shipment performed by the pure play foundries in 2025 (Figure 10.17).

The current most advanced technology nodes produced by the European semiconductor industry (18–22 nm), by excluding Intel-Ireland, comply with the current requests of edge-AI logic chips (Figure 10.18a). However, due to the future huge demand for edge-AI devices in the coming years and also for lowering their power consumption, these technology nodes would not be sufficient for fulfilling the edge-AI chips specifications in 2030–2035 (Figure 10.18b).

Accordingly, the European chip manufacturing industry will need to develop more advanced technology nodes ("X"nm, with X in one digit) to keep in the race for manufacturing edge-AI chips and remaining a significant actor in the worldwide semiconductor market. Such a development must be an utmost important project of the European Chips Act 2.0, and could be ideally achieved through a "precoopetition" between the European chip manufacturers to accelerate their development.

Due to their lower operational costs (Figure 9.33) Chinese actors could also gain huge market shares in manufacturing edge-AI, power electronic and analog chips at the expenses of European Actors.

China's progresses are direct threats to previous European strongpoints in the supply chain. In mature technology nodes (above 20 nm), China will account for 39% of world's production capacity by 2027. The mainland registers 4 million wafer starts per month (wspm, 200 mm equivalent) compared to around 3.3 million wspm of planned capacity in the rest of the world. Owing to the early development of the EV internal market – in the first half of 2025, China produced 43% of the worldwide EV – this increased capacity is concentrated in power as well as analog chips, segments where European and Japanese players are historically strong. Chinese foundries will be able to produce en masse and below global market prices, thereby outcompeting and outperforming European plants [444].

Moreover, the Chinese manufacturing capacity is expected to even increase in the forthcoming years (Figure 10.19). Chinese suppliers continue to massively invest in the mature technology nodes, when the rest of the world has rather stopped to invest in such nodes.

As a consequence, China presents the highest chip manufacturing capacity worldwide, with a focus on more mature technology nodes from 45 to 300 nm and higher (Figure 10.20).

Another example is SiC and GaN based components. These devices are critical in energy, automotive as well as defense industries (see Chapter 2). It could require the implementation of an dedicated European DRAM or DRAM-like production line in Europe. China made progress in epitaxy and related machine tools. In all these segments, Chinese companies receive state support to increase production (30% of the world's capacity in 09/2025), already leading to prices collapse and increasing their market share against their main competitors [444].

So, the move to more advanced technology nodes by European device suppliers looks mandatory to compete with the Chinese and Asian chip manufacturers in the edge-AI, power electronics, and analog sectors too.

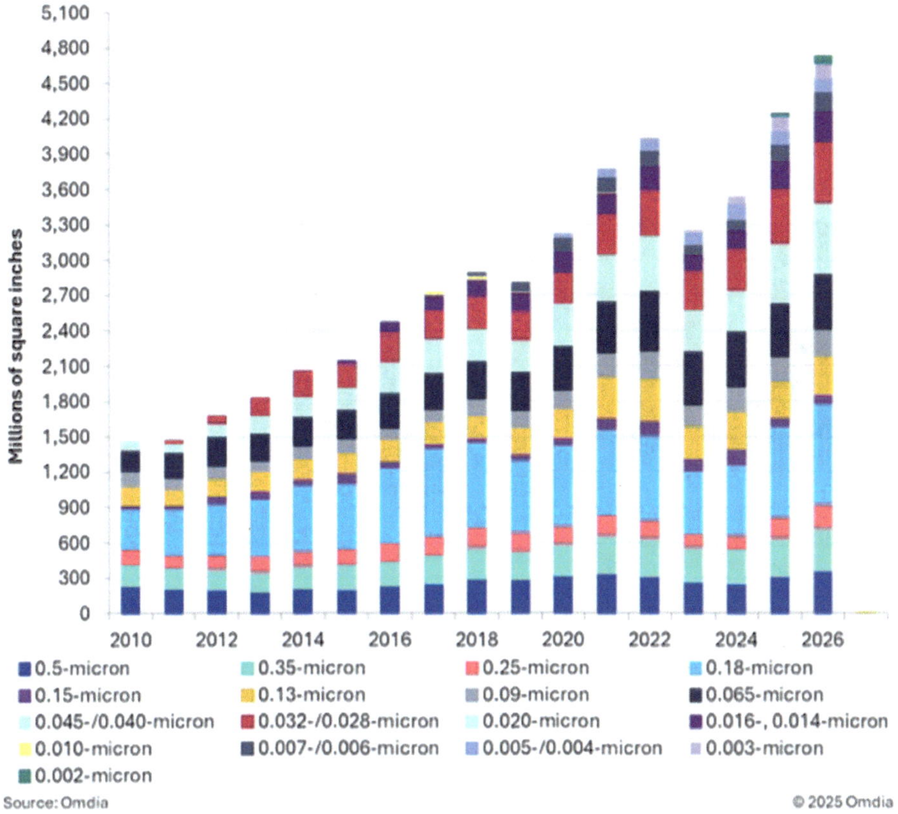

Figure 10.17: Pure play foundry shipment by technology node from 2010 to 2026 – courtesy of Omdia.

Another approach for competing with Chinese chipmakers could be to invest in more automated European semiconductor production lines. This investment would favorably concern 300 mm fabs, which are already partially automated. An increase in the automation of an existing fab will allow reducing the labor costs, which are much higher in Europe than in China (Figure 9.32a).

The recent Nexperia experience pinpoints the importance of components even produced using mature nodes. Nexperia, which produces billions of simple chips diodes for cars and electronics, has been in a tug-of-war since the Dutch government seized the company in September 2025 on economic security grounds. An Amsterdam court subsequently stripped the Chinese company Wingtech of control. Beijing retaliated by halting exports of Nexperia's packaged products in October 2025, leading to disruptions in global automotive supply chains.

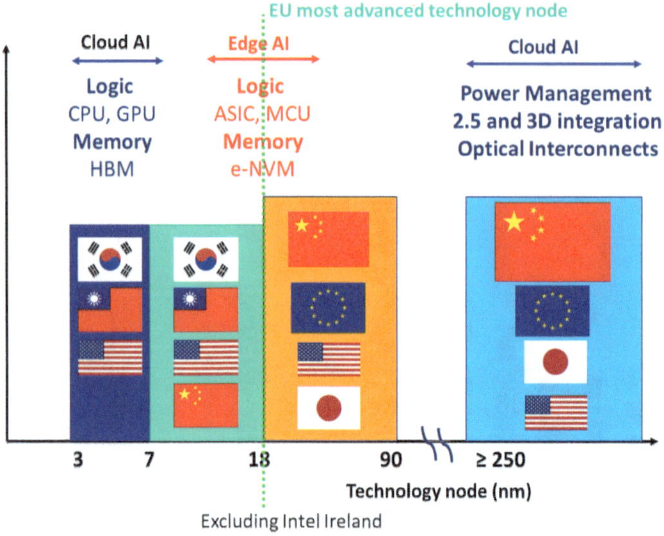

(a)

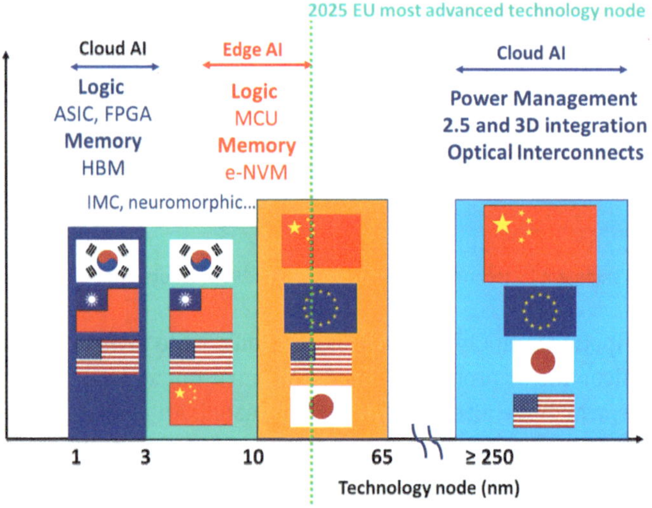

(b)

Figure 10.18: Regional distribution of the manufacturing, and evolution of AI-related chips in (a) 2025 and (b) in 2030–2035.

Just as the crisis threatened to idle Western car factories, the geopolitical winds shifted again – this time from Washington. On October 30, 2025, President Trump and President Xi Jinping struck a tactical trade truce that redounded on Nexperia. As part of the deal, the USA suspended the implementation of the export controls for 1 year.

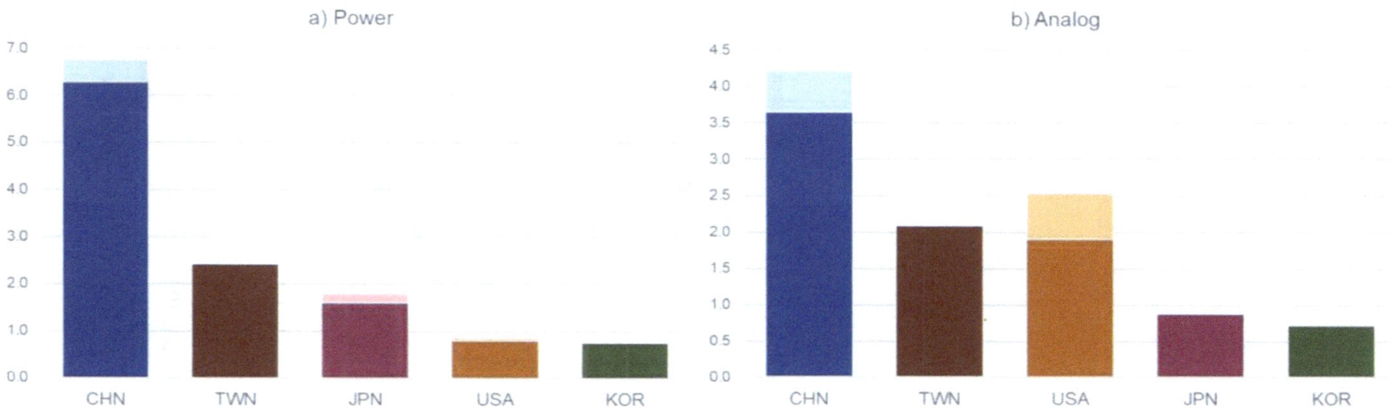

Figure 10.19: Total wafer capacity for (a) power and (b) analog of chips measured in million wspm in 200 mm equivalents. The dark shade corresponds to the capacity accounted for by the in-production fabs while the light shade reflects upcoming fabs in each of the economy [515].

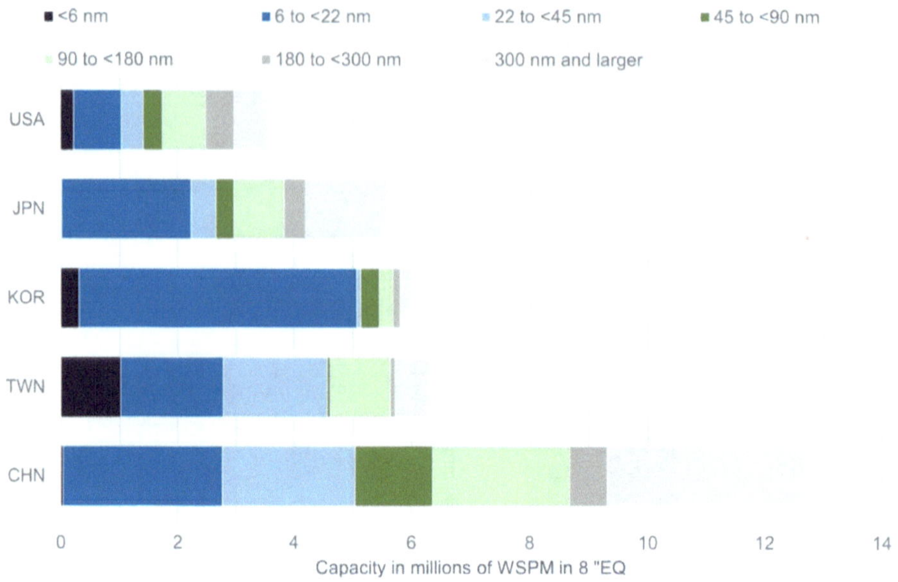

Figure 10.20: Feature size distribution in the five economies with the largest in-production capacity [515].

Simultaneously, Beijing began granting exemptions for Nexperia to resume exports [520].

This diplomatic breakthrough relieved the immediate pressure on the auto industry, but for the Netherlands and Europe, this was a devastating reversal.

The lesson from the Nexperia saga is unmistakable: in today's geopolitical economy, alignment with Washington offers no guarantee of consistency or cover. Allies can be exposed the moment US domestic politics or strategic priorities shift. For Washington, aligning allies is a strategic necessity to prevent technology leakage. For European states, however, the process increasingly resembles coercion by the USA rather than coordination. For smaller states whose prosperity depends on predictable access to global electronics markets, the volatility of the US-China technological rivalry transforms uncertainty into a structural risk.

Furthermore, many of the components used in defense applications originate from outside Europe today. This dependence on external component suppliers is not really acceptable for the European defense industry, where critical components could be withheld or delayed for strategic reasons by foreign governments. In an attempt to provide an alternative solution for Europe, in the development of advanced RF CMOS based Systems on a Chip, the e-Perfect project has been funded. This funding has been provided in response to a call from the European Defence Fund for the creation of an ITAR-free supply and value chains for advanced, radiation-hard, fine geometry, RF CMOS Technologies within Europe.

The e-Perfect project aims to formulate a strategy for the development and ownership of an advanced RF-CMOS supply chain within Europe that will address the requirement for a high-performance CMOS technology platform for space and defence applications. To achieve the overall project aims, European end users (OEMs) with different technology requirements and product ranges will engage with European design groups and technology suppliers to create components for the defence electronics industry for the coming decades.

It is the intention of e-Perfect to share the capabilities and performance of the RF-Processor and RF-Transceivers to be developed within the project with the whole European OEM community. This will ensure maximum awareness of the process technology and its capabilities through an OEM information exchange forum. This will create critical mass in Europe around the core elements of the supply chain [521].

10.2.7 Stand-alone memories

The memory market would increase at a CAGR between 10% (Figure 10.21) and 18% [522] driven by the forthcoming AI boom. The demand for HBM, which is mandatory to enhance the bandwidth between the processor and the off-chip memory will drive the memory market (Figure 10.21).

In late 2025, the global semiconductor ecosystem is experiencing an unprecedented memory chip shortage with knock-on effects for the device manufacturers and end users that could persist well into 2027. DRAM prices have surged significantly as demand from AI data centers continues to outstrip supply, creating a supply/demand imbalance.

As aforementioned, no European company currently produces the main memories: DRAM and NAND flash. Europe will have to set up partnerships with like-minded and trusted among the memory suppliers with headquarters located according to Figure 10.22.

The creation of European manufacturing line(s) of AI chips would favorably include the production of DRAMs and stacked memories such as HBMs and/or 3D-DRAMs, which are the foremost nano-electronic devices for training and inferencing AI models. Indeed, mixing the production of logic devices and memories on the same manufacturing line is rather difficult.

In order to differentiate from its competitors, the European manufacturing line(s) could develop and then produce those memories with new materials and/or new architectures or combine different memory architectures, as it has been previously described. A European fab, which will only produce logic processors as AI chips would be less meaningful for the EU sovereignty without the manufacturing of stand-alone memories on its territory.

Few European chip makers and R&D labs develop and produce emerging memories such as MRAM, ReRAM, PCRAM, and FeRAM. STT-MRAM looks currently confined in a niche market, but could replace SRAM as a memory cache. FeRAMs could act as an alternative to DRAMs and NAND flash memories.

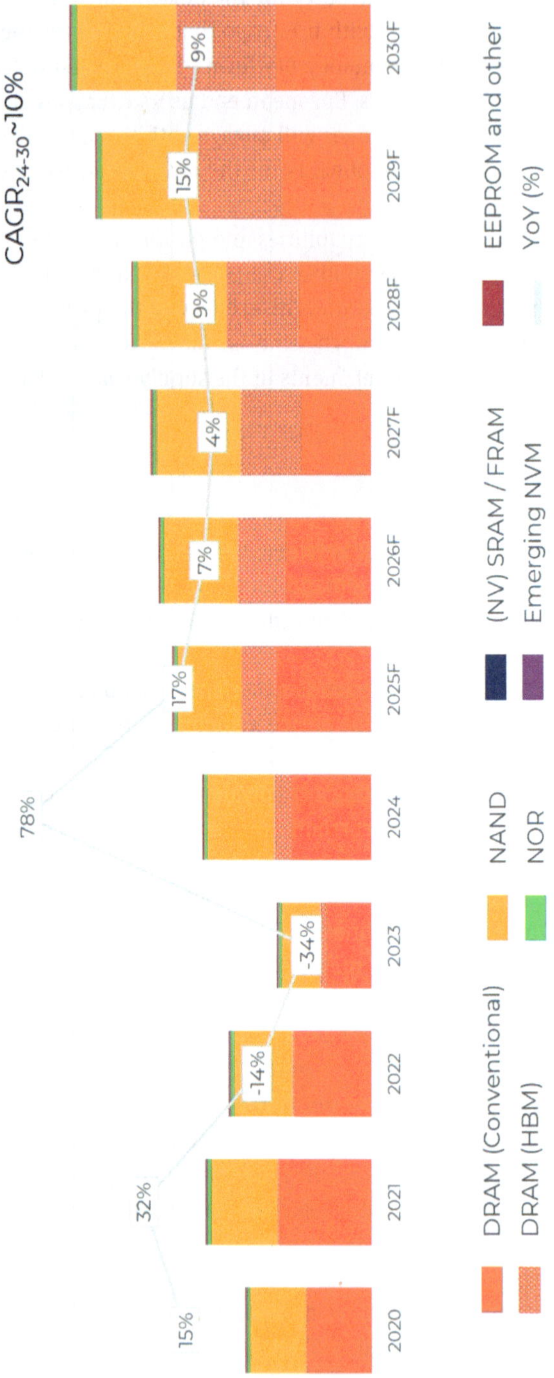

Figure 10.21: The 2020–2030 memory market revenue by technology – courtesy of Yole Status of the Memory Industry 2025.

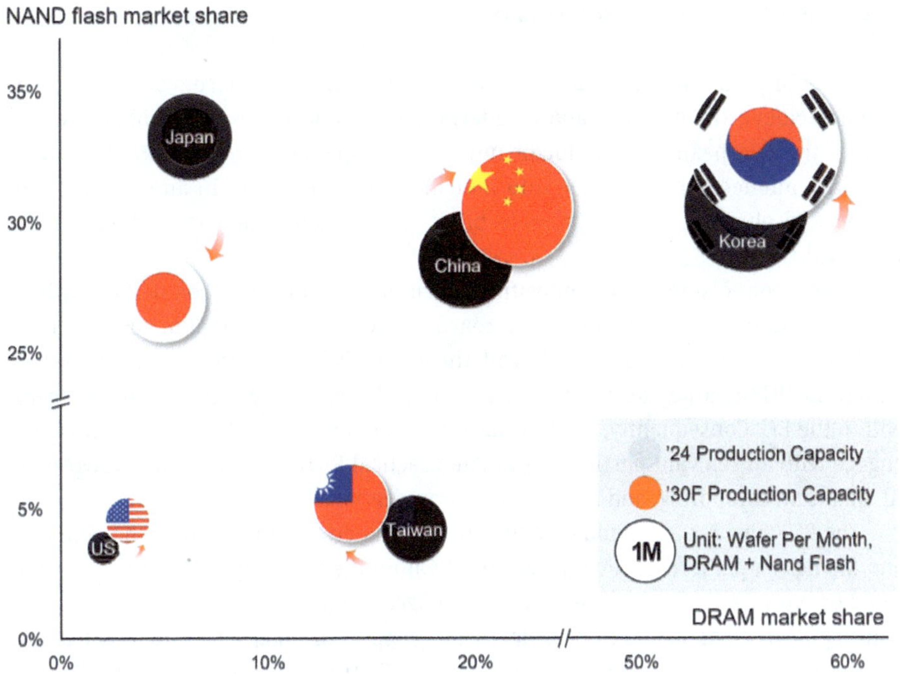

Figure 10.22: Evolution of the NAND flash memory wafer production capacity between 2024 and 2030 [388].

The development of this still-emerging non-volatile memory needs to be pursued within the long-term European strategy, due to their potentials.

10.2.8 R&D expenditures

Due to the technology and trade rivalry between the USA and China, and its volatility, Europe must possess at least minimum viable capacity across critical nodes of the semiconductor chain. Innovation is key to develop the design and technologies related to those critical nodes of the semiconductor value chain.

Despite committing large investments through the EU Chips Act, Europe is trailing both China and the USA in private funding and public support of the semiconductor industry (Figure 9.36). An EU semiconductor strategy could coordinate funding and architectural choices, addressing the availability of grants, subsidies, or R&D tax incentives for design centers active in chip design, foundries, and IDMs in selected strategic segments [523].

10.2.9 Industrialization of R&D results

While the first EU Chips act is an important preliminary step, more work needs to be done to achieve an ambitious and long-term European R&D program that keeps competitive the European semiconductor industry. Europe needs to translate its R&D results and intellectual excellence into industrial scale, as it did in automotives, airplanes, and pharmaceuticals. It would bring back its semiconductor industry in the worldwide lead.

In addition to supporting innovation, catching up with global competitors will require a simplified and very unified approach, as well as significant long-term funding.

This extended R&D approach will thrive solely due to focused partnerships among EU RTOs, a key asset of the EU, alongside the entire semiconductor sector within the EU. Consequently, the European Commission and its MS are encouraged to engage with all relevant parties to promote practical R&D that will propel progress in AI, quantum computing, and advanced communications.

The EU Chips Act 2.0 would ease the interaction between the industry the pilot lines and the design platform, which have started within the first Chips Act. This interaction could be encouraged by dedicate a specific budget line in pilot lines projects to support industry's participation and to provide specific support to smaller players. It will aim to realize demonstrators, in order to highlight the R&D results and to quickly and efficiently prepare the transfer of those results to the European semiconductor industry.

10.2.10 Conventional packaging

For cost reasons, Europe strongly depends on Far-East suppliers in conventional packaging, using wire bonding. Back-end manufacturing (ATP) is a labor-intensive value chain step that involved limited innovation, small value added and low profit margins.

China and Taiwan together host nearly 60% of global ATP capacity (including both IDM ATP facilities and OSATs). Out of 36 ATP facilities announced since 2020, 25 are projected to be in Mainland China and Taiwan. Due to lower construction and skilled labor costs, this pattern is likely to persist into the near future [489, 498].

However, the Nexperia saga highlights the importance of conventional packaging and how it could quickly disrupt the semiconductor supply chain.

The global PCB supply chain is also concentrated in China, and Chinese suppliers are growing their market share. In 2019, China accounted for more than half of the global market for chemicals used in PCB production and was the fastest-growing market followed by Taiwan and South Korea [463].

Outsourcing the conventional packaging to untrusted Asian OSATs also threatens the integrity of the European critical infrastructures and defense capabilities. The PCB industry is central to security concerns surrounding the potential for the semiconductor value chain's back-end to be exploited for espionage. In 2018 and 2021,

Bloomberg reported that an additional chip had been installed on server motherboards manufactured in China and was found to be transmitting data to China from networks in US companies and government agencies [463].

Therefore, conventional packaging should also benefit from a long-term European strategy, even if innovations in this field are rather limited. Due to the higher European wages (Figure 9.32), and since conventional packaging is labor-intensive, implementing an important assembly line in Europe is probably too risky. However, increasing the production of the assembly equipment and the essential components of conventional packaging, such as PCBs, could be parts of this long-term strategy.

10.2.11 Advanced packaging

Moore's law has shifted toward advanced packaging over the past few years. Packaging now sits much closer to the center of system performance. It no longer makes sense to talk about the architecture of advanced AI and HPC complex systems as if the package were a passive shell wrapped around the real innovation. Power delivery, thermals, interconnect density, substrate behavior, and process sequence all affect what can be built and what can be manufactured economically.

Advanced packaging technologies are erasing long-established boundaries between silicon design, packaging, and PCB layout, forcing traditionally separate domains to collaborate far more tightly.

Advanced packaging offers Europe a distinct chance to become competitive in the market for premium chips. Innovative advanced packaging techniques are essential for enhancing the performance and energy efficiency of upcoming chips, particularly those designed for AI chips.

However, over 70% of global advanced packaging capacity remains concentrated in Taiwan and South Korea, and only around 1–2% in Europe. Moreover, advanced packaging requires base dies, interposers, underfill materials, and precise thermal interfaces, which are not produced in Europe.

Only nine significant semiconductor back-end factories are left on EU soil, after decades of offshoring of production toward Asia (Figure 10.23). In addition, the EU also benefits from nine small factories from SMEs [20].

The main current European end-user market for advanced packaging is, once again, automotive, driven by trends such as ADAS, infotainment, and Electrical/Electronical architectures (central car units). While wire-bonding continues to be the dominant interconnect for automotive packages, ADAS modules are increasingly using advanced interconnect technologies such as flip chip ball grid array, wafer level fan-out, and flip-chip chip scale package (CSP) [20].

With traditional 2D scaling reaching physical and economic limits, advanced packaging is becoming the primary engine of semiconductor innovation – but packaging capacity has not kept pace. This imbalance between demand and available infrastructure suggests that packaging, rather than wafer supply, could emerge as the most significant bottleneck in high-performance computing [517].

Advanced packaging provides the raw computational fabric required for the next generation of AI. Paradoxically, the multi-die systems, especially for AI applications, themselves are becoming so large and complex that engineering (designing, testing, verifying, etc.) these devices require AI.

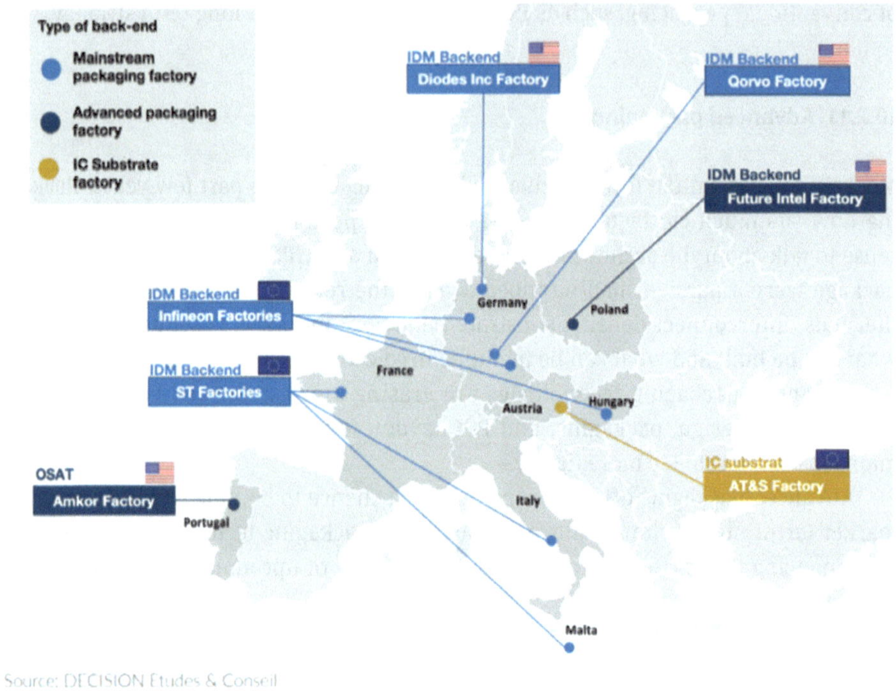

Figure 10.23: European ecosystem of back-end, advanced packaging, and IC substrate in 2023 [20] – The Polish Intel Factory has been cancelled.

The growth in multi-die design is driven by advances in packaging technologies, including 2.5D interposers, 3D memory die stacking, and 3D SoC stacking (i.e., stacking SoC dies on top of one another). Therefore, the development of the advanced packaging technologies is forecasted to exhibit an amazing growth rate in the forthcoming years. Their revenue is expected to grow from $47 billion in 2025 up to $158 billion in 2033 (Figure 10.24).

Therefore, Europe could consider building up an innovative back-end manufacturing capability with an ambition of around 5–10% market share. This may reduce reliance on foreign supply chains to safeguard against global disruptions. Joint ventures could help build these capabilities quickly, particularly if regions can be identified with a competitive cost base and access to renewable energy [523].

Taking into account that advanced packaging adds another layer of complexity, its development and its domestic production needs be fully included in the aforementioned European long-term strategy, including eventual friendly partnerships that will need to be defined for developing AI chips and advanced semiconductor technologies.

Especially, the European line(s) for manufacturing AI chips must also include large capabilities for advanced packaging. Such capabilities will allow producing new architectures or combination of different architectures, which would enhance the energy-efficiency, the memory bandwidth and the performance of AI chips.

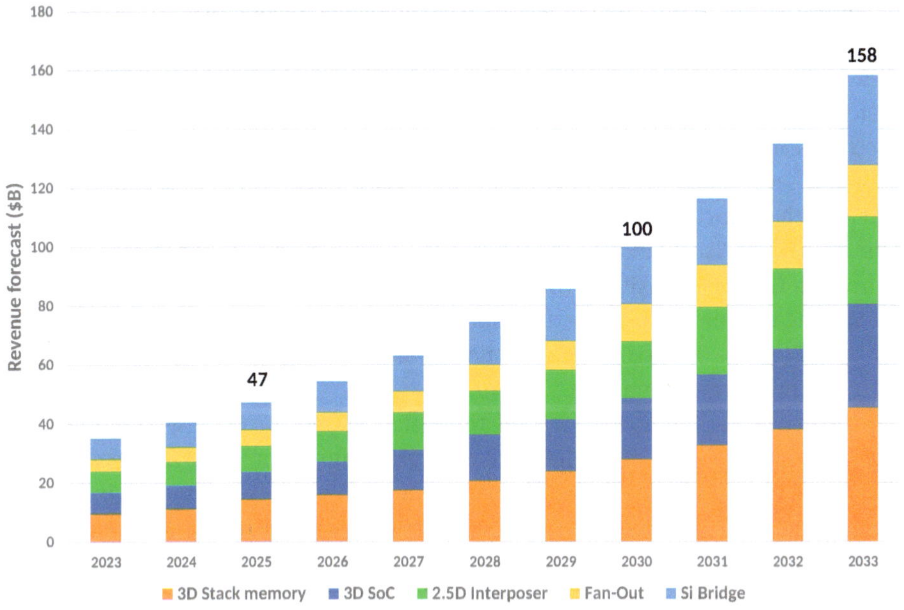

Figure 10.24: Evolution of the advanced packaging technology revenues – source: Synopsys.

10.2.12 Critical raw materials

Critical raw material (CRM) value chains are more vulnerable than ever, and entire vital industries in Europe, including semiconductor, are now at risk if supplies are not secured through strategic and urgent actions, given mounting geopolitical confrontation, resource nationalism, growing demand, and a limited supply increase.

The mobilization by the USA to build resilient CRM supply chains is unprecedented, with a total of over $40 billion of public money now mobilized in various forms, including direct equity, debt, guarantees, and fixed offtake prices via five public institutions mainly. The USA has de facto established a quasi-sovereign fund to take equity into key companies inside the USA and is extremely active also abroad, not least in Latin America, Africa, and Australia [524]. Critical materials are integrated as a key element of the Pax Silica initiative. Launched by the United States in December 2025, Pax Silica brings together a selected group of countries around the coordinated control and development of AI value chains: semiconductors, advanced computing infrastructure, energy, logistics, and critical minerals. Technological leadership is now defined along these supply chains. Those who control key nodes shape both the pace and the direction of innovation [525].

Since the future demand for chips will nearly double in the five coming years, the European access to mineral resources for manufacturing these chips must also be guaranteed. That makes the EU semiconductor ecosystem vulnerable – and building domestic capacity is difficult because of high capital and labor costs.

In order to increase its strategic resilience in the semiconductor supply chain as well as in its industry and defense sectors, and to fulfil its huge and ever-growing raw materials needs, Europe must deeply invest again in its mining industry while respecting leading sustainable, ethical, and responsible mining practices.

After several decades of mine closures, reopening several mines is planned in the frame of the Critical Raw Materials Act, which was launched by the European Commission in 2024. This initiative should ease the reopening of mines across Europe for extracting Ga from bauxite in Greece, rare earth elements in Sweden, and increase the mining volume of currently domestically extracted critical metals such as Co, Cu, Ni, and W (Figure 9.45a) in order that the extraction volume will reach at least 10% of the EU's annual consumption in 2030 [491].

Furthermore, for not relying on the risks coupled to Chinese supply chains (Figure 6.9a), European stakeholders must urgently invest in improving their knowledge of the country's subsurface in order to identify areas likely to contain mineral resources of interest, which could then be studied by more in-depth exploration by mining companies under exclusive research permits. This inventory will help to secure the supply of mineral resources and European sovereignty, thereby strengthening the resilience of its economy. This inventory will take a while and so requires a long-term strategy for Europe and its MS of more than 10 years. During this inventory, European stakeholders will also need to develop advanced and sustainable mining technologies such as biochemical extraction, and AI-driven mining for a better social acceptation of mining across Europe.

However, it is more than likely that this inventory will demonstrate that the European subsurface does not contain all the critical raw materials essential to the semiconductor value chain. Then, it will be appropriate for Europe to establish collaborations with allied or like-minded countries capable of extracting these metals.

Extracting minerals from electronic wastes could be a complementary and slightly more sustainable approach than ore mining for providing access to some critical minerals. However, their recycling rates are currently too low (Figure 6.11) and would deserve huge R&D efforts and investments before becoming economically viable.

In a similar way, the processing and refining of mineral resources need to be greatly enhanced in Europe compared to the current status (Figure 9.45b). The Chinese market shares for refining the minerals is even higher than those for extracting them (Figure 6.9b), and could lead to export restrictions like the ones on Ga, Ge, and rare earth elements.

10.2.13 Sustainability

Europe leads the world in reducing carbon and GHG emissions. Semiconductors are a key enabler of low-carbon, energy efficiency, and carbon footprint reduction solutions. They optimize energy usage in transportation, manufacturing, and consumer products. Sustainability acts as a catalyst for low-power chip design.

But, as it has been aforementioned, semiconductor manufacturing is energy, resource, and water intensive, and also uses small quantities of hazardous materials such as PFAS. For a decade, the semiconductor industry has made substantial strides toward sustainability, particularly in the areas like energy efficiency, material recycling, and renewable energy adoptions, especially in Europe. However, it can still significantly reduce its environmental impact by focusing on several key strategies. They can concern the efficiencies at the process level, the transition to low-carbon energy sources, and the development of opportunities to reduce, reuse, and recycle (in an economic viable way) materials and water across the supply chain.

In addition to environmental benefits, these sustainability efforts can bring significant advantages to the semiconductor industry. Improving material efficiency and reducing energy and water consumption can allow lowering operational costs and raw material dependencies, for instance.

Sustainability can be a strategic differentiator for the European semiconductor ecosystem. The sustainable design and manufacturing of chips can attract worldwide customers who are sensitive to the climate change and to the planet health. Customers and policymakers are scrutinizing emissions along the entire supply chain – and in many cases, semiconductor companies will account for a substantial amount of these. Achieving substantial emission reductions will require new technologies, innovative thinking, collaboration with peers and suppliers, and the complete redesign of semiconductor fabrication plants that will require major investment [498].

On the opposite, sustainability could disadvantage Europe, and its net-zero objectives by 2050, if the USA and China do not care about it. Developing sustainable design and manufacturing will require innovation-related expenses, which could increase the European chip sale price, reducing the competitiveness of the European semiconductor industry.

Despite these contradictory statements, enhancing the sustainability of its semiconductor supply chain should be integrated in the long-term European semiconductor strategy.

10.2.14 Workforce development

The worldwide technology sector currently faces a widening skills shortage. As in every region, the skill shortage is a serious challenge for the European economy and particularly for its semiconductor industry. The semiconductor relies on a highly skilled and specialized labor market. As a result, access to graduates with advanced science, technology, engineering, and mathematics (STEM) is essential to designing, manufacturing and using leading-edge semiconductors.

Due to its high-level universities, Europe is best positioned to provide highly trained experts. It is the region that trains the highest number of master and PhD graduates in STEM disciplines; 414,000 graduates annually compared to 274,000 in China and 201,000 in the USA. The European semiconductor industry will need to hire around 75,000 specialized engineering talents by 2030, which looks like a difficult task today.

Indeed, attracting, developing, and retaining talent remains a critical challenge for the European semiconductor ecosystem. This challenge partially results from the perception of the semiconductor industry by young people. Those young people, as well as their teachers, know the end-users of semiconductors, like Apple for the smartphones, but they do not know the brands name of the semiconductor chipmakers. As Europe does not have the prerogatives in matters of education, Europe and the MS could launch a wide education strategy across Europe to attract young people to STEM. This would occur at the primary and secondary schools, through a dedicated training of their teachers, for instance. It could consist of a first step within the semiconductor industry.

Future semiconductor workers will need expertise in chip design and microelectronics, materials science, and process engineering, and advanced manufacturing. To meet this demand, it will be important for universities and colleges to work alongside RTOs and industry leaders to ensure students graduate with the semiconductor industry will need. Offering hands-on experience in semiconductor fabrication, clean-room -based labs and advanced process design across Europe will be essential for preparing a workforce for the European semiconductor industry.

A targeted technology skills acquisition program could attract, develop, and retain the talent critical for advancing semiconductor technology in front- and back-end manufacturing, AI chip design and new materials development. Such a program could include a special visa for graduates and researchers in advanced electronics, enhanced subsidies for high-demand skills like electronic engineering, new EU-wide scholarships for top graduate and PhD students, and early internships and temporary contracts in public and private research centers. Such a program could be performed in dedicated universities widely spread across Europe for involving all the European MS. They could lead to a specific diploma, which will be recognized by the European semiconductor industry.

Semiconductor training of students will be performed in selected universities across Europe, with the full support of the European semiconductor industry and RTOs, which could provide hands-on experience, for instance. A European coordination entity will need to ensure that the training content is similar across the selected universities, and to provide students with the opportunity of traineeships and PhDs in the industry and the RTOs.

Workforce shortage could push companies toward automation and internal talent development. The use of AI in EDA tools could probably reduce the high number of engineers who are compulsory for designing advanced chips. Automation of front-end manufacturing fabs is on-going but could be accelerated. In contrast, back-end manufacturing still requires high human resources, which is the reason why conventional packaging had been outsourced to Asia, due to their lower wages. Automation of conventional packaging processes could be an attractive way to re-shore the conventional packaging in Europe by reducing the number of human resources.

Another approach will concern the job re- and up-skilling of current workers. Incentives to upskill and reskill employees and a robust career support system, including internships and job placement services, could also begin to bridge the technical skills gap in the European semiconductor industry.

Filling the workforce gap will represent a very tough task for the near-future. As succeeding to quickly reduce this gap is compulsory, it will require a long-term strategy, which needs to be precisely set-up between the European Commission, as the co-ordinator, MS, as the acting actors, the schools and the universities, and the European semiconductor ecosystem. Without a strong collaboration between those actors, the skill gap will remain, and the European semiconductor industry will be in a bad shape after 2030.

10.2.15 Call for Europe actions

Semiconductors are vital to economic prosperity, regional security, and technological innovation because they are ubiquitous. They underpin modern communication, energy, transportation, healthcare, and defense systems. The semiconductor industry is currently booming at the heart of global technological progress, driven by applications in AI, but also by the constant need for faster, sustainable, more versatile, and more energy-efficient applications in all sectors.

Induced by both fast AI advancements, and its huge demand for high-performance chips, and, as the consequence of the chip shortage observed during the COVID-19 pandemic and of the China–USA technology rivalry, the semiconductor industry is undergoing rapid transformation. Indeed, due to its strategic character, governments are investing heavily in domestic production to ensure supply chain resilience and technological sovereignty and security. Furthermore, export controls, restrictions on critical raw materials, and shifting trade alliances are strongly impacting the overall semiconductor community.

The strategies adopted in Washington and Beijing to effectively nationalize segments of the global semiconductor value chain send a clear signal: semiconductors must be treated as critical infrastructure, indispensable to strategic resilience and national security. The subsidies are being delivered by the USA and China in the name of economic security. Achieving strategic indispensability and a stable supply in the production of semiconductors promises economic and national security benefits because semiconductors are seen as the backbone of economic and informational exchanges in an increasingly digital, interconnected world.

The very recent Nexperia experience allows to see one of Europe's few remaining legacy chip manufacturers caught between US export controls and Chinese retaliation. This saga has strengthened the argument that Europe must possess at least minimum viable capacity across critical nodes of the semiconductor chain.

In a world where the USA and China are increasingly using technology as a tool of statecraft, countries in the middle cannot insulate themselves from coercion, even when acting rationally to protect domestic industry. The recent Nexperia experience is not a one-off failure of policy coordination – it is a preview of how Europe's advanced economies will be treated in the next phase of geo-economic competition [520].

Currently, Europe lacks fabless and end-user companies to drive demand for leading-edge AI chips. To achieve data sovereignty, Europe must develop large design houses, gather the whole design community, including OEMS, automotive and aeronautic suppliers, defense stakeholders, etc., and invest in advanced infrastructures for data, 5G, and 6G communications.

The growth of the European design community would create an internal demand for AI and HPC chips using leading-edge technologies. It would also allow designing advanced logic devices – for replacing GPUs, for instance – and memory chips using new architectures or new concepts – in-memory and neuromorphic computing – to reduce their energy consumption and to enhance their "computing" performance.

The demand for AI chips will drive the semiconductor industry in the forthcoming years, and govern the modern economy. Owing to the geopolitical shifts, the US–China technological rivalry for dominating AI, the conflicts in Ukraine and in the Middle East, the threat of a Chinese invasion in the Taiwan Strait, as well as more frequent weather events, the supply chain for AI chips has never been so uncertain as it is today.

For not depending on a single supplier or on a single country, Europe must massively invest in the design and the production of AI chips using leading-edge technologies. Europe will deeply invest in several AI giga-factories; it also needs to invest in the chips -the main constitutive elements of the data centers- for ensuring its technological sovereignty and its further prosperity.

The most advanced technology nodes for producing "made by EU" chips are currently limited to 18–22 nm – excluding Intel-Ireland. The proposed long-term strategy could include either a similar to Japan's Rapidus initiative, or a European initiative with the "support" of a trusted semiconductor for manufacturing AI chips using leading-edge technologies including logic devices, memories and advanced packaging. Such a compulsory initiative will require substantial public subsidies, similar as the ones scheduled in the US chips Act and/or the Chinese Big Fund.

Europe should also focus on energy-efficient technologies like power electronics, PICs, and advanced packaging/3D integration to remain a chip provider for European and worldwide data centers.

European manufacturers, currently focused on automotive and industrial chips, must advance beyond 18–22 nm nodes to meet future edge-AI demands. Developing sub-10 nm nodes is essential and should be another priority in the European Chips Act 2.0. Moving to sub-10 nm nodes is also compulsory for the European semiconductor industry to compete with Chinese firms, which benefit from subsidies and show lower operational costs that are gaining ground in mature nodes in the More than Moore approach.

The European semiconductor industry should also reinforce their strengths, notably in R&D organizations, equipment and their subsystems, power electronics, PICs, and their material and chemical suppliers in order to become "indispensable" in the semiconductor supply chain. To be self-sufficient in the next future, China deeply invests in the production of semiconductor equipment, and especially on lithography tools. Accordingly, this constitutes a major threat for the European tool suppliers, which are one of the strengths of the European semiconductor industry.

In order to assess the compulsory missing parts in the European semiconductor supply chain, important stakeholders involved in the defense/aerospace, communications, automotive, and from other critical industry sectors should define the priorities of their long-term strategy through efficient pan-European collaborations. The e-Perfect project (Chapter 10.2.6) funded by the European Defense Fund is a pretty nice example of collaboration between several European industry partners for ensuring the European sovereignty in the defense sector. Such an initiative must be absolutely and urgently been reproduced in the defense domain as well as in other applicative sectors such as AI, communications, automotive, industry, and health for guaranteeing a European technological sovereignty and resilience.

Reshoring parts of the supply chain – in conventional packaging – securing access to critical minerals, and recycling electronic waste – are key to sovereignty. Europe must assess which mines to reopen and invest in sustainable extraction and recycling technologies.

Technological leadership is now defined along the overall semiconductor supply chain. Those who control key nodes shape both the pace and the direction of innovation, ensuring their security and prosperity.

Even if Europe has numerous famous and worldwide-known universities, workforce shortages pose another challenge. Education and industry must collaborate to attract students to STEM and provide hands-on training in chip design and manufacturing. A coordinated European training strategy is needed, with harmonized curricula and industry-supported internships and PhDs.

Despite global competition, no region can dominate the entire supply chain. International cooperation remains essential. Europe has always been very open for global investors but now, due to the USA–China technological rivalry, Europe can see that these dependencies can be weaponized against it. That is why it is of foremost importance that Europe will not be dependent on one country or one company when it comes to some very critical fields. While building its own capabilities in sectors such as AI, and semiconductors, the EU must continue to work with like-minded partners by choosing with who and how EU could collaborate with.

Viewing the semiconductor industry in terms of strategic autonomy and strategic indispensability leads to a new and important paradigm shift: new industrial policies to acknowledge legitimate national security concerns. Therefore, Europe must unite to strengthen its semiconductor ecosystem. Being stronger and more resilient in chips means a stronger and more resilient technology sector in Europe, which translates into a stronger and more resilient Europe economically, competitively, militarily, and politically [526].

Without a unified and ambitious long-term plan like the EU Chips Act 2.0, which is pivotal for Europe's industry, Europe risks falling below 5% of global chip production by 2040. The high growing demand for AI and edge-AI chips offers a real chance to regain market share, Europe cannot afford to miss.

11 Conclusions

Semiconductors are ubiquitous and become critical technologies for the digital growth, industrial strength, and strategic independence of any country. The global semiconductor growth is driven by demand in key sectors like AI, automotive electrification, cloud/datacenter computing, IoT and green energy technologies. Nonetheless, AI becomes the key applicative sector for the semiconductor industry. AI is now compulsory for processing data in various domains such as autonomous vehicles, robotics, healthcare, telecommunications and defense. Hence, AI acts as a key enabler in economic growth, industrial competitiveness and geopolitical influence, and mainly drives the current and future exponential growth of semiconductors.

This exponential growth is already highlighted today; the 2026 semiconductor revenue is expected to reach $1.5 trillion according to recent economist forecasts, e.g., a nearly 100% increase compared to the 2025 revenue [527]. This quite impressive growth is mainly induced by the increase in the average memory selling price, owing to the huge request for DRAM and HBM for AI data centers. Memories, as the real AI bottleneck in terms of storage capacity and bandwidth, and a strategic resource, need to be produced in a massive volume to enable the development of AI data centers, as well as other applications like smartphones and PCs. This is the reason why the world's two largest memory chip companies very recently planned to invest $518 billion to build four new memory fabs in South Korea, for answering to the amazing demand for AI memories [528].

The semiconductor industry has thus entered a new era defined by AI, and the pace of change is accelerating faster than previously expected [527]. Therefore, semiconductors will underpin an "Intelligence Revolution" that extends from data centers to autonomous systems (vehicles, drones etc.) and into the physical world. Semiconductor growth is increasingly driven by the tremendous scaling characteristics of AI, which drives the expansion of the worldwide semiconductor manufacturing capacity.

However, increasing the semiconductor manufacturing capacity is not only a matter of cleanrooms and equipment set, it also relies on resources (energy, water, raw materials etc. . . .) and on skilled workforce.

The increase in semiconductor manufacturing capacity and the move to more advanced technology nodes will strongly increase the consumption of electrical energy, water and raw materials. The forecasted growing electrical energy consumption will enhance the scope 2 GHG emission and could contribute to a detrimental acceleration of climate change. Reducing the energy consumption of the process equipment and optimizing the process steps looks mandatory for lowering the carbon foot print of the semiconductor industry

In more and more countries, water becomes a scare resource. Since semiconductor manufacturing is water intensive, recycling water used from various processes re-

 | https://doi.org/10.1515/9783112247952-011

mains an emergent need. Even, if large chip manufacturers have considerably improved the water recycling rate for some years.

The availability of raw materials, such as Ga, Ge, In and rare earths is at the core of the USA-China rivalry, and induces some shortages for these quite important minerals. China has perfectly understood that controlling the semiconductor supply chain, starting from the raw materials, is a very strong geopolitical asset. The USA has just reacted with the launch of the "Pax Silica" initiative, in which one important aim consists in encouraging efforts to partner on strategic stacks of the global technology supply chain, including minerals refining and processing [529].

If the semiconductors that power everything from AI data centers, smartphones to spacecrafts and coffee machines are abundant, nowadays the engineers, scientists and technicians who design and produce them are in short supply. Recent forecasts indicate that by 2030, the global semiconductor sector will need to hire approximately 1 million additional skilled workers while also retaining its current workforce [530]. Since the semiconductor industry is at a pivotal moment with the AI explosion addressing the talent shortage is highly critical.

If semiconductor companies want to build the skilled work-force they need to support their ambitions, it is imperative they start massively invest now. That means changing the industry's reputation, and building comprehensive training, up- and reskilling programs that can create the technicians of tomorrow, today.

If chips are essential for AI development, AI can be also beneficial for semiconductor design and manufacturing. AI can reduce the time and resources to design chips, especially for advanced packing, which becomes of utmost importance for enhancing the performance of logic chips and memories. Furthermore, AI can be used for improving the detection and identification of defects during chip manufacturing, shortening the ramp-up phase to reach relevant production yield.

Chip developments never stand still. Moreover, they are the source of complex AI knowledge, and this enhanced AI capability is also the driving force behind the revolutionary chips of the coming decades. It is essentially impossible to predict where these developments will eventually plateau. This means that there is no end in sight for technological applications across the entire spectrum – from medicine to mobility, energy and defense.

References

[1] C. Wyon and M. V. de Voorde, “New Challenges for the Semiconductor Ecosystem,” IEEE Electron Devices Reviews, vol. 2, pp. 330–360, 2025, doi: 10.1109/EDR.2025.3638339.

[2] J-P. Kleinhans, J. Hess and W. Denkena, “Governments’ role in the global semiconductor value chain #1,” June 2022, https://www.interface-eu.org/storage/archive/files/governments_role_in_the_global_semiconductor_value_chain_1_0.pdf

[3] A. Ahmad, “Automotive Semiconductor Industry – Trends, Safety and Security Challenges,” 2020 8th International Conference on Reliability, Infocom Technologies and Optimization (Trends and Future Directions) (ICRITO), Noida, India, 2020, pp. 1373–1377, doi: 10.1109/ICRITO48877.2020.9197894.

[4] F. Boeuf, C. Barrera, I. Charlet et al. 2021, “Silicon Photonics Platform from Datacom to Sensing Applications,” 2021 IEEE BiCMOS and Compound Semiconductor Integrated Circuits and Technology Symposium (BCICTS), Monterey, CA, USA, 2021, pp. 1–8, doi: 10.1109/BCICTS50416.2021.9682461.

[5] S. Mowlavi, S. Giannakopoulos, A. Grabowski and L. Svensson 2024, “A Review of IC Drivers for VCSELs in Datacom Applications,” IEEE Transactions on Very Large Scale Integration (VLSI) Systems, vol. 32, no. 1, pp. 42–54, Jan. 2024, doi: 10.1109/TVLSI.2023.3326876.

[6] A.S. Augustine Fletcher and D. Nirmal 2017, “A survey of Gallium Nitride HEMT for RF and high power applications,” Superlattices and Microstructures, vol. 109, pp. 519–537, 2017, https://doi.org/10.1016/j.spmi.2017.05.042

[7] P. Vishnuram, G. Ramachandiran, B. Sudhakar et al 2021., “Induction Heating in Domestic Cooking and Industrial Melting Applications: A Systematic Review on Modelling, Converter Topologies and Control Schemes” Energies, vol. 14, 6634, 2021, https://doi.org/10.3390/en14206634

[8] Y. Yu, X. Cao, C. Li et al. 2025, “A Review of Machine Learning-Assisted Gas Sensor Arrays in Medical Diagnosis. Biosensors,” vol. 15, no 8, 548, 2025, https://doi.org/10.3390/bios15080548

[9] N. Simon, T. Stieglitz and V. Bucher 2025, “Area Selective Atomic Layer Deposition for the Use on Active Implants: An Overview of Available Process Technology” Adv. Healthcare Mater. vol. 14, 2403149, 2025, https://doi.org/10.1002/adhm.202403149

[10] T.TH. Nguyen, C.M. Nguyen, M.A. Huynh et al. 2023, “Field effect transistor based wearable biosensors for healthcare monitoring” J. Nanobiotechnol vol. 21, 411, 2023, https://doi.org/10.1186/s12951-023-02153-1.

[11] V. B. Juska, G.Maxwell, P. Estrela et al.2023, “Silicon microfabrication technologies for biology integrated advance devices and interfaces,” Biosensors and Bioelectronics, vol. 237, 115503, 2023, https://doi.org/10.1016/j.bios.2023.115503.

[12] O. Elijah, T. A. Rahman, I. Orikumhi, et al. 2018, “An Overview of Internet of Things (IoT) and Data Analytics in Agriculture: Benefits and Challenges,” in IEEE Internet of Things Journal, vol. 5, no. 5, pp. 3758–3773, Oct. 2018, doi: 10.1109/JIOT.2018.2844296.

[13] T. Miller, G. Mikiciuk, I. Durlik, et al. 2025, “The IoT and AI in Agriculture: The Time Is Now – A Systematic Review of Smart Sensing Technologies,” Sensors vol. 25, 3583, 2025, https://doi.org/10.3390/s25123583

[14] Y. Faqir, A. Qayoom, E. Erasmus et al. 2024, “A review on the application of advanced soil and plant sensors in the agriculture sector,” Computers and Electronics in Agriculture vol. 226, 109385, 2024, https://doi.org/10.1016/j.compag.2024.109385.

[15] A. Q. Huang 2017, “Power Semiconductor Devices for Smart Grid and Renewable Energy Systems,” in Proceedings of the IEEE, vol. 105, no. 11, pp. 2019–2047, Nov. 2017, doi: 10.1109/JPROC.2017.2687701.

[16] D. Siakas, G. Lampropoulos and K. Siakas 2025, “Autonomous Cyber-Physical Systems Enabling Smart Positive Energy Districts,” Appl. Sci. vol. 15, 7502, 2025, https://doi.org/10.3390/app15137502

[17] S. Fontana and F. Di Lauro 2022, “An Overview of Sensors for Long Range Missile Defense,” Sensors vol. 22, 9871, 2022, https://doi.org/10.3390/s22249871.

 | https://doi.org/10.1515/9783112247952-012

[18] "Global Semiconductor Sales Increase 9.2% Month-to-Month in May", July 6, 2026, https://www.semiconductors.org/global-semiconductor-sales-increase-9-2-month-to-month-in-may/
[19] "Gartner Says Worldwide Semiconductor Revenue Grew 21% in 2025," January 12, 2026, https://www.gartner.com/en/newsroom/press-releases/2026-01-12-gartner-says-worldwide-semiconductor-revenue-grew-21-percent-in-2025
[20] EU project ICOS N° 101092562, "Economic analysis of the EU and international semiconductor ecosystem," December 2023, https://www.decision.eu/wp-content/uploads/2024/07/Public-Version-ICOS-Economic-Analysis.pdf
[21] "Global gross domestic product (GDP) from 2000 to 2030," https://www.statista.com/statistics/268750/global-gross-domestic-product-gdp/?srsltid=AfmBOopeqSB6rZMK64gxDjO7jgsoIr9QHEqt58Bodg98PdgigRyZuUd6
[22] J. Kusters, D. Bhattacharjee, J.T. Nicholas et al. 2026, "2026 Global Semiconductor Industry Outlook," February 5, 2026, https://www.deloitte.com/us/en/insights/industry/technology/technology-media-telecom-outlooks/semiconductor-industry-outlook.html
[23] "AI drives semiconductor revenues past $1 trillion for the first time in 2026," January 15, 2026, https://omdia.tech.informa.com/pr/2026/jan/ai-drives-semiconductor-revenues-past-1-trillion-dollars-for-the-first-time-in-2026#:~:text=%E2%80%9CSemiconductor%20revenue%20growth%20in%202026,have%20historically%20influenced%20the%20market.
[24] M. Jabłoński and W. Furtak 2024, "Crisis on the Semiconductor Market: Causes, Effects, and Countermeasure Activities." Quality of the Future – the Future of Quality (QFFQ). Krakow University of Economics, 2024, pp. 70–80, https://doi.org/10.2478/9788367405843
[25] A. Ortiz-Conde 2022, "Evolution of MOSFETs toward nanoelectronics," Special Mini Colloquia (MQ) on "75th Anniversary of Transistor Invention," DL IEEE EDS Delhi Chapter, New Delhi, India, August 2022, doi: 10.13140/RG.2.2.27238.91202.
[26] G. E. Moore 1965, "Cramming more components onto integrated circuits, Reprinted from Electronics, volume 38, number 8, April 19, 1965, pp.114 ff., " in IEEE Solid-State Circuits Society Newsletter, vol. 11, no. 3, pp. 33–35, Sept. 2006, doi: 10.1109/N-SSC.2006.4785860.
[27] R. H. Dennard, F. H. Gaensslen, H. N. Yu, V. L. Rideout, E. Bassous, and A. R. LeBlanc 1974, "Design of ion-implanted MOSFET's with very small physical dimensions," IEEE J. Solid-State Circuits, vol. JSSC-9, no. 5, pp. 256–268, Oct. 1974, doi: 10.1109/JSSC.1974.1050511.
[28] P. Kogge and J. Shalf 2013, "Exascale Computing Trends: Adjusting to the "New Normal"' for Computer Architecture," Computing in Science & Engineering, vol. 15, no. 6, pp. 16–26, Nov.–Dec. 2013, doi: 10.1109/MCSE.2013.95.
[29] M. Bohr 2007, "A 30 Year Retrospective on Dennard's MOSFET Scaling Paper," in IEEE Solid-State Circuits Society Newsletter, vol. 12, no. 1, pp. 11–13, Winter 2007, doi: 10.1109/N-SSC.2007.4785534.
[30] J.-R. Léquepeys, M. Duranton, S. Bonnetier et al. 2021, "Overcoming the Data Deluge Challenges with Greener Electronics," ESSDERC 2021 – IEEE 51st European Solid-State Device Research Conference (ESSDERC), Grenoble, France, 2021, pp. 7–14, doi: 10.1109/ESSDERC53440.2021.9631774.
[31] S. Hsieh, P-Y. Lin, I-H. Lin et al. 2023, "Assessing the contribution of semiconductors to the sustainable development goals (SDGs) from 2017 to 2022," Heliyon, vol. 9, iss. 11,e21306, 2023, https://doi.org/10.1016/j.heliyon.2023.e21306
[32] B. Frieske B and S. Stieler 2022, The "Semiconductor Crisis" as a Result of the COVID-19 Pandemic and Impacts on the Automotive Industry and Its Supply Chains. World Electric Vehicle Journal vol. 13, no. 10, 189, 2022, https://doi.org/10.3390/wevj13100189.
[33] O. Villafranca, "Chip shortage cost U.S. economy billions in 2021," https://www.google.fr/search?q=Impact±Chip±shortage±on±the±US±economy&rlz=1C1GCGY_enFR1186FR1186&oq=Impact±Chip±shortage±on±the±US±economy&gs_lcrp=EgZjaHJvbWUyBggAEEUYOdIBCTIwMDMxajBqNKgCALACAA&sourceid=chrome&ie=UTF-8, accessed in September 2025

[34] ZVEI, "From chips to chances – The importance of and the economic case for supporting microelectronics," February 2025, https://www.zvei.org/fileadmin/user_upload/Presse_und_Medien/Publikationen/2025/Februar/2025-005-Studie-Von-Chips-zu-Chancen-HL-ME/ZVEI-Mikroelektronik-Studie-EN-2025-02-19.pdf
[35] J. Vervey, "Chinese Semiconductor Industrial Policy: Past and Present." Journal of International Commerce and Economics, July 2019. https://www.usitc.gov/journals/jice_home.htm.
[36] J. Corrado, "Clash or Consensus?: The Conflicting Economic and Security Imperatives of Semiconductor Supply-Chain Collaboration in the Indo-Pacific," October 2022, https://www.airuniversity.af.edu/JIPA/Display/Article/3212581/clash-or-consensus-the-conflicting-economic-and-security-imperatives-of-semicon/
[37] European Commission, "Statement by President von der Leyen on the European Chips Act." See also, https://ec.europa.eu/commission/presscorner/detail/en/statement_22_866
[38] CISCO, "What is 5G vs 4G," https://www.cisco.com/site/us/en/learn/topics/networking/5g-vs-4g.html, accessed in October 2025.
[39] Qualcomm, "Unleash the full 5G potential with mmWave," https://www.qualcomm.com/research/5g/5g-nr/mmwave, accessed in October 2025.
[40] Tech Research, 6G vs 5G: Understanding the Next Leap in Mobile Connectivity, October 2025, https://techresearchonline.com/blog/6g-vs-5g-understanding-the-next-leap-in-mobile-connectivity/
[41] J.J. Liou, M. Ziegler and F. Schwierz 2024, "Gigahertz and terahertz transistors for 5G, 6G, and beyond mobile communication systems," Appl. Phys. Rev. vol. 11, 031318, 2024, https://doi.org/10.1063/5.0213011.
[42] Z. Ren, J. Xu, X. Le et al. 2021, "Heterogeneous Wafer Bonding Technology and Thin-Film Transfer Technology-Enabling Platform for the Next Generation Applications beyond 5G," Micromachines vol. 12, no 8, 946, 2021, https://doi.org/10.3390/mi12080946.
[43] Y. Morandini 2023, "RF technology roadmap for 5G and 6G RF front end systems," in Technologies Enabling Future Mobile Connectivity and Sensing, River Publishers, 2023, pp. 1–10.
[44] K. Ben Ali, C. Roda Neve, A. Gharsallah and J.-P. Raskin, 2012, "RF SOI CMOS technology on commercial trap-rich high resistivity SOI wafer," 2012 IEEE Int. SOI Conf. (SOI), Napa, CA, USA, 2012, pp. 1–2, doi: 10.1109/SOI.2012.6404404.
[45] P. Schiefer 2025, "The Crucial Role of Semiconductors in the Software-Defined Vehicle," 2025 IEEE Int. Solid-State Circuits Conference (ISSCC), San Francisco, CA, USA, 2025, pp. 37–41, doi: 10.1109/ISSCC49661.2025.10904625.
[46] W. Xu, Y. Chen, X. Zhou et al. 2025, "Recent advancement and perspective of epitaxial growth and doping of β-Ga_2O_3 thin films for power electronics," APL Mater. vol. 13, 100601, September 2025, doi: 10.1063/5.0271220.
[47] H. A. Ignatious, H-El. Sayed and M. Khan, "An overview of sensors in Autonomous Vehicles," Procedia Computer Science vol. 198, pp. 736–741, 2022, https://doi.org/10.1016/j.procs.2021.12.315
[48] G. Liang and M. Williams, "Semiconductors: Powering the Future of Everything Efficiently from Energy Generation to Consumption," July 2021, https://www.powerelectronicsnews.com/semiconductors-powering-the-future-of-everything-efficiently-from-energy-generation-to-consumption/, accessed in April 7, 2025.
[49] Y. Zhang, F. Udrea & H. Wang 2022, "Multidimensional device architectures for efficient power electronics," Nature Electronics vol. 5, pp. 723–734, 2022, https://doi.org/10.1038/s41928-022-00860-5.
[50] A. Naulea, "How AI and IoT Can Revolutionize Energy Management: 7 Key Strategies," October 2024, https://areusdev.com/how-ai-and-iot-can-revolutionize-energy-management-7-key-strategies/, accessed in August 12, 2025.

[51] S. Lovati, “The critical role of sensors in medical devices,” March 2025, https://www.edn.com/the-critical-role-of-sensors-in-medical-devices/#:~:text=Beyond%20diagnostics%2C%20sensors%20play%20a,movements%2C%20improving%20functionality%20and%20comfort

[52] M.M.H. Shuvo, T. Titirsha, N. Amin et al., “Energy Harvesting in Implantable and Wearable Medical Devices for Enduring Precision Healthcare”. Energies vol. 15, no 20, 7495. https://doi.org/10.3390/en15207495.

[53] “Semiconductors in Healthcare Enhancing Medical Devices,” Semiconductor Review, January 2025, https://www.semiconductorrevieweurope.com/news/semiconductors-in-healthcare-enhancing-medical-devices-and-treatments-nwid-1009.html, accessed in April 9, 2025.

[54] Md. Al. Haque and H. R. Siddique 2026, “Generative artificial intelligence and large language models in smart healthcare applications: Current status and future perspectives,” Computational Biology and Chemistry, vol. 120, Part 1, 108611, 2026, https://doi.org/10.1016/j.compbiolchem.2025.108611.

[55] H. Xiong, C. Dai, X. Chen et al. 2026, “AI-Assisted Bioelectronics for Personalized Health Management,” Advanced Electronic Materials, 2026; 12e00744, https://doi.org/10.1002/aelm.202500744

[56] “World population projected to reach 9.8 billion in 2050, and 11.2 billion in 2100,” United Nations, https://www.un.org/en/desa/world-population-projected-reach-98-billion-2050-and-112-billion-2100#:~:text=World%20population%20projected%20to%20reach,billion%20in%202100%20%7C%20United%20Nations

[57] S. L. Ullo and G. R. Sinha 2020, “Advances in Smart Environment Monitoring Systems Using IoT and Sensors,” Sensors vol. 20, 3113, 2020, doi:10.3390/s20113113.

[58] M. Javaid, A. Haleem, R. P. Singh et al. 2022, “Enhancing smart farming through the applications of Agriculture 4.0 technologies,” International Journal of Intelligent Networks ,vol. 3, pp. 150–164, 2022, https://doi.org/10.1016/j.ijin.2022.09.004.

[59] Y. Zhang, G. Huang, Y. Zhao et al. 2025, “Revolutionizing Crop Breeding: Next-Generation Artificial Intelligence and Big Data-Driven Intelligent Design,” Engineering, vol. 44, pp. 245–255, 2025, https://doi.org/10.1016/j.eng.2024.11.034

[60] “Military & Defense Semiconductor Market,” Market US, July 2025, https://market.us/report/military-defense-semiconductor-market/

[61] N. Suri, M. Tortonesi, J. Michaelis et al. 2016, “Analyzing the applicability of Internet of Things to the battlefield environment,” 2016 International Conference on Military Communications and Information Systems (ICMCIS), Brussels, Belgium, 2016, pp. 1–8, doi: 10.1109/ICMCIS.2016.7496574.

[62] J. V. D’Amico, S. T. Vibbert, A. C. Watkins et al. 2023, “Analysis of Single-Event Upsets and Transients in 22 nm Fully Depleted Silicon-On-Insulator Logic,” in IEEE Transactions on Nuclear Science, vol. 70, no. 4, pp. 387–393, April 2023, doi: 10.1109/TNS.2023.3237178.

[63] P. E. Dodd, M. R. Shaneyfelt, J. R. Schwank and J. A. Felix 2010, “Current and Future Challenges in Radiation Effects on CMOS Electronics,” in IEEE Transactions on Nuclear Science, vol. 57, no. 4, pp. 1747–1763, Aug. 2010, doi: 10.1109/TNS.2010.2042613.

[64] D. F. Brown, P. Srivastava, K. Chu, W. Zhu, D. Dugas and M. Litchfield 2023, “Millimeter Wave GaN MMIC Technologies for Next-gen Defense Applications,” 2023 IEEE/MTT-S International Microwave Symposium – IMS 2023, San Diego, CA, USA, 2023, pp. 101–104, doi: 10.1109/IMS37964.2023.10188181.

[65] H. Kala, W.Pan, R.Gu, et al. 2024, “Emerging technologies for infrared sensing and imaging,” Proc. SPIE 13200, Electro-Optical and Infrared Systems: Technology and Applications XXI, 132000X (1 November 2024); https://doi.org/10.1117/12.3031838.

[66] A. U. Sokolnikov 2013, “THz Identification for Defense and Security Purposes: Identifying Materials, Substances, and Items”, January 2013, DOI: 10.1142/8729, ISBN: 978-981-4452-38-0.

[67] IDCs "Data Age 2025" whitepaper, 2018, see also "the Digitalization of the world-data Age 2025" (video) Dave Reinsel, hosted by Seagate Technology.https://www.seagate.com/files/www-content /our-story/trends/files/dataage-idc-report-final.pdf
[68] R.V.W. Putra and M. Shafique 2024, "A Design Methodology for Energy-Efficient Embedded Spiking Neural Networks," in Pasricha, S., Shafique, M. (eds) Embedded Machine Learning for Cyber-Physical, IoT, and Edge Computing. Springer, Cham. https://doi.org/10.1007/978-3-031-39932-9_2
[69] N. Zhang, Y. Wang, Y. Yan et al. 2026, "Floating-Gate Synaptic Transistors for Energy-Efficient Neuromorphic Computing," Adv. Mater. 2026, 38, e15605, DOI: 10.1002/adma.202515605.
[70] K. Yamazaki K, V-K Vo-Ho, D. Bulsara et al. 2022, "Spiking Neural Networks and Their Applications: A Review," Brain Sciences 12(7):863, 2022, https://doi.org/10.3390/brainsci12070863
[71] A. Apicella, F. Donnarumma, F. Isgrò et al. 2021, "A survey on modern trainable activation functions," Neural Networks 138, 14–32, 2021, https://doi.org/10.1016/j.neunet.2021.01.026
[72] M.E. Paoletti, J.M. Haut, J. Plaza et al. 2019, "Deep learning classifiers for hyperspectral imaging: A review," ISPRS J. of Photogrammetry and Remote Sensing 158, 279-317, 2019, https://doi.org/10.1016/j.isprsjprs.2019.09.006
[73] A. Juarez-Lora, V. H. Ponce-Ponce, H. Sossa et al. 2022, "R-STDP Spiking Neural Network Architecture for Motion Control on a Changing Friction Joint Robotic Arm," Frontiers in Neurorobotics 16:904017, DOI: 10.3389/fnbot.2022.904017.
[74] V. Sze, Y-H. Chen, T-J. Yan et al. 2017, "Efficient Processing of Deep Neural Networks: A Tutorial and Survey," Proc. IEEE 105(12) 2295–2329, doi: 10.1109/JPROC.2017.2761740.
[75] T. Soliman, S. Chatterjee, N. Laleni et al. 2023, "First demonstration of in-memory computing crossbar using multi-level Cell FeFET," Nat Commun vol. 14, art. 6348, October 2023, https://doi.org/10.1038/s41467-023-42110-y
[76] A. Mehonic and A.J. Kenyon 2022, "Brain-inspired computing needs a master plan," Nature vol. 604, pp. 255–260, April 13, 2022, doi.org/10.1038/s41586-021-04362-w.
[77] M. M. Mijwil, K. K. Hiran, R. Dosh et al. 2023, "ChatGPT and the Future of Academic Integrity in the Artificial Intelligence Era: A New Frontier," Al-Salam J. for Engineering and Technology, vol. 2, no 2, pp. 116–127, 2023, DOI: 10.55145/ajest.2023.02.02.015.
[78] Epoch AI, 'Data on AI Models'. Published online at epoch.ai. Retrieved from https://epoch.ai/data/ai-models [online resource]. Accessed 27 Feb 2026.
[79] H.S. Wong and S. Salahuddin 2015, "Memory leads the way to better computing," Nature Nanotech 10, 191–194 (2015), https://doi.org/10.1038/nnano.2015.29
[80] J. Shalf 2020, "The future of computing beyond Moore's Law," Phil. Trans. R. Soc. A. vol. 378, 20190061, 2020, https://doi.org/10.1098/rsta.2019.0061
[81] "CPU vs GPU vs TPU," January 2020, https://www.geekboots.com/story/cpu-vs-gpu-vs-tpu
[82] R.R. Vangala, "Decoding CPU vs. GPU: A Detailed Exploration of NVIDIA and AMD GPU Architectures," March 2025, https://medium.com/@rohithreddy66666/decoding-cpu-vs-gpu-a-detailed-exploration-of-nvidia-and-amd-gpu-architectures-e09ebfb594ed
[83] F. Aguirre, A. Sebastian, M. Le Gallo et al. 2024, "Hardware implementation of memristor-based artificial neural networks," Nature Commun. vol. 15, 1974, March 2024, doi.org/10.1038/s41467-024-45670-9.
[84] P. C. Chidambaram 2025, "Electron Device Challenges to Enable Individual Centric Edge AI," 2025 IEEE Int. Electron Devices Meet. (IEDM), San Francisco, CA, USA, 2025, pp. 1–4, doi: 10.1109/IEDM50572.2025.11353580.
[85] D. Ganapathy and E.J. Warner 2008, "Defining thermal design power based on real-world usage models," 2008 11th Intersociety Conf. on Thermal and Thermomechanical Phenomena in Electronic Systems, Orlando, FL, USA, 2008, pp. 1242–1246, doi: 10.1109/ITHERM.2008.4544402.
[86] B. Hanindhito, A. Fathi, D. Gourounas et al. 2025, "Technology trends in computing hardware and their impacts on high-performance scientific computing. Part I: General-purpose processors and

hardware accelerators," The International Journal of High Performance Computing Applications. 2025;0(0) 1–60, doi:10.1177/10943420251348799.

[87] R. Swaminathan 2025, "Future of AI Hardware Enabled by Heterogeneous Integration Technologies," October 8, 2025, https://eps.ieee.org/wp-content/uploads/2025/11/Oct_8_Raja_Swaminathan.pdf

[88] A. Gholami, Z. Yao, S. Kim et al. 2024, "AI and Memory Wall," IEEE Micro, 44(3) 33–39, May-June 2024, doi: 10.1109/MM.2024.3373763.

[89] J. Janukowicz, "The Importance of Memory in High-Performance Computing and AI," January 2025, https://assets.micron.com/adobe/assets/urn:aaid:aem:8a68fc66-7658-4d0a-98ef-3d70f93181a2/renditions/original/as/import_of_mem_in_hi_perf_compute_and_ai_white_paper.pdf

[90] S. Yu and T.-H. Kim, 2024, "Semiconductor Memory Technologies: State-of-the-Art and Future Trends," in Computer, vol. 57, no. 4, pp. 150–154, April 2024, doi: 10.1109/MC.2024.3363269.

[91] R. Golman, H. Cohen, G. Shapiro et al., "Cyclic Refresh Algorithm for Ensuring 100% Memory Availability in Gain-Cell embedded DRAM Macros," IEEE Transactions on Computer-Aided Design of Integrated Circuits and Systems, doi: 10.1109/TCAD.2026.3652963.

[92] F. Schirrmeister 2024, "The role of cache in AI processor design," March 22, 2024, https://www.edn.com/the-role-of-cache-in-ai-processor-design/

[93] H. Fujiwara, H. Mori, W-C. Zhao et al. 2022, "A 5-nm 254-TOPS/W 221-TOPS/mm^2 Fully-Digital Computing-in-Memory Macro Supporting Wide-Range Dynamic-Voltage-Frequency Scaling and Simultaneous MAC and Write Operations," 2022 IEEE Int. Solid-State Circuits Conf. (ISSCC), San Francisco, CA, USA, 2022, pp. 1–3, doi: 10.1109/ISSCC42614.2022.9731754.

[94] F. Zarrinfar 2016, "1-T SRAMs in high-density, portable applications," February 2016, https://semiwiki.com/eda/6046-1-t-srams-in-high-density-portable-applications

[95] T-Y. J. Chang, Y-H. Chen, K. Venkateswara Reddy et al. 2025, "A 38.1 Mb/mm^2 SRAM in a 2 nm-CMOS -Nanosheet Technology for High-Density and Energy-Efficient Compute," 2025 IEEE Int. Solid-State Circuits Conf. (ISSCC), San Francisco, CA, USA, 2025, pp. 492–494, doi: 10.1109/ISSCC49661.2025.10904759.

[96] J. Chang, Y-H. Chen, G. Chan et al. 2020, "15.1 A 5 nm 135 Mb SRAM in EUV and High-Mobility-Channel FinFET Technology with Metal Coupling and Charge-Sharing Write-Assist Circuitry Schemes for High-Density and Low-VMIN Applications," 2020 IEEE Int. Solid-State Circuits Conf. – (ISSCC), San Francisco, CA, USA, 2020, pp. 238–240, doi: 10.1109/ISSCC19947.2020.9062967.

[97] X. Wang, Y. Kim, G.H. Baek et al. 2026, "0.021-μm2 High-Density SRAM in Intel 18A RibbonFET Technology With PowerVia Backside Power Delivery," IEEE Journal of Solid-State Circuits 61(1) 259–266, doi: 10.1109/JSSC.2025.3615590.

[98] G. Desoli, N. Chawla, T. Boesch et al. 2023, "16.7 A 40-310TOPS/W SRAM-Based All-Digital Up to 4b In-Memory Computing Multi-Tiled NN Accelerator in FD-SOI 18 nm for Deep-Learning Edge Applications," 2023 IEEE Int. Solid-State Circuits Conf. (ISSCC), San Francisco, CA, USA, 2023, pp. 260–262, doi: 10.1109/ISSCC42615.2023.10067422.

[99] Z. Guo, D. Kim, S. Nalam et al. 2019, "A 23.6 Mb/mm^2 SRAM in 10 nm FinFET technology with pulsed PMOS TVC and stepped-WL for low-voltage applications," 2018 IEEE Int. Solid-State Circuits Conf. – (ISSCC), San Francisco, CA, USA, 2018, pp. 196–198, doi: 10.1109/ISSCC.2018.8310251.

[100] J. Chang, Y-H. Chen, W-M. Chan et al. 2017, "12.1 A 7 nm 256 Mb SRAM in high-k metal-gate FinFET technology with write-assist circuitry for low-VMIN applications," 2017 IEEE Int. Solid-State Circuits Conf. (ISSCC), San Francisco, CA, USA, 2017, pp. 206–207, doi: 10.1109/ISSCC.2017.7870333.

[101] J. Kulkarni, J. Keane, K-H. Koo et al. 2017, "5.6 Mb/mm^2 1R1W 8T SRAM Arrays Operating Down to 560 mV Utilizing Small-Signal Sensing With Charge Shared Bitline and Asymmetric Sense Amplifier in 14 nm FinFET CMOS Technology," IEEE J. of Solid-State Circuits 52(1) 229–239, doi: 10.1109/JSSC.2016.2607219.

[102] E. Karl, Z. Guo, J.W. Conary et al. 2016, "A 0.6 V, 1.5 GHz 84 Mb SRAM in 14 nm FinFET CMOS Technology With Capacitive Charge-Sharing Write Assist Circuitry," IEEE J. of Solid-State Circuits 51(1) 222–229, doi: 10.1109/JSSC.2015.2461592.

[103] M-F. Chang, C-F. Chen, T-H. Chang et al. 2015, "17.3 A 28 nm 256 kb 6T-SRAM with 280mV improvement in VMIN using a dual-split-control assist scheme," 2015 IEEE Int. Solid-State Circuits Conf. – (ISSCC) Digest of Technical Papers, San Francisco, CA, USA, 2015, pp. 1–3, doi: 10.1109/JSSC.2015.2461592.

[104] Y-H. Chen, W-M. Chan, W-C. Wu et al. 2014, "13.5 A 16 nm 128 Mb SRAM in high-κ metal-gate FinFET technology with write-assist circuitry for low-VMIN applications," 2014 IEEE Int. Solid-State Circuits Conf. Digest of Technical Papers (ISSCC), San Francisco, CA, USA, 2014, pp. 238–239, doi: 10.1109/ISSCC.2014.6757416.

[105] J. Chang, Y-H. Chen, H. Cheng et al. 2013, "A 20 nm 112 Mb SRAM in High-κ metal-gate with assist circuitry for low-leakage and low-VMIN applications," 2013 IEEE Int. Solid-State Circuits Conf. Digest of Technical Papers, San Francisco, CA, USA, 2013, pp. 316–317, doi: 10.1109/ISSCC.2013.6487750.

[106] H. Pilo, C.A. Adams, I. Arsovski et al. 2013, "A 64 Mb SRAM in 22 nm SOI technology featuring fine-granularity power gating and low-energy power-supply-partition techniques for 37% leakage reduction," 2013 IEEE Int. Solid-State Circuits Conf. Digest of Technical Papers, San Francisco, CA, USA, 2013, pp. 322–323, doi: 10.1109/ISSCC.2013.6487753.

[107] E. Karl, Y. Wang, Y-G. Ng et al. 2012, "A 4.6 GHz 162 Mb SRAM design in 22 nm tri-gate CMOS technology with integrated active VMIN-enhancing assist circuitry," 2012 IEEE Int. Solid-State Circuits Conf., San Francisco, CA, USA, 2012, pp. 230–232, doi: 10.1109/ISSCC.2012.6176988.

[108] K. Miyaji, T. Suzuki, S. Miyano et al. 2012, "A 6T SRAM with a carrier-injection scheme to pinpoint and repair fails that achieves 57% faster read and 31% lower read energy," 2012 IEEE Int. Solid-State Circuits Conf., San Francisco, CA, USA, 2012, pp. 232–234, doi: 10.1109/ISSCC.2012.6176989.

[109] H. Pilo, I. Arsovski, K. Batson et al. 2011, "A 64 Mb SRAM in 32 nm High-k metal-gate SOI technology with 0.7V operation enabled by stability, write-ability and read-ability enhancements," 2011 IEEE Int. Solid-State Circuits Conf., San Francisco, CA, USA, 2011, pp. 254–256, doi: 10.1109/ISSCC.2011.5746307.

[110] M. E. Sinangil, H. Mair and A. P. Chandrakasan 2011, "A 28 nm high-density 6T SRAM with optimized peripheral-assist circuits for operation down to 0.6V," 2011 IEEE Int. Solid-State Circuits Conf., San Francisco, CA, USA, 2011, pp. 260–262, doi: 10.1109/ISSCC.2011.5746310.

[111] K. Nii, M. Yabuuchi, Y. Tsukamoto et al. 2010, "A 0.5V 100MHz PD-SOI SRAM with enhanced read stability and write margin by asymmetric MOSFET and forward body bias," 2010 IEEE Int. Solid-State Circuits Conf. – (ISSCC), San Francisco, CA, USA, 2010, pp. 356–357, doi: 10.1109/ISSCC.2010.5433817.

[112] Y. Wang, U. Bhattacharya, F. Hamzaoglu et al. 2009, "A 4.0 GHz 291 Mb voltage-scalable SRAM design in 32 nm high-κ metal-gate CMOS with integrated power management," 2009 IEEE Int. Solid-State Circuits Conf. – Digest of Technical Papers, San Francisco, CA, USA, 2009, pp. 456–457,457a, doi: 10.1109/ISSCC.2009.4977505.

[113] A. Kawasumi, T. Yabe, Y. Takeyama et al. 2008, "A Single-Power-Supply 0.7 V 1 GHz 45 nm SRAM with An Asymmetrical Unit-ß-ratio Memory Cell," 2008 IEEE Int. Solid-State Circuits Conf. – Digest of Technical Papers, San Francisco, CA, USA, 2008, pp. 382–622, doi: 10.1109/ISSCC.2008.4523217.

[114] M. Yamaoka, N. Maeda, Y. Shimazaki et al. 2008, "65 nm Low-Power High-Density SRAM Operable at 1.0V under 3σ Systematic Variation Using Separate Vth Monitoring and Body Bias for NMOS and PMOS," 2008 IEEE Int. Solid-State Circuits Conf. – Digest of Technical Papers, San Francisco, CA, USA, 2008, pp. 384–622, doi: 10.1109/ISSCC.2008.4523218.

[115] W. Shim and S. Yu 2021, "Ferroelectric Field-Effect Transistor-Based 3-D NAND Architecture for Energy-Efficient on-Chip Training Accelerator," IEEE J. on Exploratory Solid-State Computational Devices and Circuits 7(1) 1–9, June 2021, doi: 10.1109/JXCDC.2021.3057856.

[116] H. Guo, Y. Li, T. Ren et al. 2026, “Non-volatile SRAM based on emerging memory: A comprehensive review and reliability challenges,” Microelectronics Reliability 177, 115995, https://doi.org/10.1016/j.microrel.2025.115995

[117] S.S. Manohar, S. Agarwal and H.K. Kapoor 2019, “Towards Optimizing Refresh Energy in embedded-DRAM Caches using Private Blocks,” GLSVLSI ‘19: Proc. of the 2019 Great Lakes Symposium on VLSI, pp. 225 –230, https://doi.org/10.1145/3299874.3317995

[118] T. Kirihata, 2010, “High Performance Embedded Dynamic Random Access Memory in Nano-Scale Technologies,” In: Iniewski, K. (eds) CMOS Processors and Memories. Analog Circuits and Signal Processing. Springer, Dordrecht. https://doi.org/10.1007/978-90-481-9216-8_10

[119] S. Ha, S. Um, S. Kim et al. 2026, “Cache-PIM: An ECC-Compatible eDRAM Processing-in-Memory for Last-Level Cache With Triple-Level Error Correction,” IEEE J. of Solid-State Circuits, 61(2) 750–762, doi: 10.1109/JSSC.2025.3568485.

[120] W. Liu, F. Wu, X. Chen et al., “Characterization Summary of Performance, Reliability, and Threshold Voltage Distribution of 3D Charge-Trap NAND Flash Memory,” ACM Transactions on Storage (TOS), vol. 18, is. 2, 16, pp; 1–25, 2022, https://doi.org/10.1145/349123

[121] L. Wang, C-H. Yang and J. Wen 2015, “Physical principles and current status of emerging non-volatile solid state memories,” Electron. Mater. Lett. 11, 505–543, https://doi.org/10.1007/s13391-015-4431-4

[122] J. Choe 2025, “Memory Technology Trends & Outlook,” The Future of Memory and Storage, August 2025, https://files.futurememorystorage.com/proceedings/2025/20250807_DRAM-304-1_Jeongdong-Choe_Short-version.pdf

[123] P. Venkatesan, L. Fernandes, S. Kang et al. 2025, “Pushing the limits of NAND technology scaling with ferroelectrics,” MRS Bulletin 50, 1094–1107, https://doi.org/10.1557/s43577-025-00991-y

[124] W. Cho, C. Jeong, J.Kim et al. 2025, “A 321-Layer 2Tb 4b/cell 3D-NAND-Flash Memory with a 75 MB/s Program Throughput,” 2025 IEEE Int. Solid-State Circuits Conf. (ISSCC), San Francisco, CA, USA, 2025, pp. 512–514, doi: 10.1109/ISSCC49661.2025.10904748.

[125] S-S. Park, J-D. Lyu, M. Kim et al., “30.1 A 28 Gb/mm^2 4XX-Layer 1Tb 3b/Cell WF-Bonding 3D-NAND Flash with 5.6 Gb/s/Pin IOs,” 2025 IEEE Int. Solid-State Circuits Conf. (ISSCC), San Francisco, CA, USA, 2025, pp. 1–3, doi: 10.1109/ISSCC49661.2025.10904543.

[126] J. Choe, “Memory Technology 2021: Trends & Challenges,” 2021 Int. Conf. on Simulation of Semiconductor Processes and Devices (SISPAD), Dallas, TX, USA, 2021, pp. 111–115, doi: 10.1109/SISPAD54002.2021.9592547.

[127] “How AI Is Transforming Data Centers and Ramping Up Power Demand,” August 29, 2025, https://www.goldmansachs.com/insights/articles/how-ai-is-transforming-data-centers-and-ramping-up-power-demand

[128] IEA (2025), Energy and AI, IEA, Paris https://www.iea.org/reports/energy-and-ai

[129] IEA (2024), World Energy Outlook 2024, IEA, Paris https://www.iea.org/reports/world-energy-outlook-2024

[130] K. M. Stiefel and J. S. Coggan, “The energy challenges of artificial superintelligence,” Front Artif Intell. vol. 24, no. 6, 1240653, doi: 10.3389/frai.2023.1240653.

[131] Guo K., Li W., Zhong K. et al. “Neural Network Accelerator Comparison” [Online]. Available: https://nicsefc.ee.tsinghua.edu.cn/projects/neural-network-accelerator/ – accessed in April 24, 2025

[132] M. Nouripayam, A. Prieto, V. K. Kishorelal and J. Rodrigues, “An Energy-Efficient Near-Memory Computing Architecture for CNN Inference at Cache Level,” 2021 28th IEEE International Conference on Electronics, Circuits, and Systems (ICECS), Dubai, United Arab Emirates, 2021, pp. 1–4, doi: 10.1109/ICECS53924.2021.9665530.

[133] U. Rueckert, “a,” In: Murmann, B., Hoefflinger, B. (eds) NANO-CHIPS 2030. The Frontiers Collection. Springer, Cham., 2020, https://doi.org/10.1007/978-3-030-18338-7_12

[134] T. P. Xiao, C. H. Bennett, B. Feinberg, S. Agarwal, M. J. Marinella, "Analog architectures for neural network acceleration based on non-volatile memory," Appl. Phys. Rev. vol. 7, no: 3, Art.: 031301, September 2020, https://doi.org/10.1063/1.5143815

[135] A. Aimar et al., "NullHop: A Flexible Convolutional Neural Network Accelerator Based on Sparse Representations of Feature Maps," in IEEE Transactions on Neural Networks and Learning Systems, vol. 30, no. 3, pp. 644–656, March 2019, doi: 10.1109/TNNLS.2018.2852335.

[136] V. Gokhale, A. Zaidy, A. X. M. Chang and E. Culurciello, "Snowflake: An efficient hardware accelerator for convolutional neural networks," 2017 IEEE International Symposium on Circuits and Systems (ISCAS), Baltimore, MD, USA, 2017, pp. 1–4, doi: 10.1109/ISCAS.2017.8050809.

[137] J. Kim, J. Kang and Y. Kim 2026, "AHCO-YOLO: An Algorithm–Hardware Co-Optimization Framework for Energy-Efficient and Real-Time Object Detection on Edge Devices," IEEE Transactions on Very Large Scale Integration (VLSI) Systems 34(02) 402–415, doi: 10.1109/TVLSI.2025.3621624.

[138] S. Datta, W. Chakraborty and M. Radosavljevic, "Toward attojoule switching energy in logic transistors," Science vol. 378, no. 6621, pp. 733–740, November 2022, doi: 10.1126/science.ade7656.

[139] J-S. Park, T. Lee, H. Lee et al., "16.3 An on-Device Generative AI Focused Neural Processing Unit in 4 nm Flagship Mobile SoC with Fan-Out Wafer-Level Package," 2025 IEEE Int. Solid-State Circuits Conf. (ISSCC), San Francisco, CA, USA, 2025, pp. 286–288, doi: 10.1109/ISSCC49661.2025.10904722.

[140] C-H. Yu, H-E. Kim, S. Shin et al., "2.4 ATOMUS: A 5 nm 32TFLOPS/128TOPS ML System-on-Chip for Latency Critical Applications," 2024 IEEE Int. Solid-State Circuits Conf. (ISSCC), San Francisco, CA, USA, 2024, pp. 42–44, doi: 10.1109/ISSCC49657.2024.10454509.

[141] P. A. Hager, B. Moons, S. Cosemans et al., "11.3 Metis AIPU: A 12 nm 15TOPS/W 209.6TOPS SoC for Cost- and Energy-Efficient Inference at the Edge," 2024 IEEE Int. Solid-State Circuits Conf. (ISSCC), San Francisco, CA, USA, 2024, pp. 212–214, doi: 10.1109/ISSCC49657.2024.10454395.

[142] H. Fujiwara, H. Mori, W-C. Zhao et al., "34.4 A 3 nm, 32.5TOPS/W, 55.0TOPS/mm^2 and 3.78 Mb/mm^2 Fully-Digital Compute-in-Memory Macro Supporting INT12 × INT12 with a Parallel-MAC Architecture and Foundry 6T-SRAM Bit Cell," 2024 IEEE Int. Solid-State Circuits Conf. (ISSCC), San Francisco, CA, USA, 2024, pp. 572–574, doi: 10.1109/ISSCC49657.2024.10454556.

[143] A. Agrawal, S.K. Lee, J. Silberman et al., "9.1 A 7 nm 4-Core AI Chip with 25.6TFLOPS Hybrid FP8 Training, 102.4TOPS INT4 Inference and Workload-Aware Throttling," 2021 IEEE Int. Solid-State Circuits Conf. (ISSCC), San Francisco, CA, USA, 2021, pp. 144–146, doi: 10.1109/ISSCC42613.2021.9365791.

[144] C-H. Lin, C-C. Cheng, Y-M. Tsai et al., "7.1 A 3.4-to-13.3TOPS/W 3.6TOPS Dual-Core Deep-Learning Accelerator for Versatile AI Applications in 7 nm 5G Smartphone SoC," 2020 IEEE Int. Solid-State Circuits Conf. – (ISSCC), San Francisco, CA, USA, 2020, pp. 134–136, doi: 10.1109/ISSCC19947.2020.9063111.

[145] D. Gil and W. M. J. Green, "1.4 The Future of Computing: Bits + Neurons + Qubits," 2020 IEEE Int. Solid-State Circuits Conf. – (ISSCC), San Francisco, CA, USA, 2020, pp. 30–39, doi: 10.1109/ISSCC19947.2020.9062918.

[146] S. Alam, C. Yakopcic, Q. Wu et al. 2024, "Survey of Deep Learning Accelerators for Edge and Emerging Computing," Electronics 13, 2988, https://doi.org/10.3390/electronics13152988

[147] C. Silvano, D. Ielmini, F. Ferrandi et al. 2025, "A Survey on Deep Learning Hardware Accelerators for Heterogeneous HPC Platforms," ACM Computing Surveys 57(11) Art. 286, https://doi.org/10.1145/3729215

[148] R. Merritt 2019, "Samsung, Toshiba Detail AI Chips," February 20, 2019, https://www.eetimes.com/samsung-toshiba-detail-ai-chips/

[149] C. Åleskog, H. Grahn and A. Borg 2022, "Recent Developments in Low-Power AI Accelerators: A Survey," Algorithms 15, 419, https://doi.org/10.3390/a15110419

[150] R Sindhu and Arunachalam V 2026, "Energy-efficient scalable memory LSTM cell with approximate feature processing units," Results in Engineering 29, 108517, https://doi.org/10.1016/j.rineng.2025.108517

[151] L. Du, Y. Du, Y. Li et al. 2028, "A Reconfigurable Streaming Deep Convolutional Neural Network Accelerator for Internet of Things," IEEE Transactions on Circuits and Systems I: Regular Papers, 65(1) 198–208, doi: 10.1109/TCSI.2017.2735490.

[152] Y. Jiao, L. Han, R. Jin et al. 2020, "7.2 A 12 nm Programmable Convolution-Efficient Neural-Processing -Unit Chip Achieving 825TOPS," 2020 IEEE Int. Solid-State Circuits Conf. – (ISSCC), San Francisco, CA, USA, 2020, pp. 136–140, doi: 10.1109/ISSCC19947.2020.9062984.

[153] S. Park, K. Bong, D. Shin et al. 2015, "4.6 A1.93TOPS/W scalable deep learning/inference processor with tetra-parallel MIMD architecture for big-data applications," 2015 IEEE Int. Solid-State Circuits Conf. – (ISSCC) Digest of Technical Papers, San Francisco, CA, USA, 2015, pp. 1–3, doi: 10.1109/ISSCC.2015.7062935.

[154] Y. Suh, J. Deng, J. Yuan et al., "Comprehensive Enhancements for AI Mobile Edge Computing from Device-Level to System and Packaging Optimization," 2025 IEEE Int. Electron Devices Meet. (IEDM), San Francisco, CA, USA, 2025, pp. 1–4, doi: 10.1109/IEDM50572.2025.11353531.

[155] M. Garcia Bardon, P. Wuytens, L.-A. Raggnarsson et al., "DTCO including Sustainability: Power-Performance-Area-Cost-Environmental score (PPACE) Analysis for Logic Technologies," 2020 IEEE International Electron Devices Meeting (IEDM), San Francisco, CA, USA, 2020, pp. 41.4.1–41.4.4, doi: 10.1109/IEDM13553.2020.9372004.

[156] "The adoption of ASML's High-NA EUV lithography tools is being delayed by major chipmakers due to their extremely high cost – about $360–400 million per," https://semiwiki.com/forum/threads/the-adoption-of-asmls-high-na-euv-lithography-tools-is-being-delayed-by-major-chipmakers-due-to-their-extremely-high-cost–about-360–400-million-per.23214/

[157] F. Hady, "AI Data Processing: Near-Memory Compute for Energy-Efficient Systems," April 12, 2022, https://medium.com/intel-tech/ai-data-processing-near-memory-compute-for-energy-efficient-systems-c13f43320811

[158] C. Liu, R. Abdul Khadar and E. Matioli, "GaN-on-Si Quasi-Vertical Power MOSFETs," IEEE Electron Device Letters, vol. 39, no. 1, pp. 71–74, Jan. 2018, doi: 10.1109/LED.2017.2779445.

[159] W. Haensch, A. Raghunathan, K. Roy et al., "Compute in-Memory with Non-Volatile Elements for Neural Networks: A Review from a Co-Design Perspective," Adv. Mater. vol. 35, 2204944, 2023, DOI: 10.1002/adma.202204944.

[160] M.-J. Park et al., "A 192-Gb 12-High 896-GB/s HBM3 DRAM with a TSV Auto-Calibration Scheme and Machine-Learning-Based Layout Optimization," 2022 IEEE International Solid-State Circuits Conference (ISSCC), San Francisco, CA, USA, 2022, pp. 444–446, doi: 10.1109/ISSCC42614.2022.9731562.

[161] J. Lim, Y. G. Kim, S. W. Chung, F. Koushanfar and J. Kong, "Near-Memory Computing With Compressed Embedding Table for Personalized Recommendation," in IEEE Transactions on Emerging Topics in Computing, vol. 12, no. 3, pp. 938–951, July-Sept. 2024, doi: 10.1109/TETC.2023.3345870.

[162] B. Kim et al., "The Breakthrough Memory Solutions for Improved Performance on LLM Inference," IEEE Micro, vol. 44, no. 3, pp. 40–48, May-June 2024, doi: 10.1109/MM.2024.3375352.

[163] A. Xu, C. Deng, J. Zhu et al. "Software-defined process-near-memory architecture using 3D hybrid bonding integration". Sci. China Inf. Sci. vol. 68, no: 112402, 2025, https://doi.org/10.1007/s11432-023-3965-1

[164] P. Mannocci, C. Zucchelli, I. Andreoli et al. 2026, "A fully integrated analogue closed-loop in-memory computing accelerator based on static random-access memory," Nature Electronics (2026), https://doi.org/10.1038/s41928-025-01549-1

[165] K. Prasad, R. Borkar and J. Mekie 2025, “Robust and High-Performance Digital In-Memory Computing in 5T Gain Cell Embedded DRAM”. In: Gupta, A., Pandey, J.G., Chaturvedi, N., Dwivedi, D. (eds) VLSI for Embedded Intelligence. VDAT 2023. Lecture Notes in Electrical Engineering, vol 1210. Springer, Singapore, https://doi.org/10.1007/978-981-97-3756-7_6

[166] H. Jiang, S. Huang and S. Yu 2025, “Compute-in-Memory Architecture,” In: Chattopadhyay, A. (eds) Handbook of Computer Architecture. Springer, Singapore, https://doi.org/10.1007/978-981-97-9314-3_62

[167] D. Kim, H. Jeong, S. Kim et al. 2025, “A Multibit ReRAM Computing-in-Memory Processor With Adaptive Decision Level Nonlinear ADC for Ultra-Low-Energy Keyword Spotting in Mobile Devices,” IEEE Transactions on Circuits and Systems I: Regular Papers 73(2) 862–874, doi: 10.1109/TCSI.2025.3603076.

[168] K. Lee, S. Cho, P.-K. Hsu et al. 2026, “Monolithically Stackable 1T1C 3-D DRAM: A Technical Survey,” IEEE Transactions on Electron Devices 73(3) 1092–1112, doi: 10.1109/TED.2026.3659106.

[169] Y. Cho, D-w. Kim, Y. Kimh et al. 2026, “Emerging Memory and Device Technologies for Hardware-Accelerated Model Training and Inference,” Advanced Electronic Materials, 2026; 0:e00796, https://doi.org/10.1002/aelm.202500796

[170] H. Wang, R. Liu, R. Dorrance et al., “A PVT Robust 8-Bit Signed Analog Compute-In-Memory Accelerator with Integrated Activation Functions for AI Applications,” 2024 IEEE Symp. on VLSI Technology and Circuits (VLSI Technology and Circuits), Honolulu, HI, USA, 2024, pp. 1–2, doi: 10.1109/VLSITechnologyandCir46783.2024.10631356.

[171] S. Kim and H-J. Yoo 2024, “An Overview of Computing-in-Memory Circuits With DRAM and NVM,” IEEE Transactions on Circuits and Systems II: Express Briefs 71(3) 1626–1631, doi: 10.1109/TCSII.2023.3333851.

[172] Z. Lin, Z. Tong, J. Zhang et al. 2022, “A review on SRAM-based computing in-memory: Circuits, functions, and applications,” J. Semicond. 43, 031401, DOI: 10.1088/1674-4926/43/3/031401.

[173] H. Jiang, X. Peng, S. Huang 2020, “CIMAT: A Compute-In-Memory Architecture for On-chip Training Based on Transpose SRAM Arrays,” IEEE Transactions on Computers 69(7) 944–954, doi: 10.1109/TC.2020.2980533.

[174] V. Iskandar, M.A.A El Ghany and D. Göhringer 2022, “Near-memory Computing on FPGAs with 3D-stacked Memories: Applications, Architectures, and Optimizations,” ACM Transactions on Reconfigurable Technology and Systems, 16(1), Art. 16, pp 1 –32, https://doi.org/10.1145/3547658

[175] Y. Li et al. 2023, “A Survey of MRAM-Centric Computing: From Near Memory to In Memory,” in IEEE Transactions on Emerging Topics in Computing, vol. 11, no. 2, pp. 318–330, April-June 2023, doi: 10.1109/TETC.2022.3214833.

[176] H. Kim, T. Yoo, T. T.-H. Kim and B. Kim, 2021, “Colonnade: A Reconfigurable SRAM-Based Digital Bit-Serial Compute-In-Memory Macro for Processing Neural Networks,” in IEEE Journal of Solid-State Circuits, vol. 56, no. 7, pp. 2221–2233, July 2021, doi: 10.1109/JSSC.2021.3061508.

[177] H. Cai, Y. Guo, B. Liu et al. 2022, “Proposal of Analog In-Memory Computing With Magnified Tunnel Magnetoresistance Ratio and Universal STT-MRAM Cell,” in IEEE Transactions on Circuits and Systems I: Regular Papers, vol. 69, no. 4, pp. 1519–1531, April 2022, doi: 10.1109/TCSI.2022.3140769.

[178] J. Gómez-Luna, I. E. Hajj, I. Fernandez, C. Giannoula, G. F. Oliveira and O. Mutlu 2022, “Benchmarking a New Paradigm: Experimental Analysis and Characterization of a Real Processing-in-Memory System,” in IEEE Access, vol. 10, pp. 52565–52608, 2022, doi: 10.1109/ACCESS.2022.3174101.

[179] S. K. Gonugondla, M. Kang and N. R. Shanbhag 2018, “A Variation-Tolerant In-Memory Machine Learning Classifier via On-Chip Training,” in IEEE Journal of Solid-State Circuits, vol. 53, no. 11, pp. 3163–3173, Nov. 2018, doi: 10.1109/JSSC.2018.2867275.

[180] J. Wang et al. 2020, "A 28-nm Compute SRAM With Bit-Serial Logic/Arithmetic Operations for Programmable In-Memory Vector Computing," in IEEE Journal of Solid-State Circuits, vol. 55, no. 1, pp. 76–86, Jan. 2020, doi: 10.1109/JSSC.2019.2939682.

[181] N. R. Shanbhag and S. K. Roy 2022, "Benchmarking In-Memory Computing Architectures," in IEEE Open Journal of the Solid-State Circuits Society, vol. 2, pp. 288–300, 2022, doi: 10.1109/OJSSCS.2022.3210152.

[182] A. Baron, A. Glukhov, E. Pérez et al. 2022, "An energy-efficient in-memory computing architecture for survival data analysis based on resistive switching memories," Frontiers in Neuroscience | Neural Technology, vol. 16, Art. 932270, https://doi.org/10.3389/fnins.2022.932270

[183] S. Yu, H. Jiang, S. Huang, X. Peng and A. Lu 2021, "Compute-in-Memory Chips for Deep Learning: Recent Trends and Prospects," in IEEE Circuits and Systems Magazine, vol. 21, no. 3, pp. 31–56, third quarter 2021, doi: 10.1109/MCAS.2021.3092533.

[184] X. Yin et al. 2025, "A Homogeneous FeFET-Based Time-Domain Compute-in-Memory Fabric for Matrix-Vector Multiplication and Associative Search," in IEEE Transactions on Computer-Aided Design of Integrated Circuits and Systems, vol. 44, no. 5, pp. 1856–1868, May 2025, doi: 10.1109/TCAD.2024.3492994.

[185] G. Desoli et al. 2023, "16.7 A 40-310TOPS/W SRAM-Based All-Digital Up to 4b In-Memory Computing Multi-Tiled NN Accelerator in FD-SOI 18 nm for Deep-Learning Edge Applications," 2023 IEEE International Solid-State Circuits Conference (ISSCC), San Francisco, CA, USA, 2023, pp. 260–262, doi: 10.1109/ISSCC42615.2023.10067422.

[186] Y-C. Chiu, W.-S. Khwa, C.-Y. Li et al. 2023, "A 22 nm 8 Mb STT-MRAM Near-Memory-Computing Macro with 8b-Precision and 46.4-160.1TOPS/W for Edge-AI Devices," 2023 IEEE Int. Solid-State Circuits Conf. (ISSCC), San Francisco, CA, USA, 2023, pp. 496–498, doi: 10.1109/ISSCC42615.2023.10067563.

[187] S. Jung, H. Lee, S. Myung et al. 2022, "A crossbar array of magnetoresistive memory devices for in-memory computing," Nature 601, 211–216, https://doi.org/10.1038/s41586-021-04196-6

[188] S. D. Spetalnick, M. Chang, S. Konno et al. 2023, "A 2.38 MCells/mm^2 9.81 -350 TOPS/W RRAM Compute-in-Memory Macro in 40 nm CMOS with Hybrid Offset/IOFF Cancellation and ICELL RBLSL Drop Mitigation," 2023 IEEE Symp. on VLSI Technology and Circuits (VLSI Technology and Circuits), Kyoto, Japan, 2023, pp. 1–2, doi: 10.23919/VLSITechnologyandCir57934.2023.10185424.

[189] P. Deaville, B. Zhang and N. Verma 2025, "Analyzing Embedded Non-Volatile Memory and Technology Optimizations for In-Memory Computing," IEEE J. of Solid-State Circuits, November 2025, pp. 1–13, doi: 10.1109/JSSC.2025.3632212.

[190] R. Mochida, K. Kouno, Y. Hayata et al. 2018, "A 4M Synapses integrated Analog ReRAM based 66.5 TOPS/W Neural-Network Processor with Cell Current Controlled Writing and Flexible Network Architecture," 2018 IEEE Symp. on VLSI Technology, Honolulu, HI, USA, 2018, pp. 175–176, doi: 10.1109/VLSIT.2018.8510676.

[191] W. Li, S. Huang, X. Sun et al. 2021, "Secure-RRAM: A 40 nm 16 kb Compute-in-Memory Macro with Reconfigurability, Sparsity Control, and Embedded Security," 2021 IEEE Custom Integrated Circuits Conference (CICC), Austin, TX, USA, 2021, pp. 1–2, doi: 10.1109/CICC51472.2021.9431558.

[192] R. Khaddam-Aljameh, M. Stanisavljevic, J. Fornt Mas et al. 2022, "HERMES-Core – A 1.59-TOPS/mm^2 PCM on 14-nm CMOS In-Memory Compute Core Using 300-ps/LSB Linearized CCO-Based ADCs," IEEE Journal of Solid-State Circuits, 57(4) 1027–1038, April 2022, doi: 10.1109/JSSC.2022.3140414.

[193] H. Jiang, S. Huang, W. Li et al. 2023, "ENNA: An Efficient Neural Network Accelerator Design Based on ADC-Free Compute-In-Memory Subarrays," IEEE Transactions on Circuits and Systems I: Regular Papers, 70(1). 353–363, Jan. 2023, doi: 10.1109/TCSI.2022.3208755.

[194] L. Wang, Y. Zhang, P. Wang et al. 2025, "A 578-TOPS/W RRAM-Based Binary Convolutional Neural Network Macro for Tiny AI Edge Devices," IEEE Transactions on Very Large Scale Integration (VLSI) Systems, 33(2) 371–383, Feb. 2025, doi: 10.1109/TVLSI.2024.3469217.

[195] Y. Li, J. Chen, L. Wang et al. 2023, "An ADC-Less RRAM-Based Computing-in-Memory Macro With Binary CNN for Efficient Edge AI," IEEE Transactions on Circuits and Systems II: Express Briefs, 70(6) 1871–1875, June 2023, doi: 10.1109/TCSII.2022.3233396.
[196] S. Kim, Z. Li, S. Um et al. 2024, "DynaPlasia: An eDRAM In-Memory Computing-Based Reconfigurable Spatial Accelerator with Triple-Mode Cell," IEEE J. of Solid-State Circuits, 59(1) 102–115, Jan. 2024, doi: 10.1109/JSSC.2023.3319962.
[197] T. Kimand Y. Chung 2024, "Logic-Compatible Embedded DRAM Architecture for Multifunctional Digital Storage and Compute-in-Memory," Applied Sciences, 14(21), 9749. https://doi.org/10.3390/app14219749
[198] S. Ha, S. Kim, D. Han et al. 2022, "A 36.2 dB High SNR and PVT/Leakage-Robust eDRAM Computing-In-Memory Macro With Segmented BL and Reference Cell Array," IEEE Transactions on Circuits and Systems II: Express Briefs, 69(5) 2433–2437, May 2022, doi: 10.1109/TCSII.2022.3159808.
[199] S. Kim, S. Um, W. Jo et al. 2024, "Scaling-CIM: eDRAM In-Memory-Computing Accelerator With Dynamic-Scaling ADC and Adaptive Analog Operation," IEEE J. of Solid-State Circuits, 59(8) 2694–2705, Aug. 2024, doi: 10.1109/JSSC.2024.3362699.
[200] X. Qiao, Y. Yang, C. Xue et al. 2024, "An eDRAM-Based Computing-in-Memory Macro With Full-Valid-Storage and Channel-Wise-Parallelism for Depthwise Neural Network," IEEE Transactions on Circuits and Systems II: Express Briefs, 71(5) 2539–2543, May 2024, doi: 10.1109/TCSII.2024.3375319.
[201] S. Whang, S. Ha, S. Um et al. 2025, "A 62.8 TOPS/W FP-INT Digital Computing-in Memory Processor with Bit-Reordered Adder Tree and Low Active Hierarchical Accumulator," 2025 IEEE Int. Symp. on Circuits and Systems (ISCAS), London, United Kingdom, 2025, pp. 1–5, doi: 10.1109/ISCAS56072.2025.11043177.
[202] S. An, H. Jeong, S. Kim et al. 2025, "A 46 TOPS/W In-/Near-Memory Computing Processor for Large Language Model with Extended Sparse Attention," 2025 IEEE Int. Symp. on Circuits and Systems (ISCAS), London, United Kingdom, 2025, pp. 1–5, doi: 10.1109/ISCAS56072.2025.11043292.
[203] W. Li, J. Zhu, Z. Zhou et al. 2025, "A 1763.9TOPS/W 520kb Scalable FeFET-Based Analog Computing-in -Memory Chip Using BEOL HZO Capacitor for On-chip Learning," 2025 IEEE Int. Electron Devices Meet. (IEDM), San Francisco, CA, USA, 2025, pp. 1–4, doi: 10.1109/IEDM50572.2025.11353707.
[204] J.M. Hung, C.X. Xue, H.Y. Kao, et al. 2021, "A four-megabit compute-in-memory macro with eight-bit precision based on CMOS and resistive random-access memory for AI edge devices," Nature Electronics 4, 921–930, https://doi.org/10.1038/s41928-021-00676-9
[205] Q. Huo, Y. Yang, Y. Wang et al. 2022, "A computing-in-memory macro based on three-dimensional resistive random-access memory," Nature Electronics 5, 469–477, https://doi.org/10.1038/s41928-022-00795-x
[206] W. Ye, Y. Liang, Y. Liu et al. 2026, "SConvNSys: Accelerating Spiking Convolutional Neural Networks With a Reconfigurable Neuromorphic Architecture for Diverse Applications" IEEE Transactions on Very Large Scale Integration (VLSI) Systems 34(02) 430–443, doi: 10.1109/TVLSI.2025.3629750.
[207] L. Yan, Z. Zhan, Z. Pan et al. 2025, "An 8 Mb Learning-Aware RRAM Compute-in-Memory Accelerator for Embodied Self-Supervised Learning," 2025 IEEE Int. Electron Devices Meet. (IEDM), San Francisco, CA, USA, 2025, pp. 1–4, doi: 10.1109/IEDM50572.2025.11353856.
[208] R. Zhu, Z. Ning, H. Shao et al. 2025, "A BEOL FeFET based Multi-bit ACiM Macro with High Accuracy and Throughput via Device-Array-System Co-Optimization for Edge LM," 2025 IEEE Int. Electron Devices Meet. (IEDM), San Francisco, CA, USA, 2025, pp. 1–4, doi: 10.1109/IEDM50572.2025.11353662.
[209] T. Bai, W. Mao, G. Wang et al. 2024, "An End-to-End In-Memory Computing System Based on a 40-nm eFlash-Based IMC SoC: Circuits, Toolchains, and Systems Co-Design Framework," in IEEE Transactions on Computer-Aided Design of Integrated Circuits and Systems 43(6) 1729–1740, doi: 10.1109/TCAD.2024.3349502.
[210] P-C. Wu, J-W. Su, Y-L. Chung et al. 2022, "A 28 nm 1 Mb Time-Domain Computing-in-Memory 6T-SRAM Macro with a 6.6 ns Latency, 1241GOPS and 37.01TOPS/W for 8b-MAC Operations for Edge-AI

Devices," 2022 IEEE Int. Solid-State Circuits Conf. (ISSCC), San Francisco, CA, USA, 2022, pp. 1–3, doi: 10.1109/ISSCC42614.2022.9731681.

[211] F. Tu, Y. Wang, Z. Wu et al. 2022, "A 28 nm 29.2TFLOPS/W BF16 and 36.5TOPS/W INT8 Reconfigurable Digital CIM Processor with Unified FP/INT Pipeline and Bitwise In-Memory Booth Multiplication for Cloud Deep Learning Acceleration," 2022 IEEE Int. Solid-State Circuits Conf. (ISSCC), San Francisco, CA, USA, 2022, pp. 1–3, doi: 10.1109/ISSCC42614.2022.9731762.

[212] Y. Jing, J. Zhou, Y. Sun et al. 2025, "37.6 A 22 nm 60.81TFLOPS/W Diffusion Accelerator with Bandwidth-Aware Memory Partition and BL-Segmented Compute-in-Memory for Efficient Multi-Task Content Generation," 2025 IEEE Int. Solid-State Circuits Conf. (ISSCC), San Francisco, CA, USA, 2025, pp. 1–3, doi: 10.1109/ISSCC49661.2025.10904573.

[213] Y. Yuan, Y. Yang, X. Wang et al. 2024, "34.6 A 28 nm 72.12TFLOPS/W Hybrid-Domain Outer-Product Based Floating-Point SRAM Computing-in-Memory Macro with Logarithm Bit-Width Residual ADC," 2024 IEEE Int. Solid-State Circuits Conf. (ISSCC), San Francisco, CA, USA, 2024, pp. 576–578, doi: 10.1109/ISSCC49657.2024.10454313.

[214] P-C. Wu, J-W. Su, L-Y. Hong et al. 2023, "A 22 nm 832Kb Hybrid-Domain Floating-Point SRAM In-Memory-Compute Macro with 16.2-70.2TFLOPS/W for High-Accuracy AI-Edge Devices," 2023 IEEE Int. Solid-State Circuits Conf. (ISSCC), San Francisco, CA, USA, 2023, pp. 126–128, doi: 10.1109/ISSCC42615.2023.10067527.

[215] Y. He, H. Diao, C. Tang et al. 2023, "7.3 A 28 nm 38-to-102-TOPS/W 8b Multiply-Less Approximate Digital SRAM Compute-In-Memory Macro for Neural-Network Inference," 2023 IEEE Int. Solid-State Circuits Conf. (ISSCC), San Francisco, CA, USA, 2023, pp. 130–132, doi: 10.1109/ISSCC42615.2023.10067305.

[216] H. Mori, W-C. Zhao, C-E. Lee et al. 2023, "A 4 nm 6163-TOPS/W/b 4790–TOPS/mm^2/b SRAM Based Digital-Computing-in-Memory Macro Supporting Bit-Width Flexibility and Simultaneous MAC and Weight Update," 2023 IEEE Int. Solid-State Circuits Conf. (ISSCC), San Francisco, CA, USA, 2023, pp. 132–134, doi: 10.1109/ISSCC42615.2023.10067555.

[217] B. Wang, C. Xue, Z. Feng et al. 2023, "A 28 nm Horizontal-Weight-Shift and Vertical-feature-Shift-Based Separate-WL 6T-SRAM Computation-in-Memory Unit-Macro for Edge Depthwise Neural-Networks," 2023 IEEE Int. Solid-State Circuits Conf. (ISSCC), San Francisco, CA, USA, 2023, pp. 134–136, doi: 10.1109/ISSCC42615.2023.10067526.

[218] S-E. Hsieh, C-H. Wei, C-X. Xue et al. 2023, "7.6 A 70.85-86.27TOPS/W PVT-Insensitive 8b Word-Wise ACIM with Post-Processing Relaxation," 2023 IEEE Int. Solid-State Circuits Conf. (ISSCC), San Francisco, CA, USA, 2023, pp. 136–138, doi: 10.1109/ISSCC42615.2023.10067335.

[219] S. Liu, P. Li, J. Zhang et al. 2023, "16.2 A 28 nm 53.8TOPS/W 8b Sparse Transformer Accelerator with In-Memory Butterfly Zero Skipper for Unstructured-Pruned NN and CIM-Based Local-Attention-Reusable Engine," 2023 IEEE Int. Solid-State Circuits Conf. (ISSCC), San Francisco, CA, USA, 2023, pp. 250–252, doi: 10.1109/ISSCC42615.2023.10067360.

[220] J. Yue, C. He, Z. Wang et al. 2023, "A 28 nm 16.9-300TOPS/W Computing-in-Memory Processor Supporting Floating-Point NN Inference/Training with Intensive-CIM Sparse-Digital Architecture," 2023 IEEE Int. Solid-State Circuits Conf. (ISSCC), San Francisco, CA, USA, 2023, pp. 1–3, doi: 10.1109/ISSCC42615.2023.10067779.

[221] F. Tu, Y. Wang, Z. Wu et al. 2023, "16.4 TensorCIM: A 28 nm 3.7 nJ/Gather and 8.3TFLOPS/W FP32 Digital-CIM Tensor Processor for MCM-CIM-Based Beyond-NN Acceleration," 2023 IEEE Int. Solid-State Circuits Conf. (ISSCC), San Francisco, CA, USA, 2023, pp. 254–256, doi: 10.1109/ISSCC42615.2023.1006728.

[222] G. Desoli, N. Chawla, T. Boesch et al. 2023, "16.7 A 40-310TOPS/W SRAM-Based All-Digital Up to 4b In-Memory Computing Multi-Tiled NN Accelerator in FD-SOI 18 nm for Deep-Learning Edge Applications," 2023 IEEE Int. Solid-State Circuits Conf. (ISSCC), San Francisco, CA, USA, 2023, pp. 260–262, doi: 10.1109/ISSCC42615.2023.10067422.

[223] S-J. Yun, J. Lee, S. Myung et al. 2025, "5-nm High-Efficiency and High-Density Digital SRAM In-Memory-Computing Macros for AI Accelerators," IEEE Solid-State Circuits Letters, 8, pp. 269–272, 2025, doi: 10.1109/LSSC.2025.3532788.

[224] H. Jia, M. Ozatay, Y. Tang et al. 2021, "15.1 A Programmable Neural-Network Inference Accelerator Based on Scalable In-Memory Computing," 2021 IEEE Int. Solid-State Circuits Conf. (ISSCC), San Francisco, CA, USA, 2021, pp. 236–238, doi: 10.1109/ISSCC42613.2021.9365788.

[225] J. Yue, X. Feng, Y. He et al. 2021, "15.2 A 2.75-to-75.9TOPS/W Computing-in-Memory NN Processor Supporting Set-Associate Block-Wise Zero Skipping and Ping-Pong CIM with Simultaneous Computation and Weight Updating," 2021 IEEE Int. Solid-State Circuits Conf. (ISSCC), San Francisco, CA, USA, 2021, pp. 238–240, doi: 10.1109/ISSCC42613.2021.9365958.

[226] Y-D. Chih, P-H. Lee, H. Fujiwara et al. 2021, "16.4 An 89TOPS/W and 16.3TOPS/mm^2 All-Digital SRAM-Based Full-Precision Compute-In Memory Macro in 22 nm for Machine-Learning Edge Applications," 2021 IEEE Int. Solid-State Circuits Conf. (ISSCC), San Francisco, CA, USA, 2021, pp. 252–254, doi: 10.1109/ISSCC42613.2021.9365766.

[227] S. Zhou and Y. Jiang 2025, "A comparison of computing-in-memory with non-volatile memory types and SRAM in DNN training," AIP Advances 15, 035317, doi: 10.1063/9.0000891.

[228] Y-C. Guo, T-S. Chang, C-S. Lin et al. 2024, "CIMR-V: An End-to-End SRAM-based CIM Accelerator with RISC-V for AI Edge Device," 2024 IEEE Int. Symp. on Circuits and Systems (ISCAS), Singapore, Singapore, 2024, pp. 1–5, doi: 10.1109/ISCAS58744.2024.10558177.

[229] L. Bao, J. Zhao, Z. Wang et al. 2025, "Hybrid Integration of IAZO/Si-based 2T0C DRAM with RRAM on CMOS Platform Enabling Bit-reconfigurable Mask for Multitask In-memory Computing," 2025 IEEE Int. Electron Devices Meet. (IEDM), San Francisco, CA, USA, 2025, pp. 1–4, doi: 10.1109/IEDM50572.2025.11353727.

[230] C. Zhang, M. Wang, Y. Mai et al. 2023, "A High-Density and Reconfigurable SRAM-Based Digital Compute-In-Memory Macro for Low-Power AI Chips," IEEE Transactions on Circuits and Systems II: Express Briefs 70(9) 3589–3593, doi: 10.1109/TCSII.2023.3276169.

[231] Z. Lin, Z. Tong, J. Zhang et al. 2022, "A review on SRAM-based computing in-memory: Circuits, functions, and applications," J. Semicond. 43, 031401, DOI 10.1088/1674-4926/43/3/031401.

[232] M. Giordano, K. Prabhu, K. Koul et al. 2021, "CHIMERA: A 0.92 TOPS, 2.2 TOPS/W Edge AI Accelerator with 2 MByte On-Chip Foundry Resistive RAM for Efficient Training and Inference," 2021 Symp. on VLSI Circuits, Kyoto, Japan, 2021, pp. 1–2, doi: 10.23919/VLSICircuits52068.2021.9492347.

[233] Y. Yuan, T. Yang, X. Wang et al. 2024, "34.6 A 28 nm 72.12TFLOPS/W Hybrid-Domain Outer-Product Based Floating-Point SRAM Computing-in-Memory Macro with Logarithm Bit-Width Residual ADC," 2024 IEEE Int. Solid-State Circuits Conf. (ISSCC), San Francisco, CA, USA, 2024, pp. 576–578, doi: 10.1109/ISSCC49657.2024.10454313.

[234] Y. He, H. Diao, C. Tang et al. 2023, "7.3 A 28 nm 38-to-102-TOPS/W 8b Multiply-Less Approximate Digital SRAM Compute-In-Memory Macro for Neural-Network Inference," 2023 IEEE Int. Solid-State Circuits Conf. (ISSCC), San Francisco, CA, USA, 2023, pp. 130–132, doi: 10.1109/ISSCC42615.2023.10067305.

[235] H. Mori, W-C. Zhao, C-E. Lee et al. 2023, "A 4 nm 6163-TOPS/W/b 4790–TOPS/mm^2/b SRAM Based Digital-Computing-in-Memory Macro Supporting Bit-Width Flexibility and Simultaneous MAC and Weight Update," 2023 IEEE Int. Solid-State Circuits Conf. (ISSCC), San Francisco, CA, USA, 2023, pp. 132–134, doi: 10.1109/ISSCC42615.2023.10067555.

[236] P. Chen, M. Wu, W. Zhao et al. 2023, "7.8 A 22 nm Delta-Sigma Computing-In-Memory (Δ∑CIM) SRAM Macro with Near-Zero-Mean Outputs and LSB-First ADCs Achieving 21.38TOPS/W for 8b-MAC Edge AI Processing," 2023 IEEE Int. Solid-State Circuits Conf. (ISSCC), San Francisco, CA, USA, 2023, pp. 140–142, doi: 10.1109/ISSCC42615.2023.10067289.

[237] P-C. Wu, J-W. Su, Y-L. Chung et al. 2022, "A 28 nm 1 Mb Time-Domain Computing-in-Memory 6T-SRAM Macro with a 6.6 ns Latency, 1241GOPS and 37.01TOPS/W for 8b-MAC Operations for Edge-AI

Devices," 2022 IEEE Int. Solid-State Circuits Conf. (ISSCC), San Francisco, CA, USA, 2022, pp. 1–3, doi: 10.1109/ISSCC42614.2022.9731681.

[238] D. Wang, C-T. Lin, G.K. Chen et al. 2022, "DIMC: 2219TOPS/W 2569F2/b Digital In-Memory Computing Macro in 28 nm Based on Approximate Arithmetic Hardware," 2022 IEEE Int. Solid-State Circuits Conf. (ISSCC), San Francisco, CA, USA, 2022, pp. 266–268, doi: 10.1109/ISSCC42614.2022.9731659.

[239] Z. Chen, X. Chen and J. Gu 2021, "15.3 A 65 nm 3T Dynamic Analog RAM-Based Computing-in-Memory Macro and CNN Accelerator with Retention Enhancement, Adaptive Analog Sparsity and 44TOPS/W System Energy Efficiency," 2021 IEEE Int. Solid-State Circuits Conf. (ISSCC), San Francisco, CA, USA, 2021, pp. 240–242, doi: 10.1109/ISSCC42613.2021.9366045.

[240] Q. Dong, M.E. Sinangil, B. Erbagci et al. 2020, "15.3 A 351TOPS/W and 372.4GOPS Compute-in-Memory SRAM Macro in 7 nm FinFET CMOS for Machine-Learning Applications," 2020 IEEE Int. Solid-State Circuits Conf. – (ISSCC), San Francisco, CA, USA, 2020, pp. 242–244, doi: 10.1109/ISSCC19947.2020.9062985.

[241] J. Wang, X. Wang, C. Eckert et al. 2019, "14.2 A Compute SRAM with Bit-Serial Integer/Floating-Point Operations for Programmable In-Memory Vector Acceleration," 2019 IEEE Int. Solid-State Circuits Conf. – (ISSCC), San Francisco, CA, USA, 2019, pp. 224–226, doi: 10.1109/ISSCC.2019.8662419.

[242] K. Ueyoshi, K. Ando, K. Hirose et al. 2018, "QUEST: A 7.49TOPS multi-purpose log-quantized DNN inference engine stacked on 96 MB 3D SRAM using inductive-coupling technology in 40 nm CMOS," 2018 IEEE Int. Solid-State Circuits Conf. – (ISSCC), San Francisco, CA, USA, 2018, pp. 216–218, doi: 10.1109/ISSCC.2018.8310261.

[243] X. Peng, S. Huang, Y. Luo et al. 2019, "DNN+NeuroSim: An End-to-End Benchmarking Framework for Compute-in-Memory Accelerators with Versatile Device Technologies," 2019 IEEE Int. Electron Devices Meet. (IEDM), San Francisco, CA, USA, 2019, pp. 32.5.1–32.5.4, doi: 10.1109/IEDM19573.2019.8993491.

[244] Q. Wu, H. Yang, Y. Liu et al. 2025, "A Novel Neuromorphic Hardware Using Area-Efficient Chain RRAM-Based Synapses and Compact Neurons with (Anti-) Integration Scheme," IEEE Transactions on Circuits and Systems I: Regular Papers 72(3) 1383–1396, doi: 10.1109/TCSI.2024.3458436.

[245] T. Kim and Y. Chung. 2024, "Logic-Compatible Embedded DRAM Architecture for Multifunctional Digital Storage and Compute-in-Memory" Applied Sciences 14, no. 21: 9749, October 2024, https://doi.org/10.3390/app14219749

[246] D. Wang, L. Xu, S. Liu et al. 2023, "TL-nvSRAM-CIM: Ultra-High-Density Three-Level ReRAM-Assisted Computing-in-nvSRAM with DC-Power Free Restore and Ternary MAC Operations," 2023 IEEE/ACM Int. Conf. on Computer Aided Design (ICCAD), San Francisco, CA, USA, 2023, pp. 1–9, doi: 10.1109/ICCAD57390.2023.10323889.

[247] W. Xu, D. Luo, M. Deng et al. 2025, "First Experimental Demonstration of Disturb-Free 3D Vertical 1T-nC-1T Ferroelectric-based KV Cache with Co-Optimization of Hybrid Analog-Digital CIM and Token-Wise Dynamic Pruning for Efficient Long-Context LLM Inference," 2025 IEEE Int. Electron Devices Meet. (IEDM), San Francisco, CA, USA, 2025, pp. 1–4, doi: 10.1109/IEDM50572.2025.11353834.

[248] M. Chang, A.S. Lele, S.D. Spetalnick et al. 2023, "A 73.53TOPS/W 14.74TOPS Heterogeneous RRAM In-Memory and SRAM Near-Memory SoC for Hybrid Frame and Event-Based Target Tracking," 2023 IEEE Int. Solid-State Circuits Conf. (ISSCC), San Francisco, CA, USA, 2023, pp. 426–428, doi: 10.1109/ISSCC42615.2023.10067544.

[249] J. He, X. Wang, K. Shao et al. 2026 "Configurable Dataflow and Adaptive Mapping Optimization for Hybrid ReRAM and SRAM Compute-in-Memory Accelerator," IEEE Transactions on Computer-Aided Design of Integrated Circuits and Systems 45(3) 1115–1128, doi: 10.1109/TCAD.2025.3596765.

[250] H. Sang, W. Xie, G. Park et al. 2024, "An 2.31uJ/Inference Ultra-Low Power Always-on Event-Driven AI-IoT SoC With Switchable nvSRAM Compute-in-Memory Macro," IEEE Transactions on Circuits and Systems II: Express Briefs 71(5) 2534–2538, doi: 10.1109/TCSII.2024.3374885.

[251] J. Song, Y. Cho, J-S. Park et al. 2019, "7.1 An 11.5TOPS/W 1024-MAC Butterfly Structure Dual-Core Sparsity-Aware Neural Processing Unit in 8 nm Flagship Mobile SoC," 2019 IEEE Int. Solid-State Circuits Conf. – (ISSCC), San Francisco, CA, USA, 2019, pp. 130–132, doi: 10.1109/ISSCC.2019.8662476.

[252] Y. He, S. Fan, X. Li et al., "34.7 A 28 nm 2.4 Mb/mm^2 6.9 – 16.3TOPS/mm^2 eDRAM-LUT-Based Digital-Computing-in-Memory Macro with In-Memory Encoding and Refreshing," 2024 IEEE Int. Solid-State Circuits Conf. (ISSCC), San Francisco, CA, USA, 2024, pp. 578–580, doi: 10.1109/ISSCC49657.2024.10454323.

[253] S. Xie, C. Ni, P. Jain et al. 2022, "Gain-Cell CIM: Leakage and Bitline Swing Aware 2T1C Gain-Cell eDRAM Compute in Memory Design with Bitline Precharge DACs and Compact Schmitt Trigger ADCs," 2022 IEEE Symp. on VLSI Technology and Circuits (VLSI Technology and Circuits), Honolulu, HI, USA, 2022, pp. 112–113, doi: 10.1109/VLSITechnologyandCir46769.2022.9830338.

[254] F. Hamzaoglu, U. Arslan, N. Bisnik et al. 2014, "13.1 A 1 Gb 2 GHz embedded DRAM in 22 nm tri-gate CMOS technology," 2014 IEEE Int. Solid-State Circuits Conf. Digest of Technical Papers (ISSCC), San Francisco, CA, USA, 2014, pp. 230–231, doi: 10.1109/ISSCC.2014.6757412.

[255] H. Jeong, S. Kim, J. Shin et al. 2026, "HYTEC: Compact and Energy-Efficient Analog-Digital Hybrid CIM With Transpose Ternary eDRAM," IEEE J. of Solid-State Circuits, doi: 10.1109/JSSC.2025.3623671.

[256] C. Yu, T. Yoo, H. Kim et al. 2021, "A Logic-Compatible eDRAM Compute-In-Memory With Embedded ADCs for Processing Neural Networks," IEEE Transactions on Circuits and Systems I: Regular Papers 68(2) 667–679, doi: 10.1109/TCSI.2020.3036209.

[257] Y. He, S. Fan, X. Li et al. 2026, "An Area-Efficient Lookup-Table -Based eDRAM Digital CIM Macro for Neural Network Inference," in IEEE Journal of Solid-State Circuits, 61(4)1697–1709, doi: 10.1109/JSSC.2025.3595012.

[258] R. Bao, H. Diao and X. Tang 2025, "A High-Density eDRAM-Based Cache Featuring Enhanced Data Retention Time and Low-Power Sense Amplifier," 2025 IEEE Int. Conf. on Integrated Circuits, Technologies and Applications (ICTA), Macao, China, 2025, pp. 43–44, doi: 10.1109/ICTA68203.2025.11329970.

[259] K. Lai, E. Yao, Z. Li et al. 2025, "A High-Density eDRAM Macro With Programmable Sense Amplifier and TG-Shifter for Logical-Instruction-Based In-Memory Computing," IEEE Transactions on Very Large Scale Integration (VLSI) Systems 33(7) 2069–2073, doi: 10.1109/TVLSI.2025.3561507.

[260] Z. Cong, Z. Qian, S. Yan et al. 2025, "A 28 nm 51.6TOPS/W 2.22 Mb/mm^2 System-in-One-Macro Computing-in-Memory Chip Utilizing Leakage-Eliminated 2T1C and Capacitor-over-Logic 1T1C eDRAM," 2025 IEEE European Solid-State Electronics Research Conf. (ESSERC), Munich, Germany, 2025, pp. 113–116, doi: 10.1109/ESSERC66193.2025.11214031.

[261] J. Song, X. Tang, B. Xu et al. 2026, "A 1131-kb/mm^2 14.0-to-53.3-TOPS/W 8-bit Analog-Assisted Digital Compute-in-Memory With Hybrid Local-Refresh eDRAM for Attention Computing," in IEEE J. of Solid-State Circuits 61(3) 1240–1253, doi: 10.1109/JSSC.2025.3580366.

[262] X. Zhang, Y-J. Jo and T. T.H. Kim 2025, "A 65-nm 55.8-TOPS/W Compact 2T eDRAM-Based Compute-in-Memory Macro With Linear Calibration," IEEE Transactions on Very Large Scale Integration (VLSI) Systems 33(5) 1477–1481, doi: 10.1109/TVLSI.2024.3520588.

[263] L. Lei, Y. Zhu, X. Zou et al. 2025, "Enabling Energy-Efficient Homomorphic Encryption Evaluation via eDRAM-Based In-Situ Computing in an Edge Processor," IEEE J. of Solid-State Circuits 60(7) 2405–2419, doi: 10.1109/JSSC.2025.3568186.

[264] C. Wyon 2025, "Material Science for Developing Emerging Non-volatile Memories," J. of Mat. Sci. Apl. Eng. 4(1), p01–08, January 2025, https://www.mkscienceset.com/articles_file/707-_article1738387673.pdf

[265] M. Adnaan, S. A. Chamazcoti, E/Karimov et al. 2025, "Benchmarking of FERAM-Based Memory System by Optimizing Ferroelectric Device Model," in IEEE Journal on Exploratory Solid-State Computational Devices and Circuits, vol. 11, pp. 99–106, 2025, doi: 10.1109/JXCDC.2025.3618883.

[266] C.Han, B. Kwak, J. Choi et al. 2025, "Synergistic Modulation of Polarization and Leakage Current in MPB-HZO Capacitors via TiO_2 Interlayer," Advanced Electronic Materials vol. 11, e00314, 2025, https://doi.org/10.1002/aelm.202500314

[267] S. Mukherjee, J. Bizindavyi, Y-C. Luo et al. 2023, "Pulse-Based Capacitive Memory Window with High Non-Destructive Read Endurance in Fully BEOL Compatible Ferroelectric Capacitors," 2023 International Electron Devices Meeting (IEDM), San Francisco, CA, USA, 2023, pp. 1–4, doi: 10.1109/IEDM45741.2023.10413879C.

[268] M. Gupta, M. Perumkunnil, K. Garello et al. 2020, "High-density SOT-MRAM technology and design specifications for the embedded domain at 5 nm node," 2020 IEEE Int. Electron Devices Meet. (IEDM), San Francisco, CA, USA, 2020, pp. 24.5.1–24.5.4, doi: 10.1109/IEDM13553.2020.9372068.

[269] D-Q. You et al. 2025, "14.1 A 22 nm 104.5TOPS/W μ-NMC-Δ-IMC Heterogeneous STT-MRAM CIM Macro for Noise-Tolerant Bayesian Neural Networks," 2025 IEEE International Solid-State Circuits Conference (ISSCC), San Francisco, CA, USA, 2025, pp. 1–3, doi: 10.1109/ISSCC49661.2025.10904540.

[270] M.G. Monteiro, A. Kumar, V. Kateel et al. 2025, "First Demonstration of Field Free Switching Perpendicular SOT-MRAM 4kb Array Using Z-Spin," 2025 IEEE Int. Electron Devices Meet. (IEDM), San Francisco, CA, USA, 2025, pp. 1–4, doi: 10.1109/IEDM50572.2025.11353687.

[271] R. Khosla and S. K. Sharma 2021, "Integration of Ferroelectric Materials: An Ultimate Solution for Next-Generation Computing and Storage Devices," ACS Appl. Electron. Mater. vol. 3, pp. 2862–2897, 2021, https://doi.org/10.1021/acsaelm.0c00851

[272] F. Oboril, R. Bishnoi, M. Ebrahimi and M. B. Tahoori 2015, "Evaluation of Hybrid Memory Technologies Using SOT-MRAM for On-Chip Cache Hierarchy," in IEEE Transactions on Computer-Aided Design of Integrated Circuits and Systems, vol. 34, no. 3, pp. 367–380, March 2015, doi: 10.1109/TCAD.2015.2391254.

[273] A. Nisar, S. Dhull, B. K. Kaushik et al. 2021, "High-performance voltage controlled multilevel MRAM cell," Semicond. Sci. Technol. vol. 36, art. 125013, November 2021, doi:10.1088/1361-6641/ac3187.

[274] I. Singh, B. Raj, M. Khosla, M. 2024, "SOT-MRAM Memories for Energy Efficient Embedded and AI Applications," In: Mehta, G., Wickramasinghe, N., Kakkar, D. (eds) Innovations in VLSI, Signal Processing and Computational Technologies. WREC 2023. Lecture Notes in Electrical Engineering, vol. 1095, pp. 13–24, Springer, Singapore. https://doi.org/10.1007/978-981-99-7077-3_2

[275] Q. Dong et al. 2019, "A 1-Mb 28-nm 1T1MTJ STT-MRAM With Single-Cap Offset-Cancelled Sense Amplifier and In Situ Self-Write-Termination," in IEEE Journal of Solid-State Circuits, vol. 54, no. 1, pp. 231–239, Jan. 2019, doi: 10.1109/JSSC.2018.2872584.

[276] P.-H. Lee et al. 2023, "33.1 A 16 nm 32 Mb Embedded STT-MRAM with a 6ns Read-Access Time, a 1M-Cycle Write Endurance, 20-Year Retention at 150°C and MTJ-OTP Solutions for Magnetic Immunity," 2023 IEEE International Solid-State Circuits Conference (ISSCC), San Francisco, CA, USA, 2023, pp. 494-496, doi: 10.1109/ISSCC42615.2023.10067837.

[277] K-F. Lin, H. Noguchi, Y-C. Shih et al. 2024, "15.9 A 16 nm 16 Mb Embedded STT-MRAM with a 20ns Write Time, a 1012 Write Endurance and Integrated Margin-Expansion Schemes," 2024 IEEE Int. Solid-State Circuits Conf. (ISSCC), San Francisco, CA, USA, 2024, pp. 292-294, doi: 10.1109/ISSCC49657.2024.10454339.

[278] W-H. Huang, T-H. Wen, J-M. Hung et al. 2023, "A Nonvolatile Al-Edge Processor with 4 MB SLC-MLC Hybrid-Mode ReRAM Compute-in-Memory Macro and 51.4-251TOPS/W," 2023 IEEE Int. Solid-State Circuits Conf. (ISSCC), San Francisco, CA, USA, 2023, pp. 15–17, doi: 10.1109/ISSCC42615.2023.10067610.

[279] D. Rossi, F. Conti, M. Eggiman et al. 2021, "4.4 A 1.3TOPS/W @ 32GOPS Fully Integrated 10-Core SoC for IoT End-Nodes with 1.7μW Cognitive Wake-Up From MRAM-Based State-Retentive Sleep Mode," 2021 IEEE Int. Solid-State Circuits Conf. (ISSCC), San Francisco, CA, USA, 2021, pp. 60–62, doi: 10.1109/ISSCC42613.2021.9365939.

[280] H. Zhong, Z. Zeng, L. Jiao et al. 2024, "First Demonstration of AFeFET-Based Capacitor-Less eDRAM Computing-in-Memory Featuring 4.84 Mb/mm^2 High Memory Density, 105 s Long Retention Time, and >1010 High Endurance," 2024 IEEE Int. Electron Devices Meet. (IEDM), San Francisco, CA, USA, 2024, pp. 1–4, doi: 10.1109/IEDM50854.2024.10873490.

[281] J. Choe 2017, "XPoint memory Comparison – Process & Architecture," Flash Memory Summit 2017, https://files.futurememorystorage.com/proceedings/2017/20170808_FR12_Choe.pdf

[282] J. Choe 2022, "Emerging Memory 2022 and Beyond," Flash Memory Summit 2022, August 2022, https://conferenceconcepts.app.box.com/s/ta1nd3pko0bpt26rzzerv9z4h6dpew0z

[283] T. Francois, J. Coignus, A. Makosiej et al. 2021, "16kbit HfO_2:Si-based 1T-1C FeRAM Arrays Demonstrating High Performance Operation and Solder Reflow Compatibility," 2021 IEEE Int. Electron Devices Meet. (IEDM), San Francisco, CA, USA, 2021, pp. 33.1.1–33.1.4, doi: 10.1109/IEDM19574.2021.9720640.

[284] H. Oh, Y. Xiang, F. Garcia Redondo et al. 2024, "Design Space Exploration of FeRAM Bit Cell for DRAM Application," IEEE Transactions on Electron Devices 71(9) 5380–5387, doi: 10.1109/TED.2024.3435630.

[285] T. S. Moise, S.R. Summerfelt, H. McAdams et al. 2002, "Demonstration of a 4 Mb, high density ferroelectric memory embedded within a 130 nm, 5 LM Cu/FSG logic process," Digest. Int. Electron Devices Meet., San Francisco, CA, USA, 2002, pp. 535–538, doi: 10.1109/IEDM.2002.1175897.

[286] N. Ramaswamy, A. Calderoni, J. Zahurak et al. 2023, "NVDRAM: A 32 Gb Dual Layer 3D Stacked Non-volatile Ferroelectric Memory with Near-DRAM Performance for Demanding AI Workloads," 2023 Int. Electron Devices Meet. (IEDM), San Francisco, CA, USA, 2023, pp. 1–4, doi: 10.1109/IEDM45741.2023.10413848.

[287] T. Moise, S. Summerfelt and John Rodriguez 2025, "A Framework for Embedded Non-Volatile Memory Development: Examples from $Pb(Zr_xTi_{1-x})O_3$ Ferroelectric Memory Development at Texas Instruments," Electronics 14(4), 818, https://doi.org/10.3390/electronics14040818

[288] K. Hoya, D. Takashima, S. Shiratake et al. 2010, "A 64-Mb Chain FeRAM With Quad BL Architecture and 200 MB/s Burst Mode," IEEE Transactions on Very Large Scale Integration (VLSI) Systems 18(12) 1745–1752, doi: 10.1109/TVLSI.2009.2034380.

[289] J.W.K. Siu, Y. Eslami, A. Sheikholeslami et al. 2003, "A current-based reference-generation scheme for 1T-1C ferroelectric random-access memories," IEEE J. of Solid-State Circuits 38(3) 541–549, doi: 10.1109/JSSC.2002.808289.

[290] R. Ogiwara, D. Takashima, S. Doumae et al. 2015, "Highly Reliable Reference Bitline Bias Designs for 64 Mb and 128 Mb Chain FeRAMs," IEEE J. of Solid-State Circuits 50(5) 1324–1331, doi: 10.1109/JSSC.2015.2405932.

[291] H.P. McAdams, R. Acklin, T. Blake et al. 2004, "A 64-Mb embedded FRAM utilizing a 130-nm 5LM Cu/FSG logic process," IEEE J. of Solid-State Circuits 39(4) 667–677, doi: 10.1109/JSSC.2004.825241.

[292] R. Ogiwara, S. Tanaka, Y. Itoh et al. 2000, "A 0.5-/spl mu/m, 3-V 1T1C, 1-Mbit FRAM with a variable reference bit-line voltage scheme using a fatigue-free reference capacitor," IEEE J. of Solid-State Circuits 35(4) 545–551, doi: 10.1109/4.839914.

[293] M-K. Choi, B-G. Jeon, N. Jang, et al. 2002, "A 0.25 /spl mu/m 3.0 V 1T1C 32 Mb nonvolatile ferroelectric RAM with address transition detector (ATD) and current forcing latch sense amplifier (CFLSA) scheme," 2002 IEEE Int. Solid-State Circuits Conf. Digest of Technical Papers (Cat. No.02CH37315), San Francisco, CA, USA, 2002, pp. 162–457 vol.1, doi: 10.1109/ISSCC.2002.992986.

[294] Y. Chung, M-K. Choi, S-K. Oh et al. 1999, "A 3.3-V 4-Mb nonvolatile ferroelectric RAM with a selectively-driven double-pulsed plate read/write-back scheme," 1999 Symp. on VLSI Circuits. Digest of Papers (IEEE Cat. No.99CH36326), Kyoto, Japan, 1999, pp. 97–98, doi: 10.1109/VLSIC.1999.797249.

[295] B-G. Jeon, M-K. Choi, Y. Song et al. 2000, "A 0.4-/spl mu/m 3.3-V 1T1C 4-Mb nonvolatile ferroelectric RAM with fixed bitline reference voltage scheme and data protection circuit," IEEE J. of Solid-State Circuits 35(11) 1690–1694, doi: 10.1109/4.881216.

[296] D. Takashima, S. Shuto, I. Kunishima et al. 1999, "A sub-40-ns chain FRAM architecture with 7-ns cell-plate-line drive," IEEE J. of Solid-State Circuits 34(11) 1557–1563, doi: 10.1109/4.799863.

[297] D. Takashima, Y. Takeuchi, T. Miyakawa et al. 2001, "A 76-mm^2 8-Mb chain ferroelectric memory," IEEE J. of Solid-State Circuits 36(11) 1713–1720, doi: 10.1109/4.962293.

[298] S-H. Song, K.C. Chun and C.H. Kim 2013, "A bit-by-bit re-writable Eflash in a generic logic process for moderate-density embedded non-volatile memory applications," Proc. of the IEEE 2013 Custom Integrated Circuits Conf., San Jose, CA, USA, 2013, pp. 1–4, doi: 10.1109/CICC.2013.6658453.

[299] M. Bavandpour et al. 2021, "3D-aCortex: an ultra-compact energy-efficient neurocomputing platform based on commercial 3D-NAND flash memories" in Neuromorph. Comput. Eng. vol.1, Art.: 014001, July 2021, doi: 10.1088/2634-4386/ac0775.

[300] D -Q. You et al. 2024, "Aimar," 2024 IEEE Symposium on VLSI Technology and Circuits (VLSI Technology and Circuits), Honolulu, HI, USA, 2024, pp. 1–2, doi:10.1109/VLSITechnologyandCir46783.2024.10631408.

[301] J; A.Diaz-Madrid, G. Domenech-Asensi, R. Ruiz-Merino et al. 2024,"A real-time and energy-efficient SRAM with mixed-signal in-memory computing near CMOS sensors". J Real-Time Image Proc vol. 21, art. 143, July 2024, https://doi.org/10.1007/s11554-024-01520-x

[302] P. Deaville, B. Zhang and N. Verma 2025, "A Fully Row/Column-Parallel MRAM in-Memory Computing Macro With Memory-Resistance Boosting and Weighted Multi-Column ADC Readout," in IEEE Journal of Solid-State Circuits, vol. 60, no. 5, pp. 1856–1866, May 2025, doi: 10.1109/JSSC.2024.3512360.

[303] S. Myung et al. 2024, "A 14 nm MRAM-Based Multi-bit Analog In-Memory Computing With Process-Variation Calibration for 72 Macros-Based Accelerator," in IEEE Solid-State Circuits Letters, vol. 7, pp. 295–298, 2024, doi: 10.1109/LSSC.2024.3465595.

[304] T. Soliman, F. Müller, T. Kirchner et al. 2020, "Ultra-Low Power Flexible Precision FeFET Based Analog In-Memory Computing," 2020 IEEE International Electron Devices Meet. (IEDM), San Francisco, CA, USA, 2020, pp. 29.2.1–29.2.4, doi: 10.1109/IEDM13553.2020.9372124.

[305] A. Shafiee, A. Nag, N. Muralimanohar et al. 2016, "ISAAC: A Convolutional Neural Network Accelerator with In-Situ Analog Arithmetic in Crossbars," 2016 ACM/IEEE 43rd Annual Int. Symp. on Computer Architecture (ISCA), Seoul, Korea (South), 2016, pp. 14–26, doi: 10.1109/ISCA.2016.12.

[306] J. Chun, E. Choi, B. Choi et al. 2024, "A 639.38GOPS Hybrid-Domain Logic-Compatible Multi-Level Embedded Flash Computing-in-Memory Macro with Spike-Edge Computation," 2024 IEEE Asian Solid-State Circuits Conf. (A-SSCC), Hiroshima, Japan, 2024, pp. 1–3, doi: 10.1109/A-SSCC60305.2024.10848.

[307] J. M. Correll, L. Jie, S. Song et al. 2025, "An 8-bit 20.7 TOPS/W Multilevel Cell ReRAM Macro With ADC-Assisted Bit-Serial Processing," IEEE J. of Solid-State Circuits 60(8) 2995–3008, doi: 10.1109/JSSC.2025.3540114.

[308] M. Webb 2025, "Future Scenarios for Emerging Memory Markets," The future of Memory and Storage, August 5-7, 2025, https://files.futurememorystorage.com/proceedings/2025/20250806_OMEM-202-1_Webb.pdf

[309] P. Mannocci, M. Farronato, N. Lepri et al. 2023, "In-memory computing with emerging memory devices: Status and outlook," APL Mach. Learn. vol. 1, 010902, 2023, https://doi.org/10.1063/5.0136403

[310] S. Schmidgall, R. Ziaei, J. Achterberg et al. 2024, "Brain-inspired learning in artificial neural networks: A review," APL Mach. Learn. vol. 2, art. 021501, May 2024, https://doi.org/10.1063/5.0186054

[311] K.S. Mohamed, "Neuromorphic Computing and Beyond – Parallel, Approximation, Near Memory, and Quantum", Springer Nature Switzerland AG 2020, https://doi.org/10.1007/978-3-030-37224-8

[312] Applied Materials "Memory", https://www.appliedmaterials.com/us/en/semiconductor/markets-and-inflections/memory.html, accessed in December 2025

[313] D. Verburg and A. Scrabeck 2025, “QLC Considerations in the Age of AI,” The Future of Memory and Storage, August 2025, https://files.futurememorystorage.com/proceedings/2025/01K5SG7CHQC65EVD8NCSAEV3CB.pdf

[314] A. Moitra, A. Bhattacharjee, Y. Li Y. et al. 2024, “When in-memory computing meets spiking neural networks – A perspective on device-circuit-system and-algorithm co-design,” Appl. Phys. Rev. vol. 11, art: 031325, September 2024, https://doi.org/10.1063/5.0211040

[315] A. Fedorova, N. Jovišiˊc, J. Vallverdù et al. 2024, “Advancing Neural Networks: Innovations and Impacts on Energy Consumption,” Advanced Electronic Materials, vol. 10, art. 2400258, November 2024, https://doi.org/10.1002/aelm.202400258

[316] S. Kundu, G. Datta, M. Pedram and P. A. Beerel 2021, “Spike-Thrift: Towards Energy-Efficient Deep Spiking Neural Networks by Limiting Spiking Activity via Attention-Guided Compression,” 2021 IEEE Winter Conference on Applications of Computer Vision (WACV), Waikoloa, HI, USA, 2021, pp. 3952–3961, doi: 10.1109/WACV48630.2021.00400.

[317] C. Choi, N. Prihatiningrum and Y. Seo 2025, “Restoring-Free Transposable eDRAM-Based Computing-In-Memory System for SNN On-Chip Unsupervised STDP Learning,” IEEE Access, 13, 197926–197941, doi: 10.1109/ACCESS.2025.3634983.

[318] Y. Jang, G. Kang, T. Kim et al. 2022, “Stochastic SOT Device Based SNN Architecture for On-Chip Unsupervised STDP Learning,” IEEE Transactions on Computers 71(9) 2022–2035, doi: 10.1109/TC.2021.3119180.

[319] H. Tang, H. Kim, H. Kim and J. Park 2019, “Spike Counts Based Low Complexity SNN Architecture With Binary Synapse,” IEEE Transactions on Biomedical Circuits and Systems, 13(6) 1664–1677, Dec. 2019, doi: 10.1109/TBCAS.2019.2945406.

[320] G. K. Chen, R. Kumar, H. E. Sumbul, et al. 2019, “A 4096-Neuron 1M-Synapse 3.8-pJ/SOP Spiking Neural Network With On-Chip STDP Learning and Sparse Weights in 10-nm FinFET CMOS,” IEEE J. of Solid-State Circuits, 54(4) 992–1002, doi: 10.1109/JSSC.2018.2884901.

[321] D. Kim, X. She, N. M. Rahmanet al. 2020, “Processing-In-Memory-Based On-Chip Learning With Spike-Time-Dependent Plasticity in 65-nm CMOS,” IEEE Solid-State Circuits Letters, 3, 278–281, doi: 10.1109/LSSC.2020.3013448.

[322] F. N. Buhler, P. Brown, J. Li et al. 2017, “A 3.43TOPS/W 48.9pJ/pixel 50.1nJ/classification 512 analog neuron sparse coding neural network with on-chip learning and classification in 40 nm CMOS,” 2017 Symp. on VLSI Circuits, Kyoto, Japan, 2017, pp. C30–C31, doi: 10.23919/VLSIC.2017.8008536.

[323] J. Park, J. Lee and D. Jeon 2019, “7.6 A 65 nm 236.5 nJ/Classification Neuromorphic Processor with 7.5% Energy Overhead On-Chip Learning Using Direct Spike-Only Feedback,” 2019 IEEE Int. Solid-State Circuits Conf. – (ISSCC), San Francisco, CA, USA, 2019, pp. 140–142, doi: 10.1109/ISSCC.2019.8662398.

[324] S. Liu, J.J. Wang, J.T. Zhou et al. 2023, “An Area- and Energy-Efficient Spiking Neural Network With Spike-Time-Dependent Plasticity Realized With SRAM Processing-in-Memory Macro and On-Chip Unsupervised Learning,” IEEE Transactions on Biomedical Circuits and Systems, 17(1) 92–104, doi: 10.1109/TBCAS.2023.3242413.

[325] H. Kim, H. Tang, W. Choi and J. Park 2020, “An Energy-Quality Scalable STDP Based Sparse Coding Processor With On-Chip Learning Capability,” IEEE Transactions on Biomedical Circuits and Systems, 14(1) 125–137, Feb. 2020, doi: 10.1109/TBCAS.2019.2963676.

[326] C. Liu, Z. Yang, X. Zhang et al. 2023, “A Low-Power Hybrid-Precision Neuromorphic Processor With INT8 Inference and INT16 Online Learning in 40-nm CMOS,” IEEE Transactions on Circuits and Systems I: Regular Papers, 70(10) 4028–4039, doi: 10.1109/TCSI.2023.3300095.

[327] M. Davies, N. Srinivasa, T-H. Lin et al. 2018, “Loihi: A Neuromorphic Manycore Processor with On-Chip Learning,” IEEE Micro 38(1) 82–99, doi: 10.1109/MM.2018.112130359.

[328] C. Frenkel, M. Lefebvre, J.-D. Legat and D. Bol, 2019, "A 0.086-mm^2 12.7-pJ/SOP 64k-Synapse 256-Neuron Online-Learning Digital Spiking Neuromorphic Processor in 28-nm CMOS," IEEE Trans;on Biomedical Circuits and Systems 13(1) 145–158, doi: 10.1109/TBCAS.2018.2880425.
[329] Y. Liu, Y. Ma, N. Shang et al. 2024, "30.2 A 22 nm 0.26 nW/Synapse Spike-Driven Spiking Neural Network Processing Unit Using Time-Step-First Dataflow and Sparsity-Adaptive In-Memory Computing," 2024 IEEE Int. Solid-State Circuits Conf. (ISSCC), San Francisco, CA, USA, 2024, pp. 484–486, doi: 10.1109/ISSCC49657.2024.10454472.
[330] C-Y. Chen, Y-S. Dai and H-C. Hong 2024, "A Neuromorphic Spiking Neural Network Using Time-to-First-Spike Coding Scheme and Analog Computing in Low-Leakage 8T SRAM," IEEE Transactions on Very Large Scale Integration (VLSI) Systems, 32(5) 848–859, doi: 10.1109/TVLSI.2024.3368849.
[331] H. Jiang, J. Lu, C. Zhang et al. 2023, "Multicore Spiking Neuromorphic Chip in 180-nm With ReRAM Synapses and Digital Neurons," IEEE J. on Emerging and Selected Topics in Circuits and Systems, 13(4) 975–985, Dec. 2023, doi: 10.1109/JETCAS.2023.3325158.
[332] J. Zhang, D. Ho, J. Zhang et al. 2023, "22.6 ANP-I: A 28 nm 1.5pJ/SOP Asynchronous Spiking Neural Network Processor Enabling Sub-0.1 μJ/Sample On-Chip Learning for Edge-AI Applications," 2023 IEEE Int. Solid-State Circuits Conf. (ISSCC), San Francisco, CA, USA, 2023, pp. 21–23, doi: 10.1109/ISSCC42615.2023.10067650.
[333] C. Frenkel and G. Indiveri 2022, "ReckOn: A 28 nm Sub-mm^2 Task-Agnostic Spiking Recurrent Neural Network Processor Enabling On-Chip Learning over Second-Long Timescales," 2022 IEEE Int. Solid-State Circuits Conf. (ISSCC), San Francisco, CA, USA, 2022, pp. 1–3, doi: 10.1109/ISSCC42614.2022.9731734.
[334] F. Akopyan, J. Sawada, A. Cassidy et al. 2015, "TrueNorth: Design and Tool Flow of a 65 mW 1 Million Neuron Programmable Neurosynaptic Chip," IEEE Transactions on Computer-Aided Design of Integrated Circuits and Systems 34(10). 1537–1557, doi: 10.1109/TCAD.2015.2474396.
[335] Z. Wang, F. Liu, N. Yang et al. 2024, "COMPASS: SRAM-Based Computing-in-Memory SNN Accelerator with Adaptive Spike Speculation," 2024 57th IEEE/ACM Int. Symp. on Microarchitecture (MICRO), Austin, TX, USA, 2024, pp. 1090–1106, doi: 10.1109/MICRO61859.2024.00083.
[336] F. Liu, S. Huang, L. Zhao et al. 2024, "LowPASS: A Low power PIM-based accelerator with Speculative Scheme for SNNs," ISLPED '24: Proceedings of the 29th ACM/IEEE Int. Symp. on Low Power Electronics and Design, August 5 – 7, 2024 CA, Newport Beach, USA, pp. 1–6, https://doi.org/10.1145/3665314.3672279
[337] J.Y. Lee, C. Nie, K. You et al. 2025, "BiNeuroRAM: Energy-Efficient ReRAM-Based PIM for Accurate Bipolar Spiking Neural Network Acceleration," 2025 62nd ACM/IEEE Design Automation Conf. (DAC), San Francisco, CA, USA, 2025, pp. 1–7, doi: 10.1109/DAC63849.2025.11132454.
[338] Cloud computing, https://digital-strategy.ec.europa.eu/en/policies/cloud-computing, accessed in January 2026
[339] D. Jang, E. Bury, R. Ritzenthaler et al. 2015, "Self-heating on bulk FinFET from 14 nm down to 7 nm node," 2015 IEEE International Electron Devices Meeting (IEDM), Washington, DC, USA, 2015, pp. 11.6.1–11.6.4, doi: 10.1109/IEDM.2015.7409678.
[340] T. Takahashi, N. Beppu, K. Chen et al. 2011, "Thermal-aware device design of nanoscale bulk/SOI FinFETs: Suppression of operation temperature and its variability," 2011 International Electron Devices Meeting, Washington, DC, USA, 2011, pp. 34.6.1–34.6.4, doi: 10.1109/IEDM.2011.6131672.
[341] S. Albawi, T. A. Mohammed and S. Al-Zawi 2017, "Understanding of a convolutional neural network," 2017 International Conference on Engineering and Technology (ICET), Antalya, Turkey, 2017, pp. 1–6, doi: 10.1109/ICEngTechnol.2017.8308186.
[342] R. Yamashita, M. Nishio, R.K.G. Do et al. 2018, "Convolutional neural networks: an overview and application in radiology," Insights Imaging vol. 9, pp. 611–629, 2018, https://doi.org/10.1007/s13244-018-0639-9

[343] T. Saito, K. Nose, T. Ogawa et al. 2025, "Trends in Embedded Volatile and Non-Volatile Memories for Scalable Edge AI Applications," 2025 IEEE Int. Electron Devices Meet. (IEDM), San Francisco, CA, USA, 2025, pp. 1–4, doi: 10.1109/IEDM50572.2025.11353725.

[344] M.M. Versace 2026, "Outlook 2026: When Intelligence Moves to the Edge," February 3, 2026, https://www.eetasia.com/outlook-2026-when-intelligence-moves-to-the-edge/?utm_source=EETA%20Article%20Alert&utm_medium=Email&utm_campaign=2026-02-03

[345] D. James 2016, "Moore's law continues into the 1x-nm era," 2016 27th Annual SEMI Advanced Semiconductor Manufacturing Conference (ASMC), Saratoga Springs, NY, USA, 2016, pp. 324–329, doi: 10.1109/ASMC.2016.7491159.

[346] M. Bhuiyan, M. Kim, H. Zhou et al. 2021, "Gate-Last I/O Transistors based on Stacked Gate-All-Around Nanosheet Architecture for Advanced Logic Technologies," 2021 IEEE Int. Electron Devices Meet. (IEDM), San Francisco, CA, USA, 2021, pp. 1–4, doi: 10.1109/IEDM19574.2021.9720507.

[347] W. P. Maszara 2018, "Logic Devices for Today and Tomorrow," ECS Trans. 85, 15, 2018, DOI 10.1149/08508.0015ecst.

[348] W. Hafez et al. 2023, "Intel Power Via Technology: Backside Power Delivery for High Density and High-Performance Computing," 2023 IEEE Symposium on VLSI Technology and Circuits (VLSI Technology and Circuits), Kyoto, Japan, 2023, pp. 1–2, doi: 10.23919/VLSITechnologyandCir57934.2023.10185208.

[349] C.J. Lockhart de la Rosa and G. Sankar Kar 2025, "Introducing 2D-material based devices in the logic scaling roadmap," January 15, 2025, https://www.imec-int.com/en/articles/introducing-2d-material-based-devices-logic-scaling-roadmap

[350] B. Potter, "How to Build a $20 Billion Semiconductor Fab," May 2024, https://www.construction-physics.com/p/how-to-build-a-20-billion-semiconductor – accessed in February 2025.

[351] "Semiconductor pricing – Chip cost drivers and trends," December 2023, https://kaizoft.com/semiconductor-pricing-a-complex-art-cost-drivers-and-trends/ – accessed in October 2024

[352] Z. Or-Bach, "Moore's Law Indeed Stopped at 28 nm," January 2024, https://www.semiconductor-digest.com/moores-law-indeed-stopped-at-28nm/ – accessed in January 14, 2024.

[353] A. Shilov, "TSMC could charge up to $45,000 for 1.6 nm wafers – rumors allege a 50% increase in pricing over prior-gen wafers," https://www.tomshardware.com/tech-industry/semiconductors/tsmc-could-charge-up-to-usd45-000-for-1-6nm-wafers-rumors-allege-a-50-percent-increase-in-pricing-over-prior-gen-wafers – accessed in June 29, 2025.

[354] J. De Vos, J. Myers, S. Van Huylenbroeck et al. 2025, "Towards SoCs leveraging 200 nm wafer-to-wafer interconnection pitch," 2025 IEEE Int. Electron Devices Meet. (IEDM), San Francisco, CA, USA, 2025, pp. 1–4, doi: 10.1109/IEDM50572.2025.11353621.

[355] H.-S. P. Wong, K. Akarvardar, D. Antoniadis et al. 2020, "A Density Metric for Semiconductor Technology [Point of View]," Proc. of the IEEE 108(4) 478–482, doi: 10.1109/JPROC.2020.2981715.

[356] O. Burkacky, T. Kim and I. Yeom 2023, "Advanced chip packaging: How manufacturers can play to win," May 24, 2023, https://www.mckinsey.com/industries/semiconductors/our-insights/advanced-chip-packaging-how-manufacturers-can-play-to-win

[357] K. Sahoo and S.S. Iyer 2026, "Heterogeneous wafer-scale integration: The new front line of semiconductor scaling," Chip Scale Review 30(1) 5–9, https://chipscalereview.com/wp-content/uploads/flipbook/49/flipBook.html

[358] Y.-J. Mii, 2024, "Semiconductor Industry Outlook and New Technology Frontiers," 2024 IEEE International Electron Devices Meeting (IEDM), San Francisco, CA, USA, 2024, pp. 1–6, doi: 10.1109/IEDM50854.2024.10873484.

[359] C. Chidambaram 2025, "Semiconductor challenges in ramping AI at the edge," Semicon West 2025, October 8, 2025, https://eps.ieee.org/wp-content/uploads/2025/11/Oct_8_PR_Chidi_Chidambaram.pdf

[360] W. Kwon, J. Lee, B. Hong et al. 2025, “Gate-All-Around Technology for Sustainable AI: A Foundation for Future Logic Architectures,” 2025 IEEE Int. Electron Devices Meet. (IEDM), San Francisco, CA, USA, 2025, pp. 1–4, doi: 10.1109/IEDM50572.2025.11353479.

[361] CP. Hung 2025, “Innovative Advanced Packaging Solutions for AI,” October 8, 2025, https://eps.ieee.org/wp-content/uploads/2025/11/Oct_8_CP_Hung.pdf

[362] J-H. Lau 2026, “Co-packaged optics: Heterogeneous integration of chiplets in switches, PICs, and EICs,” Chip Scale Review 30(1) 32–41, https://chipscalereview.com/wp-content/uploads/flipbook/49/flipBook.html

[363] X. Guo, E. Soltanian, J. Zhang et al. 2025, “Micro-transfer-printed short-wave infrared InP-on-silicon tunable laser,” Opt. Lett. 50, 1589-1592, https://doi.org/10.1364/OL.549540

[364] Y. Arafat, S. Ghosh, C. Barbier et al. 2025, “Micro Transfer Printed Semiconductor Optical Amplifier at the 2 μm waveband onto SOI,” 2025 IEEE Silicon Photonics Conf. (SiPhotonics), London, United Kingdom, 2025, pp. 1–2, doi: 10.1109/SiPhotonics64386.2025.10984987.

[365] M. Kiewiet, S. Cuyvers, M. Billet, et al. 2026 “Micro-Transfer Printed Continuous-Wave and Mode-Locked Laser Integration at 800 nm on a Silicon Nitride Platform.” Laser Photonics Rev20 no. 7: e00956. https://doi.org/10.1002/lpor.202500956.

[366] S.H. Badri, M.V. Kotlyar, R. Das et al.2025, “Compact modulators on silicon nitride waveguide platform via micro-transfer printing of thin-film lithium niobate”, Sci Rep 15, 11681, https://doi.org/10.1038/s41598-025-95397-w

[367] Z. Shi, H. Huang, R. Penty et al. 2026, “Photonic Hybrid Integration: Strategies and Promises of Advanced Additive Manufacturing.” Adv. Optical Mater.14(10) e01750, https://doi.org/10.1002/adom.202501750

[368] S. Ghosh, J. O’Callaghan, O. Moynihan et al. 2024, “Scalable transfer printing approach to heterogeneous integration of InP lasers on silicon-on-insulator waveguide platform”, Appl. Phys. Lett. 125, 081104, https://doi.org/10.1063/5.0223167

[369] J. Vardaman 2025, “AI Technology and the Markets,” October 6, 2025, https://eps.ieee.org/wp-content/uploads/2025/11/Oct_8_Vardaman_.pdf

[370] G. Cong and S. Namiki 2025, “Large-scale Silicon Photonic Integrated Circuits for Sustainable AI Infrastructure,” 2025 IEEE Int. Electron Devices Meet. (IEDM), San Francisco, CA, USA, 2025, pp. 1–4, doi: 10.1109/IEDM50572.2025.11353850.

[371] P. Kumar, D. E. Shim and A. Naeemi 2025, “Comprehensive Device to System Co-Design for SOT-MRAM at the 7 nm Node,” IEEE J. on Exploratory Solid-State Computational Devices and Circuits, 11, 139–147, doi: 10.1109/JXCDC.2025.3621279.

[372] I. Singh, B. Raj, and M. Khosla 2024, “SOT-MRAM Memories for Energy Efficient Embedded and AI Applications,” In: Mehta, G., Wickramasinghe, N., Kakkar, D. (eds) Innovations in VLSI, Signal Processing and Computational Technologies. WREC 2023. Lecture Notes in Electrical Engineering, vol 1095. Springer, Singapore, https://doi.org/10.1007/978-981-99-7077-3_2

[373] B. Kim, Y. Kim, P. Nair P et al. 2022, “Exploiting Data Compression for Adaptive Block Placement in Hybrid Caches”. Electronics. 11(2):240. https://doi.org/10.3390/electronics11020240

[374] Q. Liu, Y. Li, L. Liu et al. 2025, “High-density, nonvolatile SRAM using monolithic 3D integration of InGaZnO thin-film transistors and $Hf_{0.5}Zr_{0.5}O_2$-based ferroelectric capacitors,” Device 3(9) 100833, https://doi.org/10.1016/j.device.2025.100833

[375] J. Han, S. Kang, K. Kim et al. 2023, “Fundamental Issues in VNAND Integration Toward More Than 1K Layers,” 2023 Int. Electron Devices Meet. (IEDM), San Francisco, CA, USA, 2023, pp. 1–5, doi: 10.1109/IEDM45741.2023.10413720.

[376] T. Schenk and S. Mueller 2021, “A New Generation of Memory Devices Enabled by Ferroelectric Hafnia and Zirconia,” 2021 IEEE Int. Symp. on Applications of Ferroelectrics (ISAF), Sydney, Australia, 2021, pp. 1–11, doi: 10.1109/ISAF51943.2021.9477377.

[377] S. Varghese, "DRAM Scaling Requires New Materials Engineering Solutions," April 2021, https://www.appliedmaterials.com/us/en/blog/blog-posts/dram-scaling-requires-new-materials-engineering-solutions.html

[378] P-K. Hsu, S. Cho, J. Sharda et al. 2025, "Monolithic 3D Stackable DRAM Design with BEOL-Compatible Oxide Channel Access Transistor," 2025 IEEE International Memory Workshop (IMW), Monterey, CA, USA, 2025, pp. 1–4, doi: 10.1109/IMW61990.2025.11026931.

[379] K.S. Choi, S.H. Kim, J.W. Seo et al. 2024, "A Three Dimensional DRAM (3D DRAM) Technology for the Next Decades," 2024 IEEE Symp. on VLSI Technology and Circuits (VLSI Technology and Circuits), Honolulu, HI, USA, 2024, pp. 1–2, doi: 10.1109/VLSITechnologyandCir46783.2024.10631471.

[380] S. Hwan Kong and W. Shim 2024, "Advanced 2T0C DRAM Technologies for Processing-in-Memory – Part I: Vertical Transistor on Gate (VTG) DRAM Cell Structure," IEEE Transactions on Electron Devices 71(11) 6633–6638, doi: 10.1109/TED.2024.3447612.

[381] T.H. Noh, S. Chen, H-B.Kim et al. 2024, "First demonstration of 2T0C-FeDRAM: a-ITZO FET and double gate a-ITZO/a-IGZO FeFET with a record-long multibit retention time of >4-bit and >2000 s†," Nanoscale, 16, 16467–16476, https://doi.org/10.1039/D4NR02393E

[382] G. Hilson 2024, "Emerging Memories Look to Displace NOR, SRAM," August 2021, https://www.eetimes.com/emerging-memories-look-to-displace-nor-sram/ – accessed in August 31, 2021

[383] M. Hellenbrand, I. Teck and J. L. MacManus-Driscoll 2024, "Progress of emerging non-volatile memory technologies in industry," MRS Communications vol. 14, pp.1099–1112, 2024, https://doi.org/10.1557/s43579-024-00660-2

[384] T. Li, H. Zhong, Y. Xu et al. 2024, "REMNA: Variation- Resilient and Energy-Efficient MLC FeFET Computing-in-Memory Using NAND Flash-Like Read and Adaptive Control," ICCAD '24: Proceedings of the 43rd IEEE/ACM International Conference on Computer-Aided Design, no. 76, pp; 1–9, https://doi.org/10.1145/3676536.367671

[385] M. J. Marinella 2021, "Radiation Effects in Advanced and Emerging Nonvolatile Memories," IEEE Transactions on Nuclear Science 68(5) 546–572, doi: 10.1109/TNS.2021.3074139.

[386] Yoshikawa, A., Kochurikhin, V., Tomida, T. et al. 2024, "Growth of bulk β-Ga_2O_3 crystals from melt without precious-metal crucible by pulling from a cold container,"Scientific Reports vol. 14, 14881, 2024, https://doi.org/10.1038/s41598-024-65420-7

[387] H. W. Then et al. 2021, "Advanced Scaling of Enhancement Mode High-K Gallium Nitride-on-300mm-Si(111) Transistor and 3D Layer Transfer GaN-Silicon Finfet CMOS Integration," 2021 IEEE International Electron Devices Meeting (IEDM), San Francisco, CA, USA, 2021, pp. 11.1.1–11.1.4, doi: 10.1109/IEDM19574.2021.9720710.

[388] G. Burn, "Semiconductor and beyond," September 3, 2025, https://www.pwc.com/gx/en/industries/technology/pwc-semiconductor-and-beyond-2026-full-report.pdf

[389] V. Leitgeb, B. Kosednar-Legenstein, L. Mitterhuber et al. 2025, "Stress and Crystal Quality Analysis of GaN Layers on Cost-Effective Substrates for Vertical High-Power Applications," 2025 26th Int. Conf. on Thermal, Mechanical and Multi-Physics Simulation and Experiments in Microelectronics and Microsystems (EuroSimE), Utrecht, Netherlands, 2025, pp. 1–5, doi: 10.1109/EuroSimE65125.2025.11006600.

[390] Y. Ma, H. Chen, S. Zhang et al. 2025, "1500 V GaN-on-Si Vertical Power MOSFETs: From Quasi-Vertical to Fully-Vertical Topology," 2025 37th Int. Symp.on Power Semiconductor Devices and ICs (ISPSD), Kumamoto, Japan, 2025, pp. 345–348, doi: 10.23919/ISPSD62843.2025.11117463.

[391] Y. Ma, H. Chen, S. Zhang et al. 2025, "1200 V Fully-Vertical GaN-on-Si Power MOSFETs," IEEE Electron Device Letters, 46(9) 1513–1516, doi: 10.1109/LED.2025.3586947.

[392] A. Shilov, "TSMC could charge up to $45,000 for 1.6 nm wafers – rumors allege a 50% increase in pricing over prior-gen wafers," https://www.tomshardware.com/tech-industry/semiconductors/tsmc-could-charge-up-to-usd45-000-for-1-6nm-wafers-rumors-allege-a-50-percent-increase-in-pricing-over-prior-gen-wafers – accessed in June 29, 2025

[393] C. Liu, R. Abdul Khadar and E. Matioli 2018, "GaN-on-Si Quasi-Vertical Power MOSFETs," IEEE Electron Device Letters, 39(1) 71–74, doi: 10.1109/LED.2017.2779445.

[394] R. Zhu, H. Jiang, C. W. Tang et al. 2022, "GaN quasi-vertical trench MOSFETs grown on Si substrate with ON-current exceeding 1 A," 2022 Appl. Phys. Express 15, 121004, 2022, DOI 10.35848/1882-0786/aca26e.

[395] R. A. Khadar, C. Liu, R. Soleimanzadeh et al. 2019, "Fully Vertical GaN-on-Si power MOSFETs," IEEE Electron Device Letters, 40(3) 443–446, doi: 10.1109/LED.2019.2894177.

[396] R. A. Khadar, A. Floriduz, C. Liu et al. 2021, "Quasi-vertical GaN-on-Si reverse blocking power MOSFETs," Appl. Phys. Express 14, 046503, DOI 10.35848/1882-0786/abf054.

[397] C. Liu, R. Abdul Khadar and E. Matioli 2018, "Vertical GaN-on-Si MOSFETs With Monolithically Integrated Freewheeling Schottky Barrier Diodes," IEEE Electron Device Letters, 39(7) 1034–1037, doi: 10.1109/LED.2018.2841959.

[398] C. Gupta, C. Lund, S. H. Chan et al. 2017, "In Situ Oxide, GaN Interlayer-Based Vertical Trench MOSFET (OG-FET) on Bulk GaN substrates," IEEE Electron Device Letters, 38(3) 353–355, doi: 10.1109/LED.2017.2649599.

[399] D. Ji, C. Gupta, S. H. Chan et al. 2017, "Demonstrating >1.4 kV OG-FET performance with a novel double field-plated geometry and the successful scaling of large-area devices," 2017 IEEE International Electron Devices Meeting (IEDM), San Francisco, CA, USA, 2017, pp. 9.4.1–9.4.4, doi: 10.1109/IEDM.2017.8268359.

[400] D. Ji, C. Gupta, A. Agarwal et al. 2018, "Large-Area In-Situ Oxide, GaN Interlayer-Based Vertical Trench MOSFET (OG-FET)," IEEE Electron Device Letters, 39(5) 711–714, doi: 10.1109/LED.2018.2813312.

[401] R. Li, Y. Cao, M. Chen et al. 2016, "600 V/1.7 Ω Normally-Off GaN Vertical Trench Metal–Oxide–Semiconductor Field-Effect Transistor," IEEE Electron Device Letters, 37(11) 1466–1469, Nov. 2016, doi: 10.1109/LED.2016.2614515.

[402] T. Oka, Y. Ueno, T. Ina et al. 2014, "Vertical GaN-based trench metal oxide semiconductor field-effect transistors on a free-standing GaN substrate with blocking voltage of 1.6 kV," Appl. Phys. Express 7, 021002, 2014, DOI 10.7567/APEX.7.021002.

[403] T. Oka, T. Ina, Y. Ueno et al. 2015, "1.8 mΩ cm^2 vertical GaN-based trench metal–oxide–semiconductor field-effect transistors on a free-standing GaN substrate for 1.2-kV-class operation," Appl. Phys. Express 8, 054101, 2015, DOI 10.7567/APEX.8.054101.

[404] T. Oka, T. Ina, Y. Ueno and J. Nishii 2016, "Over 10 a operation with switching characteristics of 1.2 kV-class vertical GaN trench MOSFETs on a bulk GaN substrate," 2016 28th Int. Symp. on Power Semiconductor Devices and ICs (ISPSD), Prague, Czech Republic, 2016, pp. 459–462, doi: 10.1109/ISPSD.2016.7520877.

[405] W. He, J. Li, Z. Liao et al. 2022, "1.3 kV Vertical GaN-Based Trench MOSFETs on 4-Inch Free Standing GaN Wafer," Nanoscale Res Lett 17, 14 (2022), https://doi.org/10.1186/s11671-022-03653-z

[406] H. W. Then, S. Bader, A. Zubair et al. 2025, "GaN Chiplet Technology Based on 300mm GaN-on-Silicon," 2025 IEEE Int. Electron Devices Meet. (IEDM), San Francisco, CA, USA, 2025, pp. 1-4, doi: 10.1109/IEDM50572.2025.11353652.

[407] X. Li, Q. Wang and C. Qiul 2021, "The Effect of Different Well Implant Element on Different Pitch Size CMOS Image Sensor," 2021 China Semiconductor Technology International Conference (CSTIC), Shanghai, China, 2021, pp. 1–2, doi: 10.1109/CSTIC52283.2021.9461427.

[408] S.-G. Wuu, H.-L. Chen, H.-C. Chien, P. Enquist, R. M. Guidash and J. McCarten, 2022, "A Review of 3-Dimensional Wafer Level Stacked Backside Illuminated CMOS Image Sensor Process Technologies," IEEE Transactions on Electron Devices, 69(6) 2766–2778, doi: 10.1109/TED.2022.3152977.

[409] https://ghgprotocol.org/

[410] J-H. Mohr, G. Tembey, K.Breidenbach et al., "For Chip Makers, the Decarbonization Challenge Lies Upstream," May 2023, https://www.bcg.com/publications/2023/why-chip-makers-need-to-focus-on-the-upcoming-decarbonization-challenges
[411] M. Ruberti, "Environmental performance and trends of the world's semiconductor foundry industry," J. of Industrial Ecology, vol. 28, pp. 1183–1197, October 2024, https://doi.org/10.1111/jiec.13529
[412] J.C. Hess and M. Nowicka 2026, "Direct Emissions in Semiconductor Manufacturing Are Increasing Again – What Is Behind the Shift?," Interface, January 30, 2026, https://www.interface-eu.org/publications/semiconductor-emissions-data-2026
[413] M. Pelcat 2023, "GHG emissions of semiconductor manufacturing in 2021", Univ Rennes, INSA Rennes, CNRS, IETR – UMR 6164, F-35000 Rennes. 2023. ffhal-04112708v2
[414] O. Burkacky, M. Nikolka, S. Göke et al. 2022, "Sustainability in semiconductor operations: Toward net-zero production," May 2022, https://www.mckinsey.com/industries/semiconductors/our-insights/sustainability-in-semiconductor-operations-toward-net-zero-production
[415] D. Lay, I. Keyte, S. Tiscione et al. 2023, "A guide to PFAS in electronics," ChemSec H2020 European Project No. SA22171340, https://chemsec.org/app/uploads/2023/04/Check-your-Tech_230420.pdf
[416] W. P. Shuley and S. K. Kurinec 2025, "Addressing Energy and PFAS Chemical Consumption in Semiconductor Manufacturing: Case Study in a University Fab," IEEE Transactions on Semiconductor Manufacturing, 38(4) 925–932, Nov. 2025, doi: 10.1109/TSM.2025.3621131.
[417] S. Göke S., M. Issler, M. Patel et al. 2022, "Keeping the semiconductor industry on the path to net zero," November 2022, https://www.mckinsey.com/industries/semiconductors/our-insights/keeping-the-semiconductor-industry-on-the-path-to-net-zero
[418] P. Nagapurkar and S. Das 2022, "Economic and embodied energy analysis of integrated circuit manufacturing processes," Sustainable Computing: Informatics and Systems, vol. 35, 100771, 2022, https://doi.org/10.1016/j.suscom.2022.100771
[419] D. Kline Jr., N. Parshook, X. Ge et al. "GreenChip: A tool for evaluating holistic sustainability of modern computing systems," Sustainable Computing: Informatics and Systems, vol. 22, pp. 322–332, June 2019, https://doi.org/10.1016/j.suscom.2017.10.001
[420] Q. Wang, N. Huang, H. Cai et al. 2023, "Water strategies and practices for sustainable development in the semiconductor industry," Water Cycle, vol. 4, pp. 12–16, 2023, https://doi.org/10.1016/j.watcyc.2022.12.001
[421] European Commission: Directorate-General for Internal Market, Industry, Entrepreneurship and SMEs, "Study on the critical raw materials for the EU 2023 – Final report", Publications Office of the European Union, 2023, https://data.europa.eu/doi/10.2873/725585
[422] R. C. Berg, H. Ziemer H. et al. 2024, "Mineral Demands for Resilient Semiconductor Supply Chains," May 2024, https://www.csis.org/analysis/mineral-demands-resilient-semiconductor-supply-chains
[423] U.S. Geological Survey, Mineral commodity summaries 2025 (ver. 1.2, March 2025): U.S. Geological Survey, 212 p., 2025, https://doi.org/10.3133/mcs2025.
[424] J. Teer and M. Bertolini 2022, "The semiconductor and critical raw material ecosystem at a time of great power rivalry," October 2022, https://hcss.nl/report/reaching-breaking-point-semiconductors-critical-raw-materials-great-power-rivalry/
[425] The Global e-waste monitor 2024, November 2024, https://ewastemonitor.info/wp-content/uploads/2024/12/GEM_2024_EN_11_NOV-web.pdf
[426] L. Talens Peiro, P. Nuss, F. Mathieux and G. Blengini 2018, "Towards Recycling Indicators based on EU flows and Raw Materials System Analysis data," EUR 29435 EN, Publications Office of the European Union, Luxembourg, 2018, ISBN 978-92-79-97247-8, doi:10.2760/092885.
[427] de Oliveira R.P., Benvenuti J. and Espinosa D.C.R. 2021, "A review of the current progress in recycling technologies for gallium and rare earth elements from light-emitting diodes," Renewable

and Sustainable Energy Rev., vol. 145, art. 111090, July 2021, https://doi.org/10.1016/j.rser.2021.111090
[428] J. Howley, "Global Trends in Critical Material Recovery: 2025 and Beyond," September 2025, https://www.idtechex.com/en/research-article/global-trends-in-critical-material-recovery-2025-and-beyond/33811
[429] D. Mohamed, A. Fayad, A.-M.O. Mohamed et al. 2025, "The Role of E-Waste in Sustainable Mineral Resource Management," Waste 3, 27, 2025, https://doi.org/10.3390/waste3030027
[430] Eco Recycling Today, "E-Waste Recycling Plants: Recovering Gold from Electronics", https://www.recyclingtoday.org/blogs/news/e-waste-recycling-plants-recovering-gold-from-electronics#:~:text=The%20e%2Dscrap%20undergoes%20hydrometallurgical,cleaner%20and%20more%20sustainable%20alternative, accessed December 2025
[431] C. B. Tabelin, I. Park, T. Phengsaart et al. 2021, "Copper and critical metals production from porphyry ores and E-wastes: A review of resource availability, processing/recycling challenges, socio-environmental aspects, and sustainability issues," Resources, Conservation and Recycling 170, 105610, 2021, https://doi.org/10.1016/j.resconrec.2021.10561
[432] R. Hossain and V. Sahajwalla 2024, "Current recycling innovations to utilize e-waste in sustainable green metal manufacturing," Philos Trans A Math Phys Eng Sci. 382 (2284): 20230239, 2 December 2024, https://doi.org/10.1098/rsta.2023.0239
[433] A. N. Erkmen, R. Ulber, T. Jüstel et al. 2025, "Towards sustainable recycling of critical metals from e-waste: Bioleaching and phytomining," Resources, Conservation & Recycling 215, 108057, 2025, https://doi.org/10.1016/j.resconrec.2024.108057
[434] N.M. Charpentier, D. Xia and J-C. P. Gabriel 2024, "Printed circuit board recycling: a focus on a novel,efficient and sustainable process for spent critical metals recovery," Comptes Rendus. Chimie, GDR Prométhée – French Research Network on Hydrometallurgical Processes for Primary and Secondary Resources, 27, 5-15, 2024, doi: 10.5802/crchim.291.
[435] B. Wiseman, B. Jay et al. 2024, "Reimagining labor to close the expanding US semiconductor talent gap," August 2024, https://www.mckinsey.com/industries/semiconductors/our-insights/reimagining-labor-to-close-the-expanding-us-semiconductor-talent-gap
[436] "Build the semiconductor workforce of the future," April 2026, https://www.semiconductors.org/wp-content/uploads/2024/04/SIA-Workforce-Policy-Blueprint-2-pager-4_8_24.pdf
[437] R. Beaujeu, L. Saint-Martin et al. 2024, "Adressing the talent in the European semiconductor ecosystem," 2024, https://www.decision.eu/wp-content/uploads/2024/11/2024-ECSA-Skills-Strategy.pdf
[438] European Semiconductor Industry Association 2024, "Towards a more competitive semiconductor industry for Europe," July 2024, https://www.eusemiconductors.eu/sites/default/files/ESIA_Key%20Recommendations%202024-2029_digital_final_0.pdf
[439] K. Zhang 2025, "Taiwan's Shortage of Chipmakers: A Major Threat to the Industry's Long-term Growth," Global Taiwan Institute, March 2025, https://globaltaiwan.org/2025/03/taiwans-shortage-of-chipmakers-a-major-threat-to-the-industrys-long-term-growth/
[440] A. Patterson 2026, "Taiwan builds the Future," March 4, 2026, https://www.eetimes.com/taiwan-builds-the-future/
[441] I. Cerutti and M. Nardo 2023, "Semiconductors in the EU," Publications Office of the European Union, Luxembourg, 2023, doi:10.2760/038299, JRC133850.
[442] Semiconductor Industry Association, "State of the US Semiconductor industry 2025," July 7, 2025, https://www.semiconductors.org/wp-content/uploads/2025/07/SIA-State-of-the-Industry-Report-2025.pdf
[443] M. Alexander and T. Kirschstein 2022, "How partnerships with non-European providers and the EU's new initiative can forge a global powerhouse," August 2022,

[444] M. Duchatel and P. Sel 2026, “A New Economic Security Consensus? Semiconductor Nationalism in China and the United States, and Europe’s Choices,” March 17, 2026, https://www.institutmontaigne.org/en/expressions/semiconducteurs-la-preference-europeenne-face-la-chine-et-aux-etats-unis
[445] G. Wong 2025, “Beneficiary of China’s Semiconductor Self-Sufficiency Drive,” June 2025, https://cdn.growbeansprout.com/strapi-uploads/Beansprout_SMIC_HK_SDR_5to1_facd53363e.pdf, accessed in June 22, 2025
[446] X. You 2026, “China intensifies push to become world leader in tech and AI,” Nature 651, 859–860, https://doi.org/10.1038/d41586-026-00814-3
[447] J. Negrine 2026, “Is Japan’s economic security strategy economical?,” March 31, 2026, https://www.hinrichfoundation.com/research/article/supply-chain-security/japan-economic-security-strategy?utm_campaign=article-negrine-japan-economic-security-strategy-l1&utm_medium=email&_hsenc=p2ANqtz-9rNB9DWBZG-VEhMJCsGH9rTZMDes6TAwWRXvvCHJlg32mEBoL-3tDxItlT13b5_StZU5MZ6GhC3UFA-npX3rtAKSlaPw&_hsmi=411467458&utm_content=20260331-weekly-research-&utm_source=hinrich-thought-leadership
[448] L. Hae-Rin 2025, “Korea unveils $534 bil. plan to lead global chip race in AI era,” December 10, 2025, https://www.koreatimes.co.kr/southkorea/politics/20251210/korea-unveils-534-bil-plan-to-lead-global-chip-race-in-ai-era, accessed in December 2025
[449] “India’s Semiconductor Leap: Four Plants to be Operational by 2026, Says Ashwini Vaishnaw,” April 1, 2026, https://www.bisinfotech.com/indias-semiconductor-leap-four-plants-to-be-operational-by-2026-says-ashwini-vaishnaw/
[450] T. Devesa, J. Seong, O. White et al. 2025, “The FDI shake-up,” September 2025, https://www.mckinsey.com/~/media/mckinsey/mckinsey%20global%20institute/our%20research/the%20fdi%20shake-up%20how%20foreign%20direct%20investment%20today%20may%20shape%20industry%20and%20trade%20tomorrow/the-fdi-shake-up-how-foreign-direct-investment-today-may-shape-industry-and-trade-tomorrow.pdf?shouldIndex=false, accessed in December 2025.
[451] Maxime Market Research 2024, “European Semiconductor Market-Industry Analysis and Forecast 2030,” November 2024, https://www.maximizemarketresearch.com/market-report/european-semiconductor-market/85994/, accessed in March 2025.
[452] L. Gagda – IMARC 2025, “Europe Semiconductor Market is Growing at a CAGR of 5.7% During the Forecast Period 2025-2033,” April 2025, https://www.linkedin.com/pulse/europe-semiconductor-market-growing-cagr-57-during-forecast-gagda-jzsic/ – accessed in December 2025.
[453] EU Commission 2024,“State of the Digital Decade 2024, Annex II”, July 2, 2024, https://eur-lex.europa.eu/legal-content/EN/TXT/DOC/?uri=CELEX:52024DC0260
[454] European Commission: Joint Research Centre, P. Bonnet, A. Ciani, J. Molnar and M. Nardo 2025, “EU’s strengths and weaknesses in the global semiconductor sector,” Publications Office of the European Union, Luxembourg, 2025, https://data.europa.eu/doi/10.2760/6302476,JRC141323
[455] J. An and H. Yin 2025, “EU Policies and prospects for achieving ‘Chip autonomy’ amid global localization trend in chip supply chain,” Innovation: The European Journal of Social Science Research, September 22, 2025, DOI: 10.1080/13511610.2025.2554765.
[456] HDIN Research 2022, “Semiconductor Wafers Producers and Semiconductor Wafers Market Trend,” January 2022, https://www.hdinresearch.com/news/111#:~:text=The%20global%20semiconductor%20wafer%20market%20is%20expected%20to%20reach%20%2419.6,%2C%20Silitronic%2C%20and%20SK%20Siltron, accessed in January 2025.
[457] ESIA, “European industry remains at forefront of greenhouse gas emissions reduction,” https://www.eusemiconductors.eu/esia/public-policy/sustainability-esh/pfc-gases, accessed in January 2025.
[458] ESIA, “Water Resilience Strategy,” February 26, 2025, https://www.eusemiconductors.eu/sites/default/files/2025.02.26_ESIAposition-WaterResilienceStrategy.pdf, accessed in March 2025.

[459] A. Varas, R. Vadarajan, J. Goodrich et al., "Strengthening the Global Semiconductor Supply Chain in an Uncertain Era," SIA/BCG Report, April 2021, https://www.semiconductors.org/wp-content/uploads/2021/05/BCG-x-SIA-Strengthening-the-Global-Semiconductor-Value-Chain-April-2021_1.pdf, accessed in May 2021.

[460] Van Eck 2024, "Fabless Chip Designers: Shaping the Future of Semiconductors," August 28, 2024, https://www.vaneck.com/us/en/blogs/thematic-investing/fabless-chip-designers-shaping-the-future-of-semiconductors/, accessed in December 2025.

[461] D. Crichton, "The looming labor crisis in chip design," August 3, 2024, https://www.luxcapital.com/content/the-looming-labor-crisis-in-chip-design, accessed in December 2025.

[462] H. Bauer, O. Burkacky, P. Kenevan et al., "Semiconductor design and manufacturing: Achieving leading-edge capabilities," August 20, 2020, https://www.mckinsey.com/industries/industrials/our-insights/semiconductor-design-and-manufacturing-achieving-leading-edge-capabilities, accessed in December 2025.

[463] J-P. Kleinhans and J. Lee 2021, "China's rise in semiconductors and Europe," December 2021, https://www.interface-eu.org/publications/3187#:~:text=China%20already%20has%20a%20substantially,depend%20on%20Chinese%20designed%20chips, accessed in February 2022.

[464] The European Innovation Council, "Backing visionary entrepreneurs," February 9, 2023, https://eic.ec.europa.eu/document/download/97312423-4149-419d-9b2d-94ffec7c6a47_en?filename=Emerging%20semiconductor%20and%20Quantum%20technology%20components%20-%20Information%20Day%20-%209%20Feb%202023-Final.pdf, accessed in December 2025.

[465] R. Palma, R. Varadajan, J. Goodrich et al., "The Growing Challenge of Semiconductor Design Leadership," November 2022, https://www.semiconductors.org/wp-content/uploads/2022/11/2022_The-Growing-Challenge-of-Semiconductor-Design-Leadership_FINAL.pdf, accessed in December 2022

[466] J. Hess, W. Denkena, J-P. Kleinhans et al., "Who is funding the chips of the future?," April 2023, https://www.interface-eu.org/publications/semiconductor-startup-funding-activities, accessed in December 2025.

[467] State of European Tech 2025, "Startup Investment Trends," October 2025, https://www.stateofeuropeantech.com/chapters/startup-investment-trends, accessed in December 2025.

[468] W. Hafez, D. Abanulo, M. Abdelkader et al. 2024, "An Intel 3 Advanced FinFET Platform Technology for High Performance Computing and SOC Product Applications," 2024 IEEE Symp. on VLSI Technology and Circuits (VLSI Technology and Circuits), Honolulu, HI, USA, 2024, pp. 1–2, doi: 10.1109/VLSITechnologyandCir46783.2024.10631513.

[469] EU project ICOS N° 101092562, "Monitoring semiconductor value chains", June 30, 2025, https://icos-semiconductors.eu/uncategorized/monitoring-semiconductor-value-chain-implications-for-international-cooperation/, accessed in October 2025.

[470] European Chips Survey Report – DG Grow, July 2022, https://digital-strategy.ec.europa.eu/en/library/european-chips-survey and European Court of Auditors, special report 12/2025: "The EU's strategy for microchips – Reasonable progress in its implementation but the Chips Act is very unlikely to be sufficient to reach the overly ambitious Digital Decade target", Publications Office of the European Union, April 2025, https://www.eca.europa.eu/en/publications/sr-2025-12

[471] J. Aurik, D. Gerdermann, M. Hales et al., "Europe's urgent need to invest in leading-edge semiconductor ecosystem," February 2022, https://www.kearney.com/industry/technology/article/-/insights/europes-urgent-need-to-invest-in-a-leading-edge-semiconductor-ecosystem, accessed in January 2025.

[472] "A Closer Look at the Qimonda Bankruptcy," March 4, 2009, https://insolvency.foxrothschild.com/2009/03/a-closer-look-at-the-qimonda-bankruptcy/, accessed in December 2025.

[473] "Micron Announces Agreement to Acquire Numonyx," February 9, 2010, https://investors.micron.com/news-releases/news-release-details/micron-announces-agreement-acquire-numonyx, accessed in December 2025.

[474] S. Bertolazzi, "High Bandwidth Memory Market and Technology Trends," The future of Memory and Storage, August 5-7, 2025, https://files.futurememorystorage.com/proceedings/2025/20250807_DRAM-302-1_Bertolazzi.pdf

[475] J. Lau, "Memory Market Overview: 2025 Update," The future of Memory and Storage, August 5-7, 2025, https://files.futurememorystorage.com/proceedings/2025/20250807_BMKT-301-1_Lau.pdf

[476] B. Wiseman, H. Marcil and M. de Jong, "Semiconductors have a big opportunity – but barriers to scale remain," April 2025, https://www.mckinsey.com/industries/semiconductors/our-insights/semiconductors-have-a-big-opportunity-but-barriers-to-scale-remain

[477] "Vietnam Holds Promise as a Global Semiconductor Hub," December 15, 2023, https://arc-group.com/vietnam-semiconductor-industry/, accessed in December 2025.

[478] R. Palanisamy, "What's the Average Cost of 1 kWh Electricity around the World?," December 15, 2024, https://www.voronoiapp.com/energy/Whats-the-Average-Cost-of-1-kWh-Electricity-around-the-World–3398, accessed in December 2025.

[479] J-P. Kleinhans, R. Goujon, J. Hess et al. "Running on Ice: China's Chipmakers in a Post-October 7 World," March 2023, https://rhg.com/research/running-on-ice/, accessed in May 2025

[480] E. Flaningan, "Semiconductor companies ranked by R&D," December 5, 2023, https://x.com/EricFlaningam/status/1732070071566999567, accessed in March 2025.

[481] E. Nindl, L. Napolitano, H. Confraria et al. "The 2025 EU Industrial R&D Investment Scoreboard", Publications Office of the European Union, Luxembourg, 2025, https://data.europa.eu/doi/10.2760/7802619,JRC144638.

[482] "ASML EUV Dominance & China's Semiconductor Equipment Push," December 23, 2025, https://www.trendforce.com/insights/asml-euv

[483] S. Shivakumar and C. Wessner 2026, "China's Localization Drive in Semiconductors Gains Impetus from Allied Chip Export Controls," March 24, 2026, https://www.csis.org/analysis/chinas-localization-drive-semiconductors-gains-impetus-allied-chip-export-controls

[484] "China's SMEE Reportedly Wins RMB 110M Lithography Tool Contract Amid Domestic Push," December 2026, 2025, https://www.trendforce.com/news/2025/12/26/news-chinas-smee-reportedly-wins-rmb-110m-lithography-tool-contract-amid-domestic-push/#:~:text=According%20to%20Commercial%20Times%2C%20citing,(Shanghai%20Micro%20Electronics%20Equipment).

[485] "Chinese ASML Emergence, U.S. Semiconductor Jones Act Loom," March 16, 2026, https://www.chosun.com/english/opinion-n/2026/03/15/TRP4CLVIFZCDPG7HAV5FQSFHQY/

[486] "The United States is making efforts in EUV lithography," July 16, 2025, https://eu.36kr.com/en/p/3380929389094921

[487] "Laser Photomask Market: Next-Gen Semiconductor Demand and Industry Forecast (2025-2032)," August 2025, https://www.maximizemarketresearch.com/market-report/global-laser-photomask-market/27013/, accessed in December 2025.

[488] "ASE: Still The Global OSAT Powerhouse (Rating Downgrade)," September 17, 2024, https://seekingalpha.com/article/4721495-ase-still-the-global-osat-powerhouse, accessed in January 2025.

[489] R. Varadajan, I. Koch-Weser, C. Richard et al., "Emerging Resilience in the semiconductor supply chain," May 2024, https://www.semiconductors.org/emerging-resilience-in-the-semiconductor-supply-chain/

[490] Yole Intelligence, "Advanced packaging market set to reach $79.4 billion by 2030," August 31, 2025, https://www.yolegroup.com/press-release/advanced-packaging-market-set-to-reach-79-4-billion-by-2030/, accessed in December 2025.

[491] "Report on critical raw materials and the circular economy," Publications Office, 2018, https://data.europa.eu/doi/10.2873/167813
[492] A. Leichthammer 2025, "The EU's critical raw materials predicament: ReSourceEU to the Rescue?," December 19, 2025, https://www.delorscentre.eu/en/publications/detail/publication/the-eus-critical-raw-materials-predicament
[493] P. Dannhauer 2025, "Asean Strategies and Partnerships in the Critical Minerals Sector," Stiftung Wissenschaft und Politik. Germany. Retrieved from https://coilink.org/20.500.12592/9lz9k5d on 29 Mar 2026.
[494] K. van Wieringen, "Global Semiconductor Trends and the Future of EU Chip Capabilities," October 12, 2022, https://www.espas.eu/files/Global-Semiconductor-Trends-and-the-Future-of-EU-Chip-Capabilities-2022.pdf
[495] IndustriAll Europe, "The semiconductor industry in Europe: between geopolitics and tech race," May 2024, https://news.industriall-europe.eu/documents/upload/2024/9/638626809844508800_Semiconductors_policy_brief.pdf, accessed in November 2025
[496] S. Kerschbaumer, "Super Chips and Supply Chains: Why Europe's Chips Act Can't Fix Chokepoint Weaponisation," November 18, 2025, https://www.cife.eu/Ressources/FCK/files/publications/policy%20paper/2025/182_Kerschbaumer_Samuel_Super_Chips_and_Supply_Chains_Europe_CIFE_Policy_Paper.pdf
[497] SIA 2025 Factbook, May 27, 2025, https://www.semiconductors.org/resources/2025-sia-factbook/
[498] "Loose connections: Rethinking semiconductor supply chains," March 16,2023, https://www.wtwco.com/en-au/insights/2023/03/rethinking-supply-chains-the-semiconductor-industry
[499] "A European strategy for micro- and nanoelectronic components and systems," May 23, 2013, https://eur-lex.europa.eu/legal-content/EN/TXT/PDF/?uri=CELEX:52013DC0298&qid=1726592105386
[500] A.D.D. Pandya 2026, "More Than Just Chips: What Europe Can Learn from Taiwan's Industrial Strategy," January 2026, https://ecipe.org/publications/what-europe-can-learn-from-taiwans-industrial-strategy/
[501] A. Mahindroo, A. Suryanarayanan and M. de Jong, "Silicon squeeze: AI's impact on the semiconductor industry," April 2025, https://www.mckinsey.com/industries/semiconductors/our-insights/silicon-squeeze-ais-impact-on-the-semiconductor-industry
[502] "What is the Artificial Intelligence (AI) Chip Market Size?," December 16, 2025, https://www.precedenceresearch.com/artificial-intelligence-chip-market#:~:text=The%20global%20artificial%20intelligence%20(AI,27.88%25%20from%202026%20to%202035, accessed in December 2025.
[503] G. Burn, "State of the Semiconductor Industry," November 28, 2024, https://www.pwc.com/gx/en/industries/technology/state-of-the-semicon-industry.html
[504] "AI Chips Defining the Future: ASIC and GPU Among Leading Technologies," May 8, 2025, https://finance.yahoo.com/news/ai-chips-defining-future-asic-081200777.html, accessed in December 2025.
[505] "AI-driven growth: $1 trillion and beyond; Who, along with TSMC, Nvidia?," November 2024, https://pradeepstechpoints.wordpress.com/2024/11/20/ai-driven-growth-1-trillion-and-beyond-who-along-with-tsmc-nvidia/, accessed on September 10, 2025.
[506] K. F. Pilz, R. Rahman and J. Sanders, "The US hosts the majority of GPU cluster performance, followed by China," June 5, 2025, Published online at epoch.ai. Retrieved from: 'https://epoch.ai/data-insights/ai-supercomputers-performance-share-by-country' [online resource], accessed in December 2025.
[507] "European Cloud Providers' Local Market Share Now Holds Steady at 15%," July 24, 2025, https://www.srgresearch.com/articles/european-cloud-providers-local-market-share-now-holds-steady-at-15, accessed in December 2025.
[508] F. Bria, P. Timmers, F. Gernone et al., "EuroStack – A European Alternative for Digital Sovereignty," Bertelsmann Stiftung. Gütersloh, February 2025, https://www.euro-stack.info/docs/EuroStack_2025.pdf

[509] A. Tournesac, K. Hjartar, M. Krawina et al. 2025, “Accelerating Europe’s AI adoption: The role of sovereign AI,” December 19, 2025, https://www.mckinsey.com/industries/technology-media-and-telecommunications/our-insights/accelerating-europes-ai-adoption-the-role-of-sovereign-ai

[510] “European High Performance Computing Joint Undertaking (EuroHPC),” December 17, 2025, https://www.europarl.europa.eu/doceo/document/TA-10-2025-0333_EN.html

[511] “The New Gold Rush: The data center investment boom,” January 8, 2025, https://www.redchalk.com/the-new-gold-rush-the-data-center-investment-boom/, accessed in December 2025.

[512] B. Kastrup 2026, “Europe’s sovereign AI future doesn’t need cutting-edge chip manufacturing,” January 8, 2026, Bits & Chips

[513] “Edge AI market,” November 2024, https://www.rootsanalysis.com/edge-ai-market, accessed in December 2025.

[514] N. Agnihotri, “What are the top edge AI chips of 2025,” February 6, 2025, https://www.engineersgarage.com/what-are-the-top-edge-ai-chips-of-2025/, accessed in December 2025.

[515] J-P. Kleinhans and S. Jha 2025, “The chip landscape: Geographical distribution of wafer fabrication capacity,” December 2025, https://www.oecd.org/content/dam/oecd/en/publications/reports/2025/12/the-chip-landscape_27ef5d87/02dbd028-en.pdf

[516] C. H. X. Gnadl, “Europe’s Semiconductor Resilience: The Gaps Behind the Policy Goals,” EE Times Europe, November 2025, pp. 23-24, https://www.eetimes.com/europes-semiconductor-resilience-the-gaps-behind-the-policy-goals/

[517] B. Kapoor, K. Fernandes, S. Kumar et al., “Braving the storm: navigating an uncertain future,” June 2025, https://www.kearney.com/service/product-excellence-and-renewal-lab/state-of-semiconductors-2025

[518] “Japan’s Pursuit of a Game-Changing Technology and Ecosystem for Semiconductor,” March 1, 2024, https://www.japan.go.jp/kizuna/2024/03/technology_for_semiconductors.html, accessed in December 2025.

[519] T. Letavic 2025, “Essential Semiconductors – Innovation and Impact,” 2025 IEEE Int. Electron Devices Meet. (IEDM), San Francisco, CA, USA, 2025, pp. 1-6, doi: 10.1109/IEDM50572.2025.11353894.

[520] K. Thompson, “The Nexperia debacle: Collateral damage in the US-China crossfire,” December 2, 2025, https://www.hinrichfoundation.com/research/wp/tech/us-china-collateral-damage

[521] “Chips for Defence; an EU initiative,” January 29, 2026, https://www.design-reuse.com/news/202530016-chips-for-defence-an-eu-initiative/

[522] “Semiconductor Memory Market By Type (DRAM, Flash ROM, SRAM, MRAM, and Others), By Application,” March 2024, https://dimensionmarketresearch.com/report/semiconductor-memory-market/, accessed in December 2025.

[523] M. Giordano and M. Dusek, “Europe in the Intelligent Age: From Ideas to Action,” January 2025, https://www.weforum.org/publications/europe-in-the-intelligent-age-from-ideas-to-action/

[524] M-A. Eyl-Mazzega, V. Donnen and T.Michel 2026, “Financial Tools for Boosting Resilience of CRM Value Chains and Strategic Stockpiling,” Ifri Studies, Ifri, March 2026, https://www.ifri.org/en/studies/financial-tools-boosting-resilience-crm-value-chains-and-strategic-stockpiling

[525] E. Rossi 2026, “Pax Silica: after Finland, Italy eyes a place in the new architecture of technological power,” April 20,2026, https://decode39.com/14388/pax-silica-after-finland-italy-eyes-a-place-in-the-new-architecture-of-technological-power/

[526] J. Bish, D. Stewart, S. Beerlage et al., “A new dawn for European chips,” November 3, 2022, https://www.deloitte.com/us/en/insights/industry/technology/semiconductor-chip-shortage-supply-chain.html, accessed in December 2025.

[527] “AI demand drives 94.1% surge in semiconductor forecast for 2026”, July 30, 2026, https://omdia.tech.informa.com/pr/2026/july/ai-demand-drives-94-point-1-percent-surge-in-semiconductor-forecast-for-2026

[528] K. Park 2026, "South Korean tech giants commit over $550B to ease 'RAMageddon", June 29, 2026, https://techcrunch.com/2026/06/29/south-korean-tech-giants-commit-over-550b-to-ease-ramageddon/

[529] "Pax Silica" https://www.state.gov/pax-silica

[530] C. Simons 2025, "A Global Imperative: Confronting The Global Semiconductor Talent Shortage", January 2025, https://www.forbes.com/councils/forbestechcouncil/2025/01/07/a-global-imperative-confronting-the-global-semiconductor-talent-shortage/

About the authors

Dr. Christophe Wyon received an Engineer degree (1980) and a PhD (1984) as well as the habilitation to manage scientific research (2009) in Physics and Chemistry of Materials from the Grenoble Polytechnique National Institute, France. He is Director of Research at CEA-Leti.

He managed several laboratories in LETI involving the crystal growth of laser, scintillator and semiconductor materials, the front-end processes for advanced transistors and Flash memories, the physical and chemical characterization of leading-edge ICs, and the development of OLED micro-displays. He also managed the physical characterization and failure analysis department of advanced ICs within ST in the frame of Crolles2 Alliance, gathering ST, NXP and Freescale.

He authored and co-authored more than 80 scientific papers.

He worked during 8 years for the French Ministry of Economy for promoting the microelectronics industry in France and in Europe. He was in charge of setting-up huge national and European programs for the development of micro- and nano-electronics.

Since June 2019, he is the CEA-LETI representative in Brussels in order to promote the nanoelectronics industry in Europe. He contributes notably to the Strategic Research and Innovation Agenda of European microelectronics under the responsibility of the European Components and System community

Prof. Dr. Marcel VAN DE VOORDE studied natural and applied sciences at Belgium and at European universities which resulted in multiple academic degrees: a multidisciplinary education. He was visiting professor at various reputed universities worldwide.

His long-standing scientific and technological career spans positions at renowned European research organizations, culminating in his role as executive advisor to the CEO and President of IMEC.

He was active as chairman stroke member of multiple Research Councils and Governing Boards, worldwide e.g., CNRS (FR), CSIC (ES), CNR (IT), NIMS (JP), ESA (European space agency), NATO R&D&T etc. He also was senator and Trustee at European and World academies of sciences; honorary professor and/or doctor honors causa of universities.

He was advisor to Ministers, Directors of research Institutes, Rectors of Universities in Europe e.g., the Science Council of the French Senate and national assembly in Paris, and plaid a role in the promotion of European university education and research systems e.g., EC-ERASMUS programme. He also received honors from the Belgian King, the Luxemburg State, and multiple scientific organizations. "Nanoscience and technology" were one of his hobbies in sciences.

 | https://doi.org/10.1515/9783112247952-013

Index

 | https://doi.org/10.1515/9783112247952-014